2nd Edition

Engineering Communication

A Practical Guide to Workplace Communications for Engineers

David Ingre
Kwantlen Polytechnic University

Robert Basil
Kwantlen Polytechnic University

CENGAGE Learning™

Australia • Brazil • Mexico • Singapore • United Kingdom • United States

CENGAGE Learning

Engineering Communication: A Practical Guide to Workplace Communications for Engineers, **Second Edition**

David Ingre, Robert Basil

Product Director, Global Engineering: Timothy L. Anderson

Associate Media Content Developer: Ashley Kaupert

Product Assistant: Alexander Sham

Marketing Manager: Kristin Stine

Director, Content and Media Production: Sharon L. Smith

Senior Content Project Manager: Jennifer Risden

Production Service: SPi Global

Compositor: SPi Global

Senior Art Director: Michelle Kunkler

Internal Designer: Grannan Graphic Design, Ltd.

Cover Designer: Grannan Graphic Design, Ltd.

Cover Image: ixpert/Shutterstock.com

Interior Design Images:
Goldenarts/Shutterstock.com
antishock/Shutterstock.com
Tarchyshnik Andrei/Shutterstock.com

Intellectual Property
Analyst: Christine Myaskovsky
Project Manager: Sarah Shainwald

Text and Image Permissions Researcher: Kristiina Paul

Manufacturing Planner: Doug Wilke

Library of Congress Control Number: 2015947526

ISBN: 978-1-305-63510-4

Cengage Learning
20 Channel Center Street
Boston, MA 02210
USA

Cengage Learning is a leading provider of customized learning solutions with employees residing in nearly 40 different countries and sales in more than 125 countries around the world. Find your local representative at **www.cengage.com**.

Cengage Learning products are represented in Canada by Nelson Education Ltd.

To learn more about Cengage Learning Solutions, visit **www.cengage.com/engineering**.

Purchase any of our products at your local college store or at our preferred online store **www.cengagebrain.com**.

Printed in the United States of America
Print Number: 01 Print Year: 2015

Dedication

I dedicate my contribution to the second edition to my late parents, Pearl and Adolph Ingre, to my late sister, Anna, and to my wonderful wife, Bettie. Without their love and support, I would have been much the poorer.

—Dave Ingre

I dedicate my contribution to the second edition to my parents, Maureen and George Basil, and to my son, Dr. Miles Basil—always inspirations.

—Bob Basil

Contents

Preface

To the Instructor

Rather than attempting to be everything to everyone, this second edition of *Engineering Communication*, like the first, focuses on the most important elements of contemporary communications likely to be used by engineers in the United States and many other countries. For its approach and content, we have drawn on our combined decades of experience in the private and public sectors, including college and university teaching, communications consulting, business ownership, textbook authorship, publishing, editing, technical writing, and instructional design.

Whether you're a seasoned professional or relatively new to this field, this book will be of value to you and your students. We believe this second edition is comprehensive and challenging enough to be the sole assigned text for a college or university communications course for engineering students. At the same time, its structure is such that you can easily incorporate complementary materials or rearrange the sequence of teaching activities provided in this text.

Like the first, this second edition of *Engineering Communication* is predicated on the ongoing, dynamic analysis of context, message, audience, purpose, and product (referred to in the book as the CMAPP model). This model has met with success in college and university courses across the United States and in other countries. In *Engineering Communication*, the model provides thematic continuity as students learn to apply the approach in a variety of communication strategies, including good news and bad news, description and instructions, and persuasion, and in the creation of such products as emails, letters, proposals, reports, and summaries. Through their active participation in the book's interrelated case studies and exercises, students come to understand and develop the critical thinking and planning skills that are essential to engineering students as they strive to produce effective communications.

The text explores the still-growing distinctions between technical communications and more traditional forms of writing such as narratives and essays. While it uses the increasingly informal language of today's marketplace, it offers students an additional learning opportunity by presenting them with possibly unfamiliar, though nontechnical, vocabulary. As well, at strategic points throughout the book, students will deal with the importance of both ethics and cultural considerations in the development of effective communication.

To the Student

As a student, you might in the past have been assigned a textbook, asked to read long and often boring passages, and expected to memorize innumerable theories and apparently unrelated details. You might have gone to the exam expecting (and prepared) to recite everything. Later on you might have realized that the exam was the only time you had put that unappealing textbook to practical use.

Will your experience with the second edition of *Engineering Communication* be any different? We hope so; however, you will decide that as you use it. In any case, it does have many aspects you we believe you'll find appealing. For example:

- It doesn't try to be everything to everyone: It deals only with the types of communication that you're likely to need as an engineering student and as an engineer, and it does so *briefly* and *concisely*. (By the way, in the communication model used in the textbook, those two words don't mean the same thing.)
- It does take into account the relevant parts of the *Criteria for Accrediting Engineering Programs* document set out by the Engineering Accreditation Commission of ABET, the Accreditation Board for Engineering and Technology. At the time of this book's publication, ABET's document was available on the commission's website at http://www.abet.org/accreditation/accreditation-criteria/criteria-for-accrediting-engineering-programs-2015-2016/.)
- Instead of the traditional, academic "third-person impersonal," this book employs the "you approach" typical of today's communications. For example, we'll make use of contractions (such as "we'll"), of the second person ("you," as reader), and of the first person ("we," as writers). Though we'll define technical terms, you may encounter other vocabulary with which you aren't familiar. If so, we suggest you look up the words; that, too, is part of improving your communications skills.

The book describes and uses a dynamic analysis model of communications that has proven successful for students like you in courses around the world. You'll look in detail at this practical model, whose acronym, CMAPP, stands for its five elements: context, message, audience, purpose, and product. You might at first think that this approach is merely formulaic. You might also wonder how learning to use the model will help you to create particular types of communication such as engineering reports. Consider this analogy: One way to prioritize a set of objectives is simply by numbering them in order of importance; you can then successfully apply that "simple formula" to your homework, your daily chores, or your workplace tasks. Applying such a deceptively simple formula can thus yield a multitude of complex, sophisticated results. Here's a perhaps even simpler analogy: We can use just the 10 digits from 0 through 9 to create an infinite number of different figures.

We hope you'll find that the CMAPP approach yields a similarly rich harvest.

We won't claim that this book will teach you to be an effective communicator (its underlying premise is that *you* have to do the work). Nonetheless, we do believe it can help you learn to apply what is largely common sense to create a variety of effective real-world communications—during your engineering studies and in your career.

All the material in this textbook derives from three things: the application of commonly accepted principles of rhetoric and communications; our decades of experience in, and observation of, professional communications in the private and public sectors; and the successful testing, in post-secondary classrooms, of everything in this text.

One more thing: Although we'll define technical terms, you might encounter other vocabulary with which you aren't familiar. In fact, we're expecting that you will. That's intentional on our part. This is, after all, a *communications* textbook. And we believe that expanding your vocabulary is part of improving your technical communication skills.

We hope that you enjoy the book and find it useful.

Organization of the Text

The sequence of topics in *Engineering Communication* differs from many other Engineering Communication books.

- Chapter 1 introduces the field of Engineering Communication.
- Chapter 2 covers the CMAPP model in detail, because it serves as the basis for all that follows.
- Chapter 3 deals with the *complementary attributes* of the CMAPP model: aspects of your communication that help you be a more effective communicator.
- Chapter 4 offers suggestions regarding the use of "traditional" and electronic reference materials.
- Chapter 5 provides recommendations for organizing your thoughts—since having a clear idea of what you actually want to communicate is crucial.
- Chapter 6, new to this edition, deals with digital and social media. It examines how they have become important factors in successful professional communication, from networking to marketing, and from online project collaboration to employment search.
- In Chapter 7, you will look at visual aspects of engineering communications, because the way you show and illustrate what you write or present will help determine your success with your audience.
- Putting all these ideas together leads to Chapter 8's discussion of various types of reports—pieces of communication you are certainly going to have to create—as well as summaries and abstracts, which are often added to reports.
- It is all but inevitable that, as engineers, you will have to create mechanism and/or process descriptions, as well as written instructions; Chapter 9 deals specifically with those issues.
- In Chapter 10, you'll look, once again, in terms of the CMAPP model, at strategies used to persuade audiences.
- Among the persuasive reports that you're likely to need to produce are proposals—the focus of Chapter 11.
- Still common types of short, written communication include letters, memos, and emails, and, though now much less frequently used, faxes. Chapter 12 offers recommendations for successfully generating those items.
- Chapters 13 and 14 show how you can apply the CMAPP approach effectively to the development and delivery of professional presentations, and to the search for career employment.
- Finally, because the way you use language as an engineer is likely to affect how successful you are in your career, Chapter 15 presents a quick review of the most frequent types of language-use issues found in engineering communication—a brief "grammar compendium," if you wish.

Key Features

CASE STUDIES

At the end of Chapter 1, readers are introduced to four organizations that are featured in the book's interrelated case studies. The case studies allow students to examine the application of the CMAPP approach in the real world. The sample communications provided allow students to learn from successes and analyze common weaknesses.

DESIGNED WITH ENGINEERS IN MIND

Engineering Communication focuses specifically on the communication tools used by professional engineers. Meaningful insights and direction help students create proposals, reports, memos, letters, and job applications most appropriate for today's workplace.

EXERCISES

Carefully crafted exercises give readers opportunities to apply the chapter's concepts. Some exercises help students review what they have learned. Others require them to take on roles associated with the case study organizations. Still others require students to interact with people in the real-world marketplace.

ETHICS AND CULTURAL CONSIDERATIONS

Engineering students become aware of the importance of both ethics and cultural considerations as they develop effective communication.

USEFUL WEBSITES

A convenient list at the end of every chapter provides useful websites that relate either to the chapter's concepts or directs readers to study more about relevant communications issues. This enables students to further explore areas of interest.

New to the Second Edition

NEW CHAPTER ON DIGITAL AND SOCIAL MEDIA

Students will learn to conduct research and professionally network using Facebook, Google+, Twitter, LinkedIn, YouTube, blogs, and mobile apps. Students will learn about open-source collaboration, how to share images, and how to create social media hubs.

NEW SECTION ON BUSINESS PLANS

Because modern engineers often become part of the business planning process, this new section and related exercises emphasize key elements for success. Students will learn about executive summaries, business overviews, risk and contingency plans, financials, and how to work with sales and marketing, operations, management, and human resources.

INCREASED COVERAGE OF THE CMAPP MODEL FOR WRITING

Readers will follow this highly successful CMAPP model for written and oral communication that emphasizes context, message, audience, purpose, and product.

EXPANDED SECTION ON JOB APPLICATION PROCESS

Readers will review effective methods for career searches with this edition's new focus on the important role of LinkedIn for professional advancement.

EXPANDED MATERIAL EMPHASIZING ONLINE SEARCHES

Students will become familiar with professional, time-saving online resources useful in locating key information, including RSS feeds, such as Feedly, as well as online alerts, such as Google News Alerts.

EXPANDED TREATMENT OF REVISION AND EDITING

Readers will learn to share ideas and concepts through effective communication tools as this edition helps learners master key writing skills.

Ancillaries

MINDTAP ONLINE COURSE AND READER

This textbook is also available online through Cengage Learning's MindTap, a personalized learning program. Students who purchase the MindTap have access to the book's electronic Reader and are able to complete homework and assessment material online on their desktops, laptops, or iPads. Instructors who use a Learning Management System (such as Blackboard, Canvas, or Moodle) for tracking course content, assignments, and grading, can seamlessly access the MindTap suite of content and assessments for this course.

With MindTap, instructors can

- personalize the Learning Path to match the course syllabus by rearranging content or appending original material to the online content.
- connect a Learning Management System portal to the online course and Reader.
- customize online assessments and assignments.
- track student engagement, progress and comprehension.
- promote student success through interactivity and exercises.

Additionally, students can listen to the text through ReadSpeaker, take notes in the digital Reader, study from or create their own Flashcards, highlight content for easy reference, and check their understanding of the material through practice quizzes and gradable homework.

The MindTap for *Engineering Communication* also includes video case study examples and exercises, enabling students to examine and analyze oral forms of communication.

INSTRUCTOR COMPANION SITE

Everything you need for your course is in one place! This collection of book-specific tools is available online via www.cengage.com/login.

ACKNOWLEDGMENTS

The authors wish to acknowledge the continued love and support of Bettie Ingre and Miles Basil.

Thanks are also in order to the following reviewers who offered valuable feedback for the creation of the second edition:

Michael Berry, Montana State University

Richard Dippery, Jr., Kettering University

Diogo Santos, New Jersey Institute of Technology

Gareth Todd, Ferris State University

The authors also wish to acknowledge and thank our Global Engineering team at Cengage Learning for their dedication to this new edition: Timothy Anderson, Product Director; Mona Zeftel, Senior Content Developer; Jennifer Risden, Senior Content Project Manager; Kristin Stine, Marketing Manager; Elizabeth Murphy, Engagement Specialist; Ashley Kaupert, Associate Media Content Developer; and Teresa Versaggi and Alexander Sham, Product Assistants.

They have skillfully guided every aspect of this text's development and production to successful completion.

CHAPTER 1

Building a Foundation

1–1 The Relevance of Engineering Communication

Since this book is about engineering communication, you might think that if you don't intend to write instruction manuals or long technical reports, it won't be useful to you. But, despite its name, the field of engineering communication includes

- emails and other written communication such as letters or memos to, from, between, or among people who work in different kinds of companies, organizations, or associations—from multinational conglomerates to home-based businesses to charities such as the American Cancer Society;
- a variety of digital communication, from "professional" social media such as LinkedIn, to websites to PowerPoint slides to CAD drawings and even to some SMS texts;
- advertising and promotional material, from magazine ads to business cards;

- a host of other documents, from the annual report of a firm the size of General Motors to your own IRS tax return, and from the operating and repair manuals for specialized equipment such as PET scanners to the last parking ticket you received;
- oral communication, from a formal "sales pitch" to a large group of prospective clients to the informal explanation of your opinion at a small meeting.

Professional engineers of all kinds typically spend at least half their working time communicating—orally or in writing. They communicate with other engineers; with professionals in other fields; with nontechnical managers; with city, county, state, and federal officials; with clients and potential clients; and with the general public. The impact of engineers' communication can be enormous—the lives of the people who make use of engineers' work can depend on how well those engineers have communicated their ideas, their cautions, their specifications, or their recommendations.

The Accreditation Board for Engineering and Technology (ABET, at www.abet.org) is likely to play a significant role in your engineering career. One of the student outcomes of ABET's 2014–2015 Criteria for Accrediting Engineering Programs includes "an ability to communicate effectively." You'll need to use engineering communication throughout your academic and professional careers, and the more effectively you do so, the greater the likelihood of your success. Suppose that you and your friend both study hard in a particular course and have some valuable ideas to contribute. Now, imagine that you've learned to express your ideas (both orally and in writing) precisely and effectively, but your friend hasn't. Which of you is likely to get the better grades? Similarly, if you were an employer deciding between two applicants with roughly equivalent engineering qualifications and experience, and only one demonstrated communication skills that would represent your organization properly and impressively, which applicant would you hire?

So, as you read this book, think of it not as a complicated explanation of a subject of little interest to you, but as a set of practical guidelines that can boost your chances of success in a wide range of areas.

1–2 Characteristics of Engineering Communication

Over the past several years, engineering communication, along with other forms of technical communication, has been developing a style of its own. This is particularly true of written material. We can clarify some of the distinctions between engineering communication and what we will loosely label "traditional prose" by examining five salient features of the former: necessity for a specific audience, integration of visual elements, ease of selective access, timeliness, and structure.

1–2a NECESSITY FOR A SPECIFIC AUDIENCE

Much traditional prose is what we might term *author driven*. Someone makes a discovery, has a revelation, develops a theory, wants to share feelings with others, wishes to entertain, or simply believes that a particular story will generate profit by appealing to a large number of anonymous readers—and so begins to write, believing that there will be "an audience out there." But no one gets up one morning and says, "Today, I want to write the definitive steel strength table," or "Today, I'll fulfill my dream of writing a lengthy report on soil mechanics."

Technical communication is *audience driven*. People create it to respond to a *specific* audience's need for information. Whether you are writing a letter, a report, an email, or a set of system specifications, you always tailor it to a specific audience. In fact, if you don't already know exactly who your audience *is*, you really don't have anything to write or say. The better you know your audience, the more effectively you'll be able to communicate with them. And, of course, the converse will also be true: The less precisely you can identify your audience and its characteristics, the less effectively you'll be able to communicate with them.

1–2b INTEGRATION OF VISUAL ELEMENTS

Great literature would be just as great if it were handwritten on loose-leaf pages rather than printed in a book. By contrast, the effectiveness of a report recommending the purchase of a more powerful processor, or of a proposal for a new avionics system, often relies at least as much on visual presentation—whether in "hard" or "soft" copy—as on content. In our context, the term "visual elements" refers to everything from illustrations such as diagrams or charts to headings and type. Careful integration of ideas and their presentation is essential to effective engineering communication.

1–2c EASE OF SELECTIVE ACCESS

Literary authors normally expect you to read every word; they don't want you to skim quickly, looking only for the main points. They assume that you, as a reader, are willing to devote your full attention to the writing, from start to finish. At the same time, you recognize that your understanding of a literary text is likely conditional on your having read all of it.

By contrast, engineering and other technical communication writers assume that readers will have other demands on their time and, as a result, may need to quickly identify only the principal points in a text, perhaps returning later for a more careful reading. (Obviously, this idea will not apply the same way to all engineering communication, such as concrete analysis specifications or technical instruction manuals. Skimming such documents, even on a first reading, might not be recommended.)

If you are checking your office email, for example, you're likely to first glance quickly at the header to see who sent the message and what it's about. In checking your advertising mail, you'll tend to scan for headings or other prominent words or phrases and, based on what you find, either delete the item or put it aside for later perusal. Good professional communication allows the engineer's reader to make the choice without penalty. And it does so in large part through the judicious integration of the visual elements mentioned previously.

In short, effective engineering communication makes it easy for a reader to gain access to specific sections of a written document, to select and read only those parts important to that reader. For instance, in a formal recommendation report regarding the expansion of parking facilities at an office complex, the owner of the complex might need to read only the report's Executive Summary, located near the front of the report. The owner's accountants, in contrast, would need to study the financial tables toward the end of the report. Each of these sections will be clearly identified through headings and/or via the table of contents, for ease of selective access.

1–2d TIMELINESS

Many people believe that the world's great books are timeless; on the other hand, you know that the technicians' manuals for an Apple II or a rotary-dial telephone are so far past their "expiry dates" that they have become little more than museum curiosities. By the middle of January, the newspaper ad for Home Depot's Christmas sale has no practical value. And once you have received your merchandise, your *shopping cart* information at walmart.com has been reduced to bits and bytes of cyber-rubbish. The useful life of most engineering communication is relatively short: It is usually over as soon as the reality it addresses changes.

1–2e STRUCTURE

In high school or in another college course, you might have been taught that a paragraph must have a topic sentence, one or more supporting sentences, and a concluding sentence. You might also have been told never to begin a sentence with "and" or "but." In contemporary engineering communication, these rules often don't apply. If you can express the idea of a paragraph in one sentence, you should do so. Also, in the interest of brevity and ease of access, you may well replace topic sentences with headings or subheadings. And most engineering and technical writers accept the practice of beginning some sentences with "and" or "but."

Furthermore, although all good writers vary sentence length and structure to avoid monotony of style, most engineering writing tends to use relatively short sentences.

1-3 Ethics

The question of ethics cannot be restricted to philosophy or religion courses. It has practical applications in education, commerce, government, engineering, and other technical communication. We refer you once again to ABET's Criteria for Accrediting Engineering Programs, which also includes a required student outcome of "an understanding of professional and ethical responsibility."

1-3a DEFINITIONS

The Merriam-Webster Online Dictionary (www.merriam-webster.com) defines "ethics" as "an area of study that deals with ideas about what is good and bad behavior; a branch of philosophy dealing with what is morally right or wrong; a belief that something is very important."

There is no consensus as to the meaning and implications of "morally good and bad." (Do you and your family and friends, for example, always agree on what is meant by "moral"?) Does it mean the same to everyone in every culture—or even in every part of your own state? Has its meaning changed over time? Over the centuries, philosophers have wrestled with these questions. More recent attention has been focused on the development and meaning of medical ethics, journalistic ethics, business ethics (which some people facetiously consider an oxymoron), and, of course, engineering ethics. Can we derive from a confusing, but important, maze of ideas any practical guidance for those who engage in engineering communication?

1-3b CODES OF ETHICS

The importance of ethics in the conduct of business and technology is broadly accepted. In fact, not only professional associations, but also many business organizations publish codes of ethical conduct. Several years ago, for example, a group of international business leaders, primarily from the United States, Europe, and Japan, developed what they called the Caux Round Table Principles for Business (see www.cauxroundtable.org), a document that is effectively a code of ethics, and includes the following:

Principle 3 - BUILD TRUST BY GOING BEYOND THE LETTER OF THE LAW

- A responsible business recognizes that some business behaviors, although legal, can nevertheless have adverse consequences for stakeholders.
- A responsible business therefore adheres to the spirit and intent behind the law, as well as the letter of the law, which requires conduct that goes beyond minimum legal obligations.
- A responsible business always operates with candor, truthfulness, and transparency, and keeps its promise.

Today almost all engineering associations and societies subscribe to a code of ethics. Many, such as the American Society of Civil Engineers (ASCE, at www.asce.org), the oldest professional engineering association in the United States, have chosen to adhere to ABET's Code of Ethics of Engineers. The second of ASCE's Fundamental Principles requires that "Engineers uphold and advance the integrity, honor and dignity of the engineering profession by . . . being honest and impartial, and servicing with fidelity the public, their employers and clients." The third of its Fundamental Canons states that "Engineers shall issue public statements only in an objective and truthful manner."

In 2010 the Institute of Electrical and Electronic Engineers (IEEE, at www.ieee.org), whose Internet home page professes it to be "The world's largest professional association for the advancement of technology," approved a code of ethics that includes the mandate "to disclose promptly factors that might endanger the public or the environment" and "to be honest and realistic in stating claims or estimates based on available data."

1–3c APPLICATIONS

However we define ethics, most of us have a sense of what we consider right and wrong, and most of us would agree that most of the time we should try to apply the Golden Rule: "Do unto others as you would have them do unto you." Most successful people seem to believe that treating others as justly as possible is one of the cornerstones of their success. How might this approach be reflected in engineering communication? Here are a few examples.

Honesty

Many of us know someone who lied on a resume and so got the job. We often forget that in most cases the lie is eventually discovered and the person loses not only the job but also his or her reputation. Telling someone that "the check is in the mail" when it isn't, or "I already emailed you the assignment" when you didn't, may bring short-term benefits; but, if you persist in this kind of deception, you'll soon come to be thought of as dishonest and will find it difficult to maintain professional relationships. Being dishonest with others is illegal in some cases; for most people, it is always unethical.

Accuracy

An extremely important criterion for engineers' communication is accuracy. If your document or presentation contains inaccuracies, and if someone in your audience notices the errors, your entire message—and likely any future ones—is compromised. Whenever you communicate, your credibility is on the line; regaining lost credibility is very difficult indeed. If the lack of accuracy in what you communicate is intentional, the issue becomes one of honesty and thus of ethics.

Emphasis

Whether you're trying to sell a product or express an opinion, you should certainly try to present your information in the best possible light. You might do this, for example, by emphasizing the advantages of your design rather than its high cost. Similarly, in describing your preferred option, you might use more forceful vocabulary than you use in describing the alternatives. Here the question of ethics tends to be one of degree. Only on purely technical matters is a technical audience likely to be swayed by neutral facts alone. Consequently, when you wish to persuade—and engineering communication often involves persuasion—you can sometimes emphasize some points over others, but you can never neglect to mention salient facts, even when they render the case you are making weaker.

For example, a client for whom author Bob Basil was once preparing a business plan, proposing to build wind farms on the coast of Finland, wanted to describe *only* the potential economic benefits of the project, including its location on a very windy coast. Bob, however, persuaded the client to accept that it would be unethical—and, ultimately, counter-productive—to omit mention of the opposition to wind farms that has previously been voiced by some coastal communities in Finland. (Other such projects in the area had killed some migrating birds.)

Creating Impressions

Whenever it conveys information, language also creates impressions. Consciously or unconsciously, audiences respond to the emotional impact of language—what we call its *connotation*—as well as to its objective meaning—what we call its *denotation*. Thus, the way you phrase a communication will invariably influence how your audience responds to it. (Think of the importance of spin doctors in political circles.) Admittedly, the importance of precise phrasing—as well as grammar, spelling, and punctuation—is frequently ignored in *texting*, in which a focus on speed and convenience often results in misunderstanding and discontent. Here is an example of very different impressions created by saying "the same thing" in different ways. Suppose that you are a manager in a company and your record shows that three-quarters of your decisions have proved to be good ones. To describe this batting average, your boss could write, in your annual performance review, that you were right 75 percent of the time *or* that you were wrong 25 percent of the time. Which phrase would you prefer to see recorded in your personnel file?

When it comes to creating impressions, the ethical path may be indistinct. In most cases, technical communication requires you to be as objective as possible, even though the nature of language itself makes it all but impossible to present communication that is entirely free of connotation.

1-3d CONCLUSIONS

As an ethical communicator, you have to be willing to put the needs of your audience before your own interests. Twisting language to camouflage an unsavory truth may produce the results you want at that moment; but, you should not ignore the potential repercussions. Rarely are ethical people comfortable with the rationalization that the end justifies the means. Whenever you are communicating, try asking yourself how you would react if you were the audience. Would you feel that you were being treated fairly and respectfully? If you can't honestly answer yes, it's likely that as a communicator you haven't given sufficient weight to the ethical aspects of what you're doing.

1-4 Cultural Impact

Working as a professional in the United States means that you will be working with people from a wide variety of cultural and religious backgrounds. Indeed, more and more American engineers work in other countries. This means that, to be successful in your career, you need to become sensitive to the different ways that communication is commonly effected and viewed in different cultures. In short, you will need to consider the possible pitfalls of intercultural communication.

1-4a CULTURAL PREFERENCES

Just as different people respond differently to one another, different groups tend to react differently to certain types of behavior. For example, in the United States the use of first names in newly established business relationships has become very common. Presumably, this behavior is designed to promote an impression of friendliness and conviviality. It is not, however, universally accepted. In a number of countries, such as Canada and Great Britain, for example, many people believe that business relationships should be more formal, and that the appropriate form of address—even among people who know each other quite well—is a last name preceded by an honorific such as "Ms." or "Mr." Until you're aware of your audience's preferences, therefore, you probably run less risk of being thought impolite if you err on the side of formality rather than casualness.

Another characteristic of American business dealings is the value placed on directness. Thus, brevity and concision are seen as desirable qualities in most professional communication. In some other cultures, these same qualities would be regarded as brusque, curt, or abrasive. In Japan, for example, tradition still tends to dictate a much more roundabout approach, in which ideas are conveyed through implication rather than stated explicitly.

An analogous issue could be represented by your planning a meeting with your subordinates. Presumably, you're not likely to call a meeting for Thanksgiving or Christmas Day. However, evangelical Christians, orthodox Jews, highly observant Muslims, or baptized Sikhs might be very uncomfortable if you scheduled the meeting for a day that was one of their important religious holidays. (One of our students once created an online calendar showing those holidays celebrated by the many and various local religious communities. The audience for this calendar were human-resources professionals who need to be sensitive to employee needs. Not a week went by without one or two important holidays.) Your lack of cultural awareness might be perceived as prejudice—however inadvertent on your part. Again, remember that people from different backgrounds may react to what you say or write differently from the way you intended.

Spelling and vocabulary should also be considered. When writing to a client in England, for example, which should you use—American or British spelling ("center" or "centre," for instance)? Again, consider both your audience and your purpose. You may make a point of using American spelling (e.g., "favor," "honor") when you want to draw attention to the fact that your firm is American; conversely, you might choose British spelling ("favour," "honour") if you don't think that fact is really relevant and therefore don't want to make it conspicuous. You should also be aware of differences in vocabulary. Again, if you are writing to someone in England, you will be better understood if you use the words "lift," "boot," and "biscuit," rather than "elevator," "trunk" (of a car), and "cookie." Nor should you be shocked if someone from England offers to "knock you up" at 7:00 a.m. It's simply that person's way of proposing a wake-up call.

Although humor plays at best a minor role in engineering communication, you should exercise care in its use. The same caution applies to references to religion and politics. North Americans tend to approach such allusions with what for some cultures is inappropriate informality and familiarity. Unless you are certain of your audience's reaction, you should avoid potentially offensive references in your communication—in other words, stick to the facts, and try to be aware of cultural differences.

The term "personal space" is often used to refer to the physical distance we like to maintain between ourselves and those with whom we are speaking. Different cultures have different norms. Many Europeans and South Americans, for example, prefer much less personal space than most Americans are comfortable with. When you see two people from different cultures talking, you might observe one advancing to decrease the personal space and the other backing up in an effort to increase it. Although the issue of personal space does not have a direct bearing on written communication, it is a good idea to be aware of the cultural differences involved.

A related issue is eye contact. Most Americans believe that looking someone in the eye while conversing is an indication of honesty and forthrightness. In some cultures, however, such behavior may be viewed as disrespectful, rude, or presumptuous. Although you should make eye contact with your audience during a presentation, you should not assume that your audience's failure to reciprocate implies shiftiness or deceit. Here again, knowledge of your audience will help you communicate more effectively and develop better professional relationships.

1–4b CULTURAL REFERENTS

People, ideas, and things that form part of the popular culture and become ingrained in our thinking and in our language can be referred to as *cultural referents*. As a communicator, you must consider whether your audience will understand the cultural referents that you take for granted. For example, an audience that had never been exposed to the Grade 1 primers common in America in the 1950s and 1960s (and you, yourself, are likely to be unfamiliar with them) would probably not know that a "Dick and Jane approach" is one that is overly simplistic. The diverse ethnic fabric of our society means that you must pay attention to the cultural referents you use, and consider whether they are relevant to your intended audience. If your audience doesn't relate to them as you do, you are not communicating effectively.

1–5 The Issue of Gender

To be effective communicators, engineers also need to be aware of linguistic change. Such change is dramatically evident in the increased use of nonsexist language in recent years. The article "Sexism in Our Language" (see Figure 1.1), which appeared in 1990 in a business periodical that targeted SOHO (small office/home office) operators, addressed the issue. As you read it, you might want to think about how much—or how little—has actually changed in some 25 years.

FIGURE 1.1: Sexism in Our Language

In recent years, people have been focusing greater attention on a topic commonly known as "sexism." For distressingly good reason, sociologists, social reformers, activists, politicians, and others have studied and commented on the ways our society distinguishes between men and women.

Did you notice the order of the last three words of the last sentence? We have been taught that it is proper form to say "ladies and gentlemen" rather than "gentlemen and ladies," and most of us would more naturally write "men and women" than "women and men." Both practices are socially learned. Are both sexist? To what extent do societal distinctions between the sexes exist in the language itself? Can we change them? What happens to the language when we do?

These are highly complex questions. They are also very loaded ones. Few people are neutral; once they have begun to look at the issue in any depth, fewer still expect quick solutions. What we can do here is look at a few examples of systemic gender discrimination in English, and point out some of the strategies for dealing with them. Since a great deal of the problem seems to reside in the words "man" and "woman" and their compounds, it might be worthwhile seeing where these words come from.

The word "man" can mean "man" as opposed to "woman," as well as "man" as opposed to "other animal species." "Man" derives from the old Anglo-Saxon word *mann*, which was originally applied to both sexes but eventually took on its modern meanings. "Woman" comes to us from another Anglo-Saxon word, *wifman*, literally "wife [or "woman"] man." (Its plural, *wimmen*, explains our modern pronunciation of "women.") Thus, it would seem that defining women in terms of men is in a sense entrenched in the language.

Another social issue that has affected English is employment. Until fairly recently, a number of fields were considered "men's work"; others (and, more often than not, those of lesser status) were seen as "women's work." So we get words in the language like "policeman," "fireman," and "chairman," contrasting sharply with others like "charwoman" (which is always strongly pejorative) and "midwife" (which derives from Old English, meaning "with woman").

Two other things to remember: First, English has a number of words that don't show gender; that is, they don't tell us whether we're talking about a man or a woman. These are words like "friend," "colleague," and "person." Second, English has words that show gender in the singular but not in the plural. Among them are the pronouns "he" and "she" (as opposed to "they") and their derivatives ("him," "his," "her," "hers," as opposed to "their" and "theirs").

When we speak (and often, when we write), we may not want to specify whether we're talking about a woman or a man. To avoid being forced to by the language, we sometimes take some slightly clumsy detours. For example, you may hear a sentence such as, "A colleague of mine always carries a spare key in their pocket." We find more problems with two sentences together: "I was talking to a colleague last night. They tell me there's a new contract coming." Although everyone understands what these sentences mean, the language is still often considered substandard: we are not (yet) really supposed to use a plural to refer to a singular.

One way around the difficulties in such sentences is to rephrase. For example: "A colleague of mine always carries a spare key." This revised version gives all the information that's likely necessary and avoids the problem of using a plural to refer to a singular. The other example would be a bit more formal (and acceptable in writing) if it were changed to "A colleague to whom I was talking last night tells me a new contract is coming."

Particularly in the public sector, people have been recommending ways to remove sexist language. Basically, they have tried three strategies. First is the one mentioned in the last paragraph: rephrase your sentences to avoid the problem.

Second is the use of an initial note of some kind, such as: "All references to the masculine in this document should be interpreted as applying equally to the feminine." This "solution," in fact, may be worse than the problem. Not only is it irritatingly bureaucratic, but it emphasizes the very distinction it is supposed to diminish.

Third is the substitution of words or the creation of entirely new terms. As far as the pronouns are concerned, we see slightly peculiar forms such as "he/she," "him/her," and so on. (I have also seen "(s)he," although never "her(m).") This route generates a visually curious document and causes obvious problems if you try to read it aloud.

Words like "ombudsperson" and "chairperson" or "chair" have become commonplace. Other creative substitutions include "flight attendant" instead of "stewardess" or "steward"; "server" rather than "waiter" or "waitress"; and "police officer" instead of "policeman" or "policewoman." As well, there is a tendency for some pairs of words to coalesce. Thus, "actor" and "singer" are commonly used for both men and women, to the exclusion of "actress" and—almost never seen now—"songstress." As our society changes, we're likely to see more and more such attempts to remove sexism from the language.

Adapted from *Hats Off to English* by David Ingre, published by David Ingre Written Image Services, 1990. © David Ingre.

The many years since the article's publication have seen continued efforts to replace sexist language with non-gender-specific terminology. Many books on writing now feature lists of gender-neutral words or phrases that can be used to replace gender-specific vocabulary. For example:

INSTEAD OF	USE
businessman/businesswoman	businessperson
businessmen/businesswomen	businesspeople
foreman	supervisor
mailman	mail carrier
man-made	synthetic
manpower	personnel

As well, the use of "they" when referring to a singular antecedent has gained almost total acceptance, even in professional writing. Consider this example: "When a geologist returns from an exploration site, they send rock samples to a certified lab." In the past this sentence would have been "fixed" in one of two ways: "a geologist returns" would have been changed to "geologists return," or "they" would have been transformed to the unwieldy "her or she."

In business letters, the use of salutations such as "Dear Personnel Manager," as an alternative to "Dear Sir" or "Dear Madam," has likewise become almost universal.

Whether you are dealing with issues of cultural difference or gender bias, as an effective communicator you are required to know what current acceptable practice is. This, in turn, means constructing your message with careful consideration of your context and your audience.

1-6 Correct Language

What we consider "correct" in terms of language use in English is likewise largely dependent on culture. Differences in acceptable vocabulary (and occasionally in structure) exist among the varieties of standard English spoken in the United States, Canada, England, South Africa, Jamaica, India, and the many other places where it is a common first (and often second) language. Further, acceptable usage within a particular language community changes not only over time, but very much according to context.

The following should help you determine what is correct for engineering communication today.

1-6a DEFINITION

We'll define correct language as "what the majority of educated, native speakers tend to use in a particular context."

1-6b EXPLANATION

We specify native speakers because, when you learn your first language, you acquire a "feel" for what is acceptable. You have an innate sense of how people would or would not say something. Acquiring this "feel" for a second language is certainly possible; however, it is usually a painstaking and time-consuming process. If you are a native English speaker, you will likely find it easier to acquire good technical communication skills in English. Interestingly, however, we have noticed that many nonnative speakers have a better command of traditional English grammar, something that seems to be taught less and less, at least in American elementary and secondary schools. Nonetheless, it is the native speakers who really define their language.

We mention *educated* because there is usually a correlation between the amount of formal education you've been lucky enough to obtain and the breadth of your exposure to different people, ideas, and vocabulary. So far as linguists know, all languages change over time. Simplistically speaking, that explains how Latin eventually produced Spanish, French, Italian, Portuguese, and Romanian, and, analogously, why we can identify a Boston or a Nashville accent. Moreover, diachronic linguistics (basically, the study of how language changes over time) has suggested that in most cultures, it has been the more educated classes who have had most to do with how the language they use changes. (The United States, by the way, might be an unusual exception. According to the noted American linguist and author, H. L. Mencken, American speech seems to have been more influenced by the cultures of black slaves, poor recent immigrant communities, and impoverished whites, than by the country's educated classes who have adopted such speech.) At any rate, our definition refers to vocabulary and grammar currently used by the majority of educated native speakers.

We talk about the *majority* because not everyone uses language the same way. In fact, every individual possesses what's called an *idiolect*, a personalized version of the language whose details differ from those of every other speaker. It is their idiolect, in effect, that generates the *style* that we can attribute to particular writers, from Harriet Beecher Stowe to Frederick Douglass to Walt Whitman, and from Isaac Asimov to Maya Angelou to Barbara Kingsolver. Obviously, though, there are significant commonalties that we recognize and accept; otherwise, we wouldn't be able to communicate with each other.

The word "tend" is in the definition because we can express ourselves in a multitude of different ways to deal with specific circumstances. Overall, however, we do show tendencies. For example, contemporary English tends to put adjectives before nouns. However, think of common expressions such as "oysters Rockefeller" or "court-martial" or "code red" or "president pro tem." All of them put the adjective (including the last one, which is, in fact, two Latin words) *after* the noun. As well, we tend to indicate size with the words "big" or "large," rather than "great," which usually suggests *importance*. Think of the difference between a *big building* and a *great building*, and of how we refer to the "*Great* Seal of the United States."

Finally, we stipulate a particular context because we use language differently depending on the context. Imagine the following scenario. You are a manager in your company, and over the weekend you have been preparing for a presentation you'll be giving on Monday morning to the executive board of Acme Inc., whom you hope to get as a client. It's now 6:15 a.m. on that Monday morning. Having finished your breakfast, you're sitting at your kitchen table, reviewing your notes and the handouts you've prepared for your 9:00 a.m. presentation. Suddenly, your three-year-old son, unexpectedly out of bed, runs into the kitchen, smiles adoringly, cries "Hi!" . . . and spills orange juice over everything on the table. Now, the language you're going to (try not to) use with your little boy is automatically different from the language you'll soon use to the Acme executives to explain the unusual coloring and sticky substance on some of their handouts. Notice: What would be correct language with your three-year-old would be incorrect with the executives, and vice versa. Analogously, note that what is entirely correct usage when talking excitedly with friends over a beer in a tavern would be quite inappropriate when interviewing for a promotion—and vice versa.

1-6c ADVICE

Effective engineering communication requires you to remain current with what is acceptable, and what is not. Once again, this means learning about, and paying attention to, your context and your audience—in effect, applying the communication model you will examine in detail in the next chapter.

CASE STUDIES

In the subsequent chapters of this book, you will look at examples of technical communication involving one or more of four fictional organizations that are, nonetheless, based on real companies and institutions.

Each case study is introduced in a section titled *Situation*, which presents an example of flawed communication produced by people within the organizations under consideration. The section titled *Issues to Think About* raises questions deriving from an examination of the case study. Some cases conclude with a section called *Revision*, which lets you examine a better-constructed version of the flawed communication shown earlier.

Here are brief profiles of the four organizations.

1a Grandstone Technical Institute (GTI)

Located in Randolph, Vermont, GTI offers two-year diplomas and bachelor degrees in fields such as computer engineering technology and civil engineering, as well as internship programs in areas such as automotive technology, electronics, and welding. Because of GTI's reputation for excellence, its graduates are often welcomed by companies across the country.

1b Ann Arbor University (AAU)

Situated in Ann Arbor, Michigan, AAU offers both undergraduate and graduate degrees, principally in arts, law, and engineering. One result of the university's successful fundraising campaigns was the construction of a new Student Association Building. The Ann Arbor University Student Association Executive (AAUSAEx) remains very active in promoting cooperation among student associations across the country.

1c Accelerated Enterprises Ltd. (AEL)

Some 20 years ago, two young Lansing, Michigan engineers, Sarah Cohen and Frank Nabata, opened a small consulting practice that they optimistically named Accelerated Enterprises Ltd. Over the next two decades, business prospered. Cohen and Nabata (now married to each other) took in a number of associates and eventually opened branches in New York, Detroit, Dallas, and San Francisco. With the expansion came broadened expertise, and AEL is now involved in chemical, civil, and geological engineering, as well as in software design and development. While the original partners still oversee the firm's overall strategy from their now-spacious offices in Lansing, they believe in allowing their branch associates significant latitude in running the local operations. As well, AEL tries to maintain good working relationships with several colleges and universities, viewing them as important contributors to the country's economic growth, and as likely sources for future AEL consultants, employees, and indeed clients as well.

1d Radisson Automobiles Inc. (RAI)

Headquartered in Dallas, and with locations in Atlanta, Boston, Chicago, Denver, Los Angeles, and Seattle, RAI is a highly successful automobile dealership chain.

Over 30 years ago, Maurice Radisson, an energetic, successful entrepreneur from Baton Rouge, moved to Dallas, where he opened a car sales business that he named after himself. Over the years, RAI prospered, and, about 10 years ago, Maurice turned the stewardship of the business over to his eldest son, Griffin, who held an automotive technology diploma from GTI and an MBA from AAU. As president and chief executive officer, Griffin built his father's Dallas dealership into a wealthy and respected cross-country network. Though still known as a somewhat authoritarian hands-on manager, Griffin is a firm believer in being a good corporate citizen and in maintaining relationships with colleges and universities.

EXERCISES

1.1 Give four or more examples of what could be classified as engineering communication.

1.2 Describe at least one real-world situation in which engineering communication skills would serve you well.

1.3 Briefly explain the significance of the following characteristics of effective engineering communication:

a. necessity for a specific audience
b. integration of visual elements
c. ease of selective access
d. timeliness
e. structure

1.4 Briefly discuss the importance of ethics in technical communication, referring specifically to the issues of honesty, accuracy, emphasis, and the creation of impressions.

1.5 Discuss how ethics played a role in an example from your own experience.

1.6 Talk to someone in business or government who is involved in hiring new employees, and report back to your class on what that person thinks are the most important ethical issues in his or her workplace.

1.7 Give two or more examples of cultural referents that you take for granted, and suggest alternatives for an audience that might not understand them.

1.8 Provide two or more examples of gender-specific words or phrases not mentioned in this chapter; beside them, list gender-neutral alternatives.

1.9 Imagine that you work for a medium-sized company with about 500 employees. The company's management believes that employees could benefit from some diversity training to better deal with the different types of employees in the workplace. You have been asked to serve on a committee to provide suggestions for overcoming barriers to communication that seem to derive from the cultural diversity in your workplace. What suggestions would you offer to the other committee members? Consider also how the diversity training should be delivered. Should employees be required to attend seminars, or will a memo or newsletter do the job? Include your suggestions for distributing the information along with your specific ideas about overcoming communication barriers.

USEFUL WEBSITES

Note: For ease of reading, we have removed the "http://www." from the URLs below. However, some browsers might still require that you insert it to precede some URLs so as not to receive an error message.

■ An informative example of a field-specific code of ethics is that of the National Society of Professional Engineers.
Visit: nspe.org/resources/ethics/code-ethics

■ Another such set of guidelines is the "Software Engineering Code of Ethics and Professional Practice," set out by the IEEE and Association for Computing Machinery.
Visit: computer.org/cms/Computer.org/Publications/code-of-ethics.pdf

■ The American Society of Civil Engineers (asce.org/) likewise subscribes to "The Engineering Code of Ethics" set out by ABET.
Visit: courses.cs.vt.edu/~cs3604/lib/WorldCodes/ASCE.html

■ The Online Ethics Center for Engineering and Science of the National Academy of Engineering offers a good set of resources relating to ethics and engineering.
Visit: onlineethics.org/

■ The home page of the Society for Technical Communication (STC) offers a broad variety of links to STC resources around the world, as well as information on educational resources, seminars and conferences, and employment opportunities.
Visit: stc.org

■ The University of Maryland's Office of Diversity and Inclusion offers links to a variety of multicultural and diversity resources.
Visit: diversity.umd.edu/resources.html

CHAPTER 2

The CMAPP Analysis

If you wanted to find out the effect on Galveston of a two-foot rise in the world's sea level, you wouldn't try to melt the polar ice cap and then visit the Gulf of Mexico; you'd try to find a computer model that would predict the likely consequences. If you're an engineer working on a new passenger airliner, you don't start fabrication before running computer models. Analogously, when we want to look at how engineers communicate effectively, we can use a communication model.

2–1 Transactional Communication Models

Various communication models have been developed over the years. Figure 2.1 shows a simple transactional model, so called to reflect the two-way nature of communication.

The model, which in principle works for all types of oral and written communication, has the following characteristics:

1. The originator of the communication (the sender) conveys (transmits) a message of some sort to someone else (the receiver).

FIGURE 2.1: A Simple Transactional Model

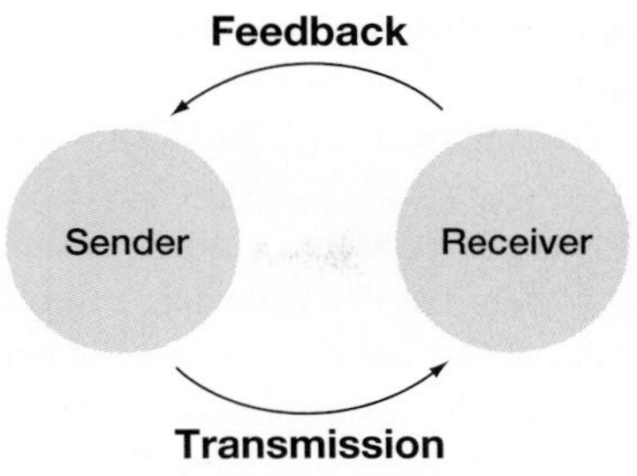

2. It is a linear model: Communication always proceeds in one direction—from sender to receiver. Note that if the original receiver responds to the original sender, their roles are reversed, but the "direction" of communication remains the same.
3. The transmission vehicle might be face-to-face speech, correspondence, telephone, email, and so forth.
4. The receiver's reaction (e.g., body language, verbal or written response)—the feedback—can have an effect on the sender, who may then modify any further communication accordingly.

As an example, think of a face-to-face conversation you have had with a friend. As sender, you mention what you think is a funny comment made by another student, named Maria. (Note that the basic transmission vehicle here is the sound waves that carry your voice.) As you refer to her, you see your friend's (the receiver's) face begin to cloud over, and you remember that your friend and Maria strongly dislike each other. This feedback makes you decide to start talking about something else. The model thus demonstrates an ongoing transaction between sender and receiver, conditioned by both the type and effectiveness of the transmission and the impact of the feedback.

A more complex transactional model appears in Figure 2.2.

FIGURE 2.2: A More Complex Transactional Model

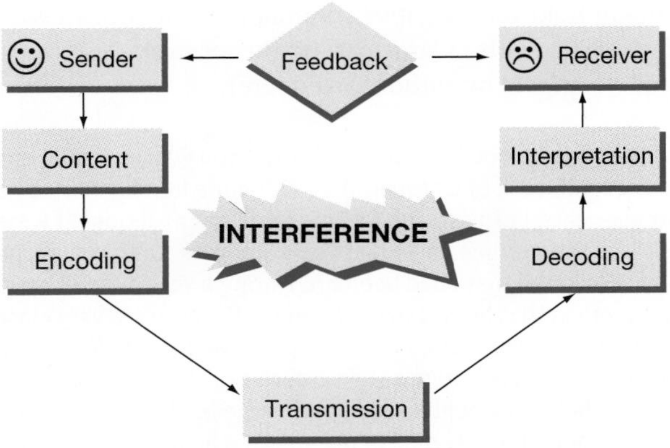

Let's suppose that you're going to be the sender. The first thing you need to do is decide exactly what it is you want to communicate. And that, in fact, is often the hardest task. All too often, we sit down at a keyboard and start typing, thinking something like, "I have to write to so-and-so about something to do with something about that kind of thing, approximately, I think . . ." Naturally, we don't do very well. A quip in common circulation years ago was, "Caution: Be sure to engage brain before putting mouth into motion." Whether you want to communicate orally or in writing, you need to know—before you begin—precisely what it is that you want to say or to write. That's what this model calls *Content*.

The next thing you need to do is *encode* that content—in effect, put it into language. In some circumstances, that might mean creating your content in English and/or another language. However, in whichever language you *are* using, you'll need to choose the right words, the right expressions, and the right grammatical structures.

Now, you'll need to *transmit* that *encoded* content. In the transactional model, we talk about *transmission channels* or *transmission vehicles*. They're not meant to be "rocket science." In a face-to-face conversation between two speakers of the same language, the principal transmission vehicle is sound (we say "principal" because gesture, for example, can also be involved). That means an *oral* (speech)/*aural* (hearing) process. Since no one's oral mechanism is *perfect* and since no one's aural mechanism is *perfect*, some "bits" get lost in transmission. We can refer to that phenomenon as *interference*. People who study interference say that from 5%–15% of what the sender transmits can be lost.

We usually talk about *external* interference and *internal* interference. Examples of the former might include someone noisily entering the hall while a professor is speaking, several people coughing repeatedly, the power suddenly going out, or the electronic distortion common over smartphones and Skype connections. Examples of the latter might include distractions such as you're sleepy or hungry or more interested in your next class than in this one, you have a headache, or you're worried about your sick pet.

Whatever parts of the transmission remain intact now have to be *decoded* by the receiver. Does the receiver understand the language being used—English or Spanish, for example? Does the receiver know the meaning of the words being used?

Then, the receiver has to *interpret* the *decoded* content. This can be problematic because, in any language, words and expressions mean different things to different people. We usually talk about two kinds of meaning: *denotation* and *connotation*. The former might be described as the simple, empirical, dictionary definition. The latter includes this but also encompasses the images, feelings, and responses elicited on the part of the receiver. Let's illustrate with the common English word "home." Its denotation might be "the principal residence of one or more individuals." Its connotation, though, depends on you. If you were lucky and had a happy childhood, "home" might have very positive connotations: happiness, security, comfort, laughter, and so forth. If you had an unhappy childhood, though, "home" might hold very negative connotations for you: stress, anger, perhaps even danger. Interpretation of words depends in large part on the receiver.

Here is a true story (told by author Dave Ingre).

> This happened a very long time ago, and yet, it still evokes strong emotions in me. At the end of the second week of October of 1991, I quite literally buried my best friend. Even after all these years, I can see and feel my reactions as though it were yesterday. We had been friends for about nine and a half years. For most of that time, he had been my best friend, and I could tell that I had been his. During a very difficult couple of years in my own life, it seemed that he was my only friend. So, we were very close. Some two years before he died, I learned that he had a progressive disease. At the time, no one really knew what caused it; and, there's still no cure. So, I knew that it was going to be just a matter of time before I would lose him. When he finally succumbed, he weighed about 80 pounds. He quite literally died in my arms. I'd helped to dig his grave, and I helped to lay him in it. I'm not embarrassed to say that shortly after, I began to sob.
>
> Now, I should tell you that, three or four years before he died, when he was still hale, hearty, and active, he weighed about 85 pounds, was just over three feet high, was brown with black markings, had small floppy ears, and a little stub of a tail that seemed to go berserk when he saw me come back from work.

You're right: Dave was talking about his dog. He was a big brindle boxer named Captain; and everything Dave told you is true. But notice. As you were reading the story, you might well have been thinking, "Why would he be telling us this as part of an engineering communication course?" And, once you realized Dave was talking about his dog, if you happened to really like dogs (as Dave obviously does), you might have thought something like, "Well, that is just so sad." If you *don't* happen to like dogs much, you might have reacted by thinking, "Sure, but it's just a *dog*!"

There's nothing wrong with either reaction, of course. What's important, though, is the following. For the story, Dave chose words and expressions that tend to have high connotation. (Go back and look.) You all read the same ones, but you responded differently—because of *who you are*. Pretty obviously, then, the receiver's *interpretation* plays a very important role in the effectiveness of communication.

The process we've just looked at helps to explain the gap you might notice between the marvelous idea you want to communicate (the *smiley face* in Figure 2.2) and the rather different reaction you get from the receiver (the *sad face* in that figure). One of the things that *can* help to narrow that gap is *feedback*. For whatever reason, people communicate more comfortably and more effectively when there's the opportunity for constant feedback. Try the following. The next time you're on the phone with someone you know well, just listen. I'm sure you think that you *do* listen. But here we mean, do nothing *but* listen: no "uh huh . . . yep . . . oh, sure . . . great . . . awesome . . . OK . . . ," and so forth. Listen attentively but *silently*. Likely within 10–20 seconds (a *very* short time), you'll hear, "Hello?" The person on the other end of the line wasn't getting any feedback.

You should remember that when you send written communication, what the receiver gets is all the receiver has. There's no opportunity for immediate feedback—and that makes for more difficult and, usually, more ineffective communication.

Finally, understand that, in many ways, language is insufficiently precise for really accurate communication—even for engineers. You really cannot preclude *intentional misinterpretation*. However, as a practitioner of good engineering communication, part of your job is to try to lessen the likelihood of *unintentional* misunderstanding. In other words, you try to narrow the gap between the *happy face* and the *sad face*.

Here is a perhaps simpler example. In an exam, you have to answer a complex question. While you have a clear *idea* of what you mean, you have to find the right way to express it. You have to *encode* your idea in language that is logical, clear, and concise. External interference while you are encoding might include the coughing of other students and the hum of the fluorescent lighting. Internal interference could come from your nervousness during the exam or from your fatigue from having been up all night studying.

The transmission vehicle is the exam paper—the composition you are writing. An example of external interference at this stage would be your pen running out of ink.

Your professor—the receiver—will have to *decode* what you have written—that is, interpret your words and assess your knowledge. During this process, there might be external interference: from other people's nearby conversations or from the difficulty of deciphering your rushed handwriting. Internal interference might stem from your professor's irritation at the poor quality of the papers already marked.

In this example, feedback cannot be immediate: You will receive it only when you get your exam back. When that happens, your dissatisfaction with your mark might interfere with your understanding of your professor's comments. Finally, the late arrival of several students might annoy your professor and thus interfere with the delivery of his or her subsequent feedback.

2–2 The CMAPP Communication Model

Transactional models appear to work for most types of communication. By contrast, the model shown in Figure 2.3 below was designed specifically for engineering and other technical communication. This so-called *CMAPP model* (or *approach*) does not include the terms "sender," "transmission," or "receiver." It is meant to reflect the deceptively simple nature of real-world engineering communication.

FIGURE 2.3:
The CMAPP Model

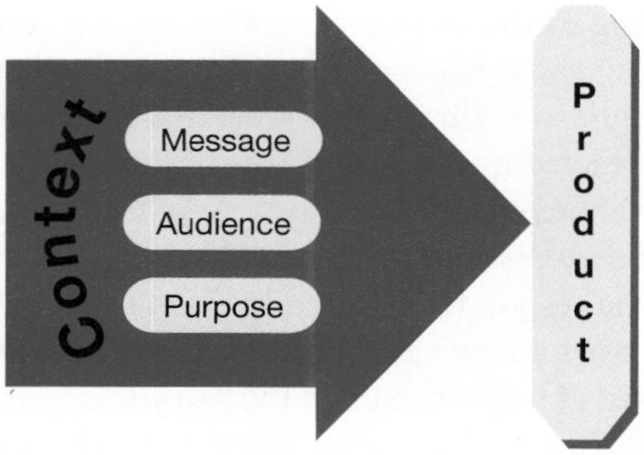

The CMAPP approach incorporates the following ideas:

- The situations (the Context) in which people find themselves affect their communication.
- The words they use (the Message) are affected by the person or group with whom they are communicating (the Audience).
- The Message is affected by the communicator's reason for communicating, his or her expectations (the Purpose), as well as the overall situation.
- The physical form of the communication (the Product) affects the way in which the communication is formulated and received.

All these ideas affect each other all the time. Thus, while the Transactional models are *linear*, the CMAPP model is *dynamic*.

Do you remember the quip about "getting your brain in gear"? Well, that's a crucial concept in applying the CMAPP model, which says that the first step in creating effective engineering or other technical communication is conducting a *CMAPP analysis*. In a moment, we'll explain that.

2–2a DEFINITIONS

Following are brief explanations of the CMAPP model's terminology.

Context refers to the surrounding situation. It may include, but isn't limited to

- personal relationships—both with and among your audience or audiences;
- time and place;
- circumstances that may influence the people and the communication involved;
- external and internal interference (in terms of the transactional model) that might have an impact on the success of one's communication.

Message refers to the content of the communication—that is, the main points and the significant details of the message you wish to communicate. Further, you may have a *primary* message—the principal idea you have in mind—and, potentially, a *secondary* message—ideas that, in a sense, might appear in parentheses.

Audience is similar to the receiver in the transactional models. Here, however, it can include

- a *primary* audience: the person or people for whom you are specifically designing your communication and to whom you will deliver it—whether in writing or orally;
- a *secondary* audience: other people whom you will be taking into definite consideration while developing—and delivering—your communication; for example, the manager

who needs to receive a copy of everyone's memos, or your teacher when you are delivering a presentation to your classmates. Note that this means you must know quite a bit about your secondary audience, and that you will be taking that audience into account to the extent that you may well modify your message to your primary audience.

Purpose refers to "why" you are communicating. The concept includes

a. your motive or motives—overt and/or covert—for communicating;

b. the anticipated reaction on the part of the audience;

c. your desired response from your audience.

In the CMAPP model, your *primary purpose* (note that you might have a *secondary* purpose as well) must be one of four: descriptive, informative, instructive, persuasive.

- If your purpose is to *describe*, you want your audience, in their mind's eye, to see something they haven't seen before, or to see something more clearly than before. Think, for example, of a set of blueprints.
- If your purpose is to *inform*, you want your audience simply to know something new—with, in fact, no subsequent reaction on their part. Think, for instance, of a direct answer to an audience's question.
- If your purpose is to *instruct*, you want your audience to know how to do something new. Think, for example, of a user guide.
- And, if your purpose is to *persuade*, you want to effect some kind of change in your audience's attitude, belief, or behavior. Think, for instance, of a proposal that must convince your audience to hire your company to manufacture certain electrical components.

There will always be overlap: You can't describe without informing; you can't persuade without describing; and so on. However, you need to identify your primary purpose as *one* of the four, so that your communication has a clear goal.

Product is the "mode" or the "platform" of the communication. Usually referring to its physical form (e.g., a document, whether on paper or in electronic format), it can sometimes specify the *job* the item does (e.g., summarizing another product). The choice of product will affect and be affected by the context, message, audience, and purpose. Thus, the term would include, but not be limited to

- memos
- letters
- reports
- emails
- text messages
- telephone conversations
- face-to-face conversations
- tweets
- social media status updates
- summaries

2-2b ANALYSIS TIPS

You should use a CMAPP analysis to plan your document or presentation. This means that before you begin drafting your CMAPP *product*, you ask yourself questions—and answer them—about all the CMAPP components. Make sure that you have adequately defined the particular context, your primary message, your primary audience, and your primary purpose. Examine each of the elements closely. As you find one component affecting others, modify each accordingly until your product is ready for delivery.

Do you still remember the quip about engaging brain before putting mouth into motion? You can look at the CMAPP approach along similar lines: Before you begin to communicate, effect a CMAPP analysis. It will allow you to identify exactly what you want to convey and how to convey it. We'd recommend that, to begin with, you conduct the analysis in writing. Over time, however, you'll find that you do the analysis instinctively—just as you might now make multicourse meals for a variety of guests without returning to original, written-out recipes.

Generally, there is no predefined set of necessary CMAPP questions. The ones you arrive at will be determined by the interrelationship of context, message, audience, purpose, and product. There is one exception—this: "What does the primary audience need and/or want to know?" Your answer to that question should be, in effect, what you list under *Message*.

Here are examples of questions you might ask regarding each CMAPP component.

Context

- What is the underlying or surrounding situation?
- What are the physical conditions (lighting, noise, and so forth)?
- How will the context affect how my audience responds to my message or to me?
- What kind of relationship do I have with my audience?
- What other relationships might have an impact?

Message

- What, exactly, am I trying to communicate?
- What are my main points?
- What significant details should I include?
- Is my message self-contained, or is it the beginning, middle, or final segment of a longer communication?
- If I have more than one message, what is my secondary one?
- What sequence of information have I used to organize my message?
- How technical should my message be? (The answer to this question should reflect the level of technical expertise possessed by your *Audience*; see below.)

Audience

- Whom, specifically, am I trying to communicate with?
- What does my primary audience already know?
- What does that audience need and/or want to know? (This, remember, is the one question that you must *always* ask. And, its answer should be a close reflection of the main points and significant details that you listed under *Message*.)

- Who might a secondary audience be?
- What assumptions have I made about my audience(s)?
- How much specialized (technical) training does my primary audience possess?
- How will my primary audience benefit from my message?

Purpose

- Overall, am I trying to inform, persuade, instruct, or describe?
- Why do I want to communicate?
- Why do I want to communicate at this particular time?
- Why would my audience need or want this message?
- What do I want to achieve through this message?
- Was my message explicitly requested and how might that be relevant?
- What deadlines are involved? How have I identified and dealt with them?

Product

- Should I be writing, phoning, visiting, and so forth?
- How is the product I've chosen (e.g., letter, memo, report, visit, presentation) appropriate for this context, audience, message, and purpose?
- How do the wording and format of my product reflect the image I want to present?

2–2c EXAMPLE

Consider the following scenario. You have completed a project for a client, who has emailed you to indicate that you will shortly be receiving a check rather than an electronic funds transfer for the work you did. You do then receive an envelope; however, it does not contain the check—only a note that says the check is enclosed. You are annoyed. What do you do? According to the CMAPP approach, you would conduct a brief analysis before you actually do anything. Here's an example:

Context

Have you dealt with this client before? If so, has he or she previously paid by check or by EFT? Is the payment on time, or was the client late? Is your relationship with that client a good or a bad one? If more than one person signed the note, how will you determine to whom you should direct your response?

Message

What, specifically, should you say? What information are you providing to ensure your audience will have everything necessary to understand and respond? Should you mention your annoyance? (Think of the context.) How much detail should you include? What kind of language should you use—very simple or sophisticated? How technical should it be? (Think of your primary audience's level of technicality.)

Audience

Are you communicating with a single individual, a company, a large bureaucracy? Will that audience know (or care) who you are? Do you have reason to believe your audience is competent to deal with the situation? Are you trying to communicate with the person who forgot to enclose the check or with that person's boss?

Purpose

It is important to note that a general purpose of almost all workplace communication is to foster and maintain relationships; the converse of that is "to not screw up." It is very rare in the professional world when one cannot care about one's future relations with a person. That said, the terms "primary purpose" and "secondary purpose" have a "technical" meaning within the CMAPP model. In that light, you need to ask yourself: Is your primary purpose descriptive, informative, instructive, or persuasive? Are you communicating simply to get your check as quickly as possible, to voice your irritation while still seeking to sustain your relationship with the company, to obtain some kind of redress, or what? Do you expect an immediate response—and what do *you* understand by "immediate"?

Product

Should you telephone? If so, would voice mail be satisfactory if you cannot reach the person you had in mind? How about texting the client? Would that be appropriate in this context and for this audience? Should you pay a visit instead? What distance would you be prepared to go—a couple of blocks, a few miles? Would a more formal written communication be more effective? If so, should you send a simple email, or should you compose a letter and send it as an email attachment? Should you perhaps print a letter and then send it by "snail mail"? Which CMAPP product is most likely to fulfill your purpose?

2-2d INTERRELATIONSHIPS AMONG CMAPP ELEMENTS

Whereas most traditional models seem to be linear in the sense that they progress from A to B to C (and then, often, back to A), the CMAPP model is dynamic: Each element continually affects all the others. In the missing-check example, the CMAPP dynamic addressed the following interactions:

- Knowing more about your audience helped determine your context.
- That knowledge about your context helped you identify the particular audience.
- Knowing your audience helped you identify and refine your purpose.
- As you refined your purpose, you got a better idea of the most appropriate product.
- Your conception of the product was also dependent on your audience, which in turn affected your message, which itself affected . . . and so forth.

Any change in one element has a ripple effect on all the others.

Incidentally, this dynamic aspect also means that you could have examined the components in any order—MPACP, for example.

You may have noticed that CMAPP does not make *explicit* reference to one of the prime components of transactional models: feedback. The concept of feedback, however, remains fundamental in CMAPP analysis. Modifications to any element of context, message, audience, purpose, or product have an inevitable impact on the other elements. These shifts are, in effect, the manifestations of feedback. For example, your message affects your audience in a particular way, which alters the context, which has an impact on both your audience's reaction and your own response to that reaction, and so forth. Communication, just like real life, can get complicated.

2-2e APPLICATIONS

Is the CMAPP model likely to be of any practical use to you? As a student, you'll be communicating with your professors, with other students, with groups to which you belong, with potential employers, with businesses, banks, government, and so on. Conducting a CMAPP analysis for each of these communications will undoubtedly help you obtain what you want. Obviously, we're not suggesting that you conduct an explicit CMAPP analysis for every instance of your daily communication. As we implied earlier, over time, you *internalize* a CMAPP analysis—it becomes an instinctive, habitual *approach* to your communication. In a sense, you *automatically* "put your brain in gear" before beginning your communication.

The benefits of understanding and using the CMAPP model could be even more valuable in the workplace where you'll be dealing with colleagues, your boss, clients and potential clients, suppliers, and consumers. It has been shown time and time again that professional success is much more likely if you communicate well. As just one example, in its Job Outlook 2013, NACE, the National Association of Colleges and Employers, points out that almost 75 percent of employers are seeking in potential new hires good written communication skills, and over 67 percent seek good oral communication skills.

The CMAPP communication model is designed to help you develop those skills.

Finally, over our many years of teaching the CMAPP model, we have heard from a number of students who have told us that the model has proved to be very useful indeed in fostering and maintaining not only *business*, but *social*—and even *family*—relationships. We will always face communication situations, and the better we deal with them, the more satisfied and successful we are likely to be.

CASE STUDIES

2a AAU Student Association Function

SITUATION

During the construction of the Student Association Building (SAB) in 2015, the Ann Arbor University Student Association (AAUSA) continued to rent from the university the following premises:

- office space in Arbor Hall (the main administration building) for the 10 members of the AAUSA Executive (AAUSAEx);
- a "student space" (a former classroom) on the main floor of Lindbergh Square (the Engineering building), containing several Internet terminals, two foosball tables, a Ping-Pong table, an ATM, food and drink vending machines, and several tables and chairs;
- a small editorial office in the Ring Lardner Building (used primarily for Department of Arts and Sciences faculty offices) for the editor and associate editor of the student newspaper, *The AAU Reporter*.

When it wishes to put on functions such as concerts, dances, or parties, AAUSAEx normally arranges to rent either Fuller Hall, the university's main auditorium, or one of two large conference rooms named after corporate benefactors Accelerated Enterprises Ltd. (AEL) and Radisson Automobiles Inc. (RAI).

At its November 9, 2015, meeting, after an often-heated debate, AAUSAEx finally decided to celebrate the Thursday, November 26 Thanksgiving holiday on the following day—the Friday—with a concert. Late on Sunday night, November 15, Jack Lee, the AAUSAEx Functions Coordinator, logged on to the Room Booking Web page and completed an online AAU room booking form. He also confirmed bookings with (and sent checks to) two popular local bands, Hodgepodge and Really Me. As usual, he contracted with a local company, House Specialties, to produce and distribute posters and flyers advertising the event and to update the AAUSAEx Facebook page and Twitter feed. He was obliged to pay extra because the work was a rush job.

Checking the AAUSAEx inbox around 7:00 p.m. on Monday, November 23, only four days before the Thanksgiving concert, Jack was shocked to find his booking form has been "rejected." He was unable to identify a contact person in the brief email containing the rejection. In a bit of a panic, and angry at AAU administration, he tried to contact Dorothy Palliser, the AAUSAEx president, but discovered that she was in the middle of a three-day Engineering field trip in Flint. Realizing that he had to take immediate action, Jack left a voice mail (since he was unable to reach her personally) for Kulwinder Atwal, the administration room booking clerk. (A transcript of Jack's message is shown in Figure 2.4.) Since he had once booked the RAI Conference Room online and later been informed there was no record of that reservation, Jack decided, at 9:00 a.m. on Tuesday morning, to hand-deliver to the Room Booking Office a printed memo (Figure 2.5). Finally, he texted Dorothy (Figure 2.6), hoping that she was still checking her smartphone.

FIGURE 2.4: Transcript of Lee's Voice Mail to Atwal

Hi! It's Jack from AAUSAEx.

How come our booking was rejected? I don't think your Web page said there was going to be a problem, and we've already got Hodgepodge and Really Me coming, and House Specialties has got some stuff up already.

I hope this was like just a mistake, 'cause we've got to have the room 'cause we're really in deep already.

Can you call me and tell me what's wrong with it?

AAUSAEx Memo

Arbor Hall 243 | **Ann Arbor University Student Association Executive** | **Local 4334**

From: Jack Lee, Functions Coordinator
To: Room Booking, Administraton
cc.
Date: Tuesday
Subject: Concert

I booked Fuller Hall for our Thanksgiving concert and already hired the bands and got the publicity arranged. Now you've rejected our booking without saying why, and we don't have time to get our arrangements changed. I've already left voice mail for Kulwinder but can't get hold of her to do anything about it. The bands want to set up at 6 p.m., and we want to arrange with the cafe to get food and refreshments. Would you please remake our booking?

If you have any questions, don't hesitate to call me. Otherwise, we'll just assume you made another mistake.

FIGURE 2.5: Jack Lee's Memo to AAU Administration

FIGURE 2.6: Jack Lee's Text to Dorothy Palliser

FWIW Admin scrwd up-no Thxgvg bookng but I fixed it. Bands etc OK. TTYL in class Mon

ISSUES TO THINK ABOUT

In considering this scenario in terms of CMAPP, you have to take into account the following interconnected communications involving AAU and AAUSAEx: the original booking form completed by Jack, Jack's voice mail to Kulwinder, Jack's memo to the Room Booking Office, and Jack's text to Dorothy.

Booking Form

Not having seen the original booking form, we don't know if it was completed incorrectly—a possible reason for its rejection. But we do know that it was submitted less than two weeks before the planned concert—another possible explanation for its rejection, since Fuller Hall might have already been reserved for another function. This might lead you to question whether AAU's Room Booking page didn't automatically prevent "double bookings," or whether Jack simply didn't pay attention. We also know very little about the relationship between AAU administration and AAUSAEx. Had there been problems before? Had the administration been unnecessarily rigid in its requirements? Had AAUSAEx been lax in submitting rental payments? Had there been incidences of damage to university property? Do you think that any of these details of context might have had an impact?

What can you say about the content of this voice mail? Was it specific? Did it contain all the information that Kulwinder would have needed in order to look into the matter? For example, would she know that Jack is the AAUSAEx functions coordinator? Shouldn't he have given his last name, too? Why didn't he leave a number and a time for her to reach him?

Was the message polite? Had Jack given sufficient thought to his audience? What might Kulwinder's reaction to the message have been? Do you think that Jack's purpose was clear to Kulwinder? More importantly, do you think that Jack himself had a clear idea of his purpose? Did he simply want, as his message states, Kulwinder to call him back and tell him what was wrong with his booking form? Or did he really want Kulwinder to find a venue for the concert?

What about Jack's choice of CMAPP product? Was leaving a voice mail the best way for him to communicate? Could he not have tried to reach someone else? Did he take likely delays into account when he decided to leave the voice mail? So, do you think the voice mail was an effective product in this context—why or why not?

Memo to Room Booking Office

This time, let's consider the product first. Is a handwritten memo acceptable, or should Jack have word-processed it? (Hint: Is it a personal or professional communication? How might the distinction be important in this context?)

Did you notice the spelling mistake in the **To** line? What do you think of what he gave as the *Date*? Should he have initialed or signed the memo? Does it make any difference that the **From** line appears above the **To** line? Why might any of these things matter? Here are some other questions you might ask yourself about this memo.

- Did Jack's message contain all the necessary information? What significant information might be lacking?
- What do you think of his **Subject** line? Would Jack's wording allow the Room Booking Office to find the relevant information easily? How might this issue influence the context?
- Did the memo contain one request or several? Were these requests clearly expressed?
- To whom is the memo addressed? Why did Jack leave a voice mail for a particular person (Kulwinder) but address his memo to "Room Booking"? From what you know of Jack so far, do you think he knew whether AAU's administration even had a section called "Room Booking"? Why might this be important?

- Does the first sentence of the last paragraph imply that, if "Room Booking" did not have questions, Jack did not want them to call?
- How might the likely audience respond to the last sentence? Who do you think that audience is? Do you think Jack knew who his audience was likely to be?

Here are some questions about that text.

- How does the fact that Dorothy is another student (possibly a friend of Jack's) affect the context?
- Should he have shown a "cc" to Dorothy in his memo to "Room Booking?" (Note that memos will be discussed in detail in Chapter 12.)
- Should he have emailed her a copy of that memo instead of just texting her—why or why not?
- What do you think Jack's purpose (in CMAPP terms) *should* be, and what do you think it *seems* to be?
- How do you think Dorothy will react to Jack's text?
- Does it provide her with useful information?
- You'll remember that words always carry two kinds of meaning—denotation and connotation. (Note that a single word or expression can have *several* connotations for a single audience.) Each kind of meaning has an impact. What might be the connotation of Jack's words "scrwd up" and "I fixed"? And what do you think those choices say about him?
- Was a text message the right CMAPP product for Jack to use? If so, why? If not, what should he have used instead?

Revision

Examples of what Jack might have done in his voice mail to Kulwinder Atwal and his memo to the Room Booking Office are shown in Figures 2.7 and 2.8. Note that these versions are not the only correct possibilities; you might want to create other versions yourself.

FIGURE 2.7: Revised Version of Jack Lee's Voice Mail to Kulwinder Atwal (Figure 2.4)

This is a message for Kulwinder Atwal, the Room Booking Clerk, from Jack Lee, the Functions Coordinator for the Student Association. It's now 7:30 p.m., Monday, November 23.

Since it is now well after office hours, I am leaving you this voice mail that I hope you will receive when you come in tomorrow morning. I have just seen the rejection notice for Ann Arbor University Student Association Executive's booking request for Fuller Hall for the evening of Friday, November 27, for the student Thanksgiving celebration dance.

I do apologize for any problem we might have caused when submitting the booking request, but wonder if there's now any way you can help us. I'm afraid that we had presumed there would be no problem booking the hall. We've already contracted with two bands, and have had our posters printed showing Fuller Hall as the venue. Would it still be possible for us to use Fuller Hall for the dance? If not, would you be able to get an alternate location for us; we could always put up a small notice on each of the posters, giving the change of room.

I'd really appreciate anything you can do. Please get in touch with me as soon as possible. My local at the Student Association office is 4334. Or you can call or text me on my cell at 734-555-8888.

Incidentally, I hope you won't mind, but, just in case you aren't in tomorrow morning, I'm also now leaving a memo at the main Administration desk so that someone else might be able to help us.

Thank you very much.

Bye.

AAUSAEx Memo

Dominion 243 **Ann Arbor University Student Association Executive** **Local 4334**

From: *Jack Lee, Functions Coordinator* J.L.
To: *Room Booking Office, Administration Building*
cc. *Dorothy Palliser, President, AAUSAEx*
Date: *Tuesday, November 24, 2015*
Subject: *Booking for Sassoon Fuller for Friday, November 27, 2015*

I have just seen the rejection notice for AAUSAEx's online booking request for Fuller Hall for the evening of Friday, November 27, for the student Thanksgiving celebration dance. Since I could not reach Kulwinder Atwal by phone around 7:30 p.m. yesterday, I left her a voice mail asking for assistance, and indicated that I would be sending you this memo as well.

On behalf of the AAUSAEx, I apologize for any problem with the initial booking request. Would you nonetheless be able to help us now?

While conducting our preparations for the Student Thanksgiving Dance, I'm afraid that we presumed there would be no problem booking the hall. We have already contracted with two bands, and have had our posters printed showing Fuller Hall as the location. Would it still be possible for us to use Fuller Hall? If not, would you be able to arrange an alternate location for us? We'd be more than willing to affix a small notice on each of the posters, indicating the change of venue.

I'd be very grateful for your assistance. Please contact me as soon as possible at the AAUSAEx office, local 4334, or on my cell at 734-555-8888.

Thank you

FIGURE 2.8: Revised Version of Jack Lee's Memo to AAU Administration (Figure 2.5)

2b Reorganization Notice at RAI

SITUATION

At RAI's head office in Dallas, Griffin Radisson, the president and CEO, has become concerned about what he sees as an increasing decentralization of authority at RAI. He believes that RAI's growing network of dealerships has diminished his own control over daily operations and overall company direction.

Recently, Radisson persuaded the board of directors to impose in three months' time an organizational change that will have an impact on all seven RAI dealerships. As of January 1, 2016, each dealership vice-president will report directly to Radisson rather than to the board. Furthermore, each dealership's general manager will report not to the respective dealership vice-president but to the head office company manager, Celine Roberts.

The board members were reluctant to endorse this change. They foresaw considerable opposition on the part of dealer vice-presidents and general managers over what might be perceived as a sudden lack of trust in their abilities. Over the last several years, all dealerships have been reporting increased profits and a decline in the number of customer complaints. Local dealership management have thus had good reason to believe that they have been instrumental in RAI's continued success.

The board also argued that the three-month deadline did not leave sufficient time to work out all the inevitable administrative complications. But Radisson, who remains RAI's principal shareholder, prevailed. Now the dealerships must be notified of the coming reorganization. Radisson has decided to impose his will on the board once again and communicate the news personally, by emailing a formal memo to the VPs and GMs. His memo to the dealerships appears in Figure 2.9. His attachments are shown in Figures 2.10 and 2.11.

Radisson Automobiles Inc.
Where all the cars are stars

7500 Lemmon Ave
Dallas TX 77802
(979) 555-0099
Fax: (979) 555-0091

Head Office Memorandum

November 12, 2015
To: Dealership Vice-Presidents and General Managers
From: Griffin Radisson, CEO GR
Re: Reorganization
cc. Board of Directors. C. Roberts

As you know, I believe in being concise and straightforward in all my dealings, with both clients and company personnel. Therefore, I am announcing to you now the reorganization of Radisson Automobiles.

As of January 1, 2016, the *Current Organization Chart* (copy attached) will be superseded by the *Revised Organization Chart* (copy attached).

The Board of Directors and I recognize that administrative details must still be worked out. Nonetheless, we are confident that the new organization will be to the benefit of the Company, a goal we are sure you all share.

Celine Roberts is looking forward to receiving 2016 operational plans from the dealership general managers. I am likewise looking forward to receiving 2016 business plans from the vice-presidents.

I know I can count on your cooperation.

Attachments

FIGURE 2.9: Radisson's Memo re Reorganization

ISSUES TO THINK ABOUT

In deciding whether Griffin Radisson conducted an effective CMAPP analysis before communicating, consider the issues outlined below.

Context

Radisson's position and forceful personality served him well in persuading reluctant board members to endorse his plan. Although we know little about his relationship with the dealership vice-presidents, we can assume that they will not be happy with a reorganization that will have a dramatic impact on two sets of relationships: those between Celine Roberts and the general managers, and those between the general managers and their local vice-presidents.

Another complication is geography. The physical distance between most of these individuals means that face-to-face communications in the future are likely to be few.

Product

What is your assessment of Radisson's choice of product? Since the communication is internal (within an organization), a memo rather than a letter would be appropriate. (See Chapter 12 for details regarding letters and memos.) On the other hand, what might have been the effect on the relationships had Radisson first held a conference call and then followed up in writing? Should he have used some other CMAPP product(s) for this communication?

FIGURE 2.10: Radisson Automobiles Current Organization Chart

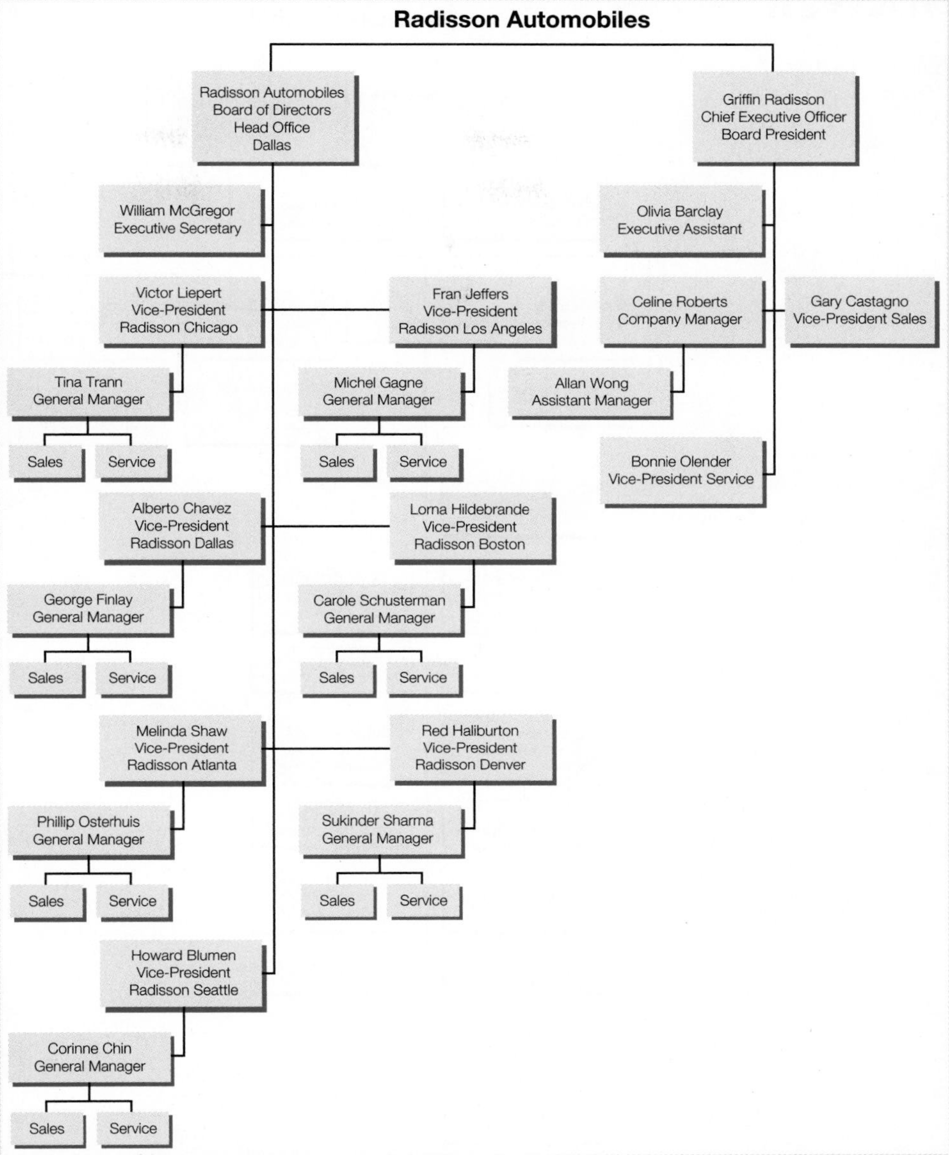

Message

What do you think is Radisson's primary message? Can you identify any secondary message? What kind of impression is conveyed by the tone of the memo? (Hint: Think about connotation as well as denotation.) What might the message say to the audience about "Radisson the man"?

Do you think the two attachments help or hinder Radisson's message? Will his audience understand them easily or do you think they will need further explanation? If further explanation is required, what form should it take?

Audience

The memo is addressed to dealership vice-presidents and general managers, with copies going to board members and the company manager. How many different audiences do you think Radisson

FIGURE 2.11: Radisson Automobiles Revised Organization Chart

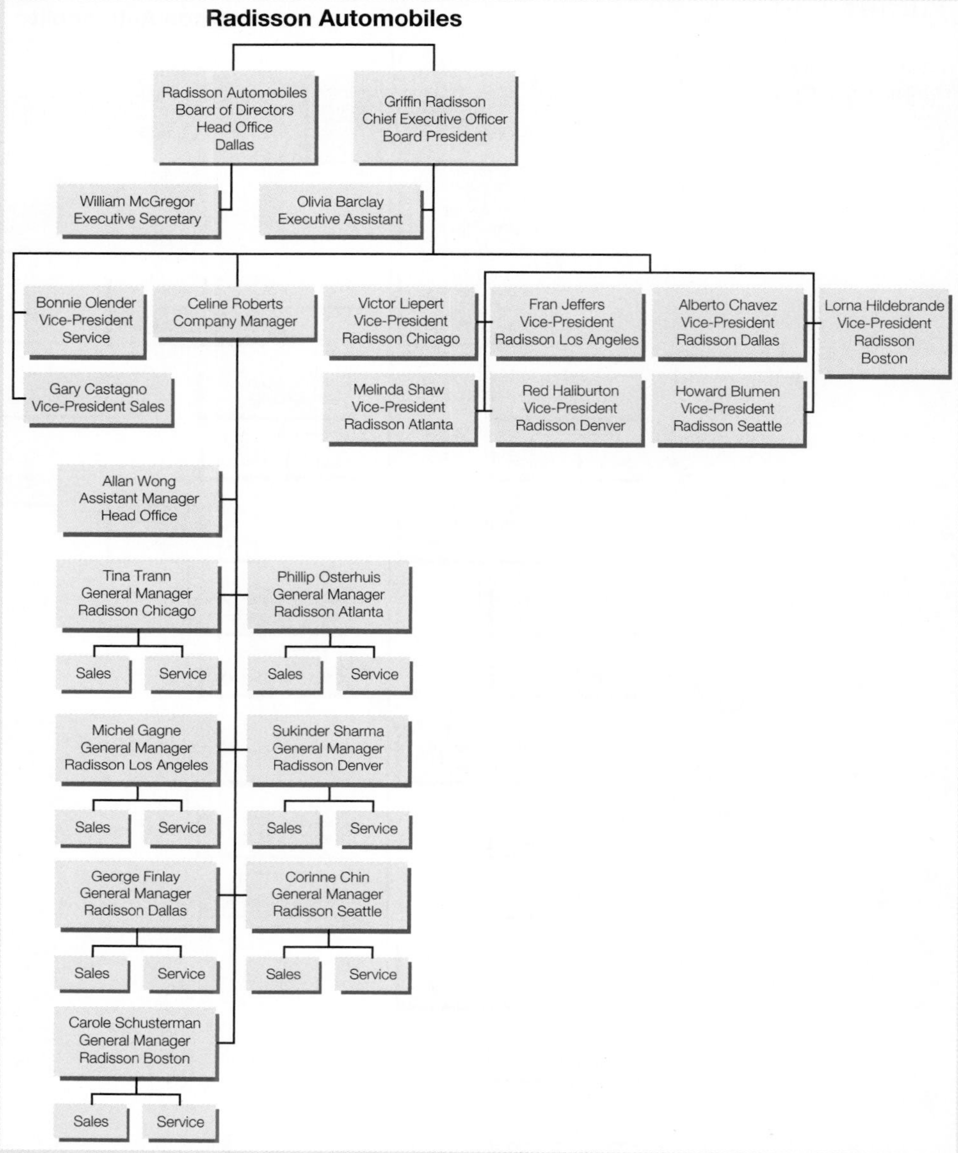

was likely considering? Which audience do you think is—or should be—the primary one? Should all audiences really receive the same message or messages? What differences in interpretation might there be on the part of these audiences?

Purpose

Does Radisson have more than one purpose? What do you think he is trying to do? Which of the four CMAPP primary purposes do you think is involved here? What do you think he is actually doing? What do you think the various audiences will identify as his purpose or purposes? Based on their perception of purpose, how do you think they will view the attached organization charts in terms of perceived purpose? Do you think that Radisson's message succeeded in enhancing his relationship with his audiences? Or, conversely, do you think he "screwed up"?

EXERCISES

2.1 Briefly define the following terms:

a. CMAPP
b. context
c. external interference
d. feedback
e. internal interference
f. message
g. primary audience
h. product
i. purpose
j. secondary audience
k. technical communication
l. transactional model

2.2 Below are two different scenarios. In scenario A you tell a student at another institution about your financial situation this term. In scenario B you communicate this information to a loan officer at your bank. For each scenario, provide answers to the CMAPP questions that follow. Invent any necessary details.

Scenario A: Fellow Student

Context

1. What is your relationship with the student?
2. Why would the student be interested in your message?
3. What is the student's own financial situation?

Message

4. What specific information will the student need or want in order to respond as you would like?
5. What details should you provide?
6. Should you exclude any details? If so, what details and why?

Audience

7. Is the student female or male? How might this make a difference?
8. What might the student know already?
9. What would he or she want to know?
10. What do you know about the student's financial situation?

Purpose

11. Why are you telling the student about your financial situation?
12. What kind of reaction do you expect?
13. What do you want the student to do with the information?

Product

14. What product do you intend to use?
15. Why have you chosen to use that product?

Scenario B: Loan Officer at Your Bank

Context

1. What is your relationship with the loan officer?
2. Why might the loan officer even be interested in your message?
3. What kind of reputation do students have in the banking community?
4. How long have you been dealing with this branch and/or this person?
5. How might your history with the bank affect how your message is received?

Message

6. What specific information and explanations will the loan officer want?
7. What information should you provide?
8. What information might you want to withhold?
9. Should you withhold this information? Why or why not?

Audience

10. What is the sex and age of the loan officer? How might this make a difference?
11. Is the loan officer the only audience?
12. If there is a secondary audience, who is it?
13. What would the loan officer know already?
14. What would the loan officer want or need to know?
15. Is anyone else likely to see or use the information you provide?

Purpose

16. Why are you telling the loan officer about your situation?
17. What do you want him or her to do with the information?
18. Do you have a secondary purpose? If so, what is it?
19. What might the loan officer's purpose be?
20. Is the loan officer's purpose of any relevance to your situation?

Product

21. What product do you intend to use?
22. Why have you chosen to use that product?

2.3 The computer system you purchased from Acme High-Tech Products has given you considerable trouble. Angry, you dash off an email (Figure 2.12) to Acme's sales manager, R. B. Kim. You don't actually hit "Send," however. Instead, you wait until you are calm, then decide to email to Kim a letter that adheres to the principles of good technical communication. Your first step is to conduct a CMAPP analysis. For the purposes of this exercise, do not write the new letter, but simply answer the CMAPP questions that follow.

CMAPP Questions Regarding the Letter

Context

1. What has prompted you to write a new letter?
2. How would you describe the relationship between you and your audience?
3. What would you be prepared to do if your letter fails to meet its intended objective?

Message

4. What are your main points?
5. If you have any secondary points, what are they?

FIGURE 2.12: Initial Version of Letter to R. B. Kim

Last month, we purchased a computer system from you, actually sold to us by your Mr. Webley, who came to our home after our phone conversation two days earlier. We spoke with him for a couple hours, gave him coffee, and eventually decided to buy the PB-2 "User-Friendly PC Package" that he showed us in the many pages of advertising material he had brought.

When the merchandise arrived two days later, the deliveryman, who said he always replaced Mr. Webley on Fridays, really didn't seem to know what he was doing. He kept looking things up in the manuals that came with the package, and had a continual worried look on his face while trying to make little jokes about the complex technology we have in our lives now. Eventually, he seemed to get it all working, and left. It was already 8 p.m., about three hours since he'd arrived.

The next morning, I had some spreadsheet work to do. As I'd told Mr. Webley before he suggested the PB-2 package, I'd had quite a bit of experience with Excel 2007 at the office. So we'd bought a copy of Office from Acme, as part of the package, along with a bunch of DVDs, including Quicken and WebEasy Pro.

Unfortunately, I got one system error after another (I think that's what they were), never did get my work finished, and, to top it off, found the printer included in the package wouldn't accept the commands I typed into the Print window.

No one was at the service number Mr. Webley had left me when I called that afternoon—a Saturday—nor the following day. On Monday, someone answered, but got quite rude when I told my story and demanded that Acme either arrange for real support for its customers or pick up your lousy equipment and give me my money back! After that call, I gave up trying to contact anyone and wrote this . . . on my iPad, not on your shoddy computer!

I better hear from you in the next couple of days, or you'll be hearing from my lawyer, who's getting a copy of this letter, too.

Audience

6. Who is your audience?
7. If you have a secondary audience, who is it?
8. What do you think your primary audience knows?
9. What do you think that audience needs/wants to know?

Purpose

10. What is your purpose? In other words, what is your intended objective mentioned under Context?
11. If you have a secondary purpose, what is it? (Hint: Think longer-term, particularly with regard to questions 3 and 7 above.)

Product

12. What product might you have chosen instead of a letter?
13. What advantages and disadvantages might that alternative product have engendered?
14. Why did you eventually choose a letter? (Hint: You might have more than one reason, particularly related, perhaps, to a long-term purpose.)
15. What kind of tone will you use?
16. How technical should your letter be?

2.4 Among RAI's products is the Minotaur, a popular luxury car manufactured by Global Vehicles. (Think of Global as a "competitor" to GM or Ford.) Having recently discovered a possible manufacturing defect in some Minotaur models, Global composed a notification (Figure 2.13) to deal with the problem. Based on your analysis of that notification, provide answers to the CMAPP questions that follow it.

FIGURE 2.13: Global Notification

A possible defect relating to the potential failure of a seal in the mounting of the left tie-rod-end assembly may exist in your vehicle. In the event of disconnection, this seal (configured as a supportive component to the front rack assembly) could result in at least partial loss of bilateral vehicle directional control, particularly during unusually significant intentional deceleration. In addition, your vehicle may require adjustment service to the front hub bearing where it comes into contact with the hub carrier beside the right caliper. Should this bearing be improperly aligned, unusually rapid acceleration during excessive directional change might result in a similar loss of bilateral vehicle directional control. In certain circumstances, either of these potential failures has the capability of causing unexpectedly severe vehicle handling. You are therefore advised to consider the benefits of contacting your reseller to arrange for relevant adjustment or replacement without personal expenditure on your part.

Audience

1. Who appears to be the primary audience for the notification? What words in the proposed notification make you think so?
2. What secondary audience or audiences might the notification's author be considering? (Hint: Look at the use of language and consider the purpose.)

Context

3. What relationship(s) might exist between Global Vehicles and the audience(s)?
4. How might the relationship(s) change as a result of the notification?
5. What kind of reaction to its notification might Global expect?
6. What impact might the situation have on RAI?

Message

7. What can you say about the language used in the notification?
8. What are the main points supposedly being made?
9. Is that information clear and concise? Give examples to justify your response.
10. What do you think the primary audience actually needs and wants to know?

Purpose

11. What is the apparent primary purpose of the notification? Identify one of the four CMAPP purposes and explain your answer in detail.
12. What secondary purpose might be involved?
13. How well do you think Global's current notification fulfills its purpose(s)?

Product

14. What *primary* product should Global, in fact, use, and why?
15. What would make that product effective?
16. Should Global use more than one product? If so, what additional products should it use and why?

2.5 Think of a real-world situation, preferably one having to do with a business or other organization that requires a response from you. Based on your scenario, construct and answer relevant CMAPP questions.

2.6 As a sales representative for your company, you want to send a message to a very important customer. This particular customer is a good friend of your company's president—they play golf together often. Also, this customer consistently comes to you and your company to place large orders for supplies. In your opinion, the two of you have had a very good relationship. However, the last time you paid him a visit "just to make sure everything is going well," he seemed very impatient, actually telling you that he "really had to get on with his work." Now, two days after your visit, your company has started a sales campaign on many of the products your customer purchases from you. The prices are unusually low, but the sale will last only two weeks. Would you contact this customer? If so, how—by letter, by phone, by email, with a personal visit? Justify your answer.

USEFUL WEBSITES

Note: For ease of reading, we have removed the "http://www." from the URLs below. However, some browsers might still require that you insert it to precede some URLs so as not to receive an error message.

- About.com's "Audience Analysis Checklist" provides useful reminders about conducting audience analysis.
  Visit: grammar.about.com/od/rhetoricstyle/a/Audience-Analysis-Checklist.htm

- About.com's "Choosing the Best Words: Denotations and Connotations" offers a brief overview of these two types of meaning.
  Visit: http://grammar.about.com/od/words/a/connotations.htm

- Daniel Chandler gives a dated but still excellent overview of "The Transmission Model of Communication."
  Visit: http://visual-memory.co.uk/daniel/Documents/short/trans.html

- Max Wideman, of Simon Fraser University in British Columbia, Canada, provides a detailed discussion of a variety of communication models. While now several years old, the material remains useful.
  Visit: maxwideman.com/issacons4/iac1432/index.htm

- Purdue University's Online Writing Lab (OWL) offers a good overview of audience analysis.
  Visit: owl.english.purdue.edu/owl/resource/629/1

- The University of Washington offers an overview of audience analysis.
  Visit: faculty.washington.edu/ezent/aaaa.htm

- "The Writing and Speaking Guidelines for Engineering and Science" at Penn State is an excellent portal page for assistance in producing written communication. It is edited by Michael Alley of Penn State's College of Engineering.
  Visit: writing.engr.psu.edu/

CHAPTER 3

Complementary Attributes of the CMAPP Model

The CMAPP model requires you to undertake a dynamic analysis of context, message, audience, purpose, and product. Before you create your communication, you pose appropriate questions and construct answers for them, regarding all the CMAPP components. Remember that you must *always* ask "what does my primary audience need/want to know?" Your answer should reflect the main points and significant details comprising your primary message. Otherwise, of course, every instance of communication will dictate its own distinct set of questions and answers. When creating that message, you should apply the CMAPP complementary attributes represented by these five abbreviations (two of which are acronyms—see below): 5WH, ABC, CAP, CFF, and KISS.

3–1 Abbreviations and Acronyms

You'll have noticed that we referred to both *abbreviations* and *acronyms*. What's the difference?

The online American Heritage College Dictionary of the English Language (ahdictionary.com) defines the former as "A shortened form of a word or phrase used chiefly in writing to represent the complete form, such as *Mass.* for *Massachusetts* or *USMC* for *United States Marine Corps*." Thus, Sept.,

Encl., and Attach. would be the respective abbreviations for September, Enclosure, and Attachment. Traditionally, abbreviations were characterized by the use of periods, as in the honorifics (defined by the same dictionary as "title[s], phrase[s], or grammatical form[s] conveying respect, used especially when addressing a social superior") Mr., Ms., and Mrs. It has now become acceptable to omit the period in many abbreviations, with Mr, Ms, and Mrs becoming common. Other examples include HIV (Human Immunodeficiency Virus), IRS (Internal Revenue Service), mm (millimeters), VFW (Veterans of Foreign Wars), and YMCA (Young Men's Christian Association), as well as the U.S. Postal Service's abbreviations for the names of the states, from Alabama to New Hampshire to Wyoming.

The same dictionary defines acronym as "[a] word formed by combining the initial letters of a multipart name, such as *NATO* from *N*orth *A*tlantic *T*reaty *O*rganization or by combining the initial letters or parts of a series of words, such as *radar* from *ra*dio *d*etecting *a*nd *r*anging." Thus, you might say that while all acronyms are abbreviations, not all abbreviations are acronyms. Technically speaking (as engineers are trained to do . . .), to be considered an acronym, the abbreviation has to be commonly pronounced as a word, a restriction confirmed by the American Heritage Dictionary's *Usage Note*. Some acronyms retain their characteristic capital letters; others more commonly appear in lowercase. Examples would include ASCII (American Standard Code for Information Interchange), DARPA (Defense Advanced Research Projects Agency), laser (light amplification by the stimulated emission of radiation), modem (modulate and demodulate), NORAD (North American Aerospace Defense Command), sonar (sound navigation and ranging), scuba (self-contained underwater breathing apparatus), and VISTA (Volunteers in Service to America).

You'll recall that the elements of the CMAPP model can be examined in any order. However, the sequence C-M-A-P-P offers an easily spoken *acronym*, while PPMCA, for example, would be an abbreviation, but relatively unpronounceable.

3-2 Complementary CMAPP Attributes

Now let's turn our attention to each of the five groupings of the CMAPP complementary attributes.

3-2a 5WH

The abbreviation 5WH refers to the six questions that are often associated with objective journalism: who, what, when, where, why, and how. Since these questions seek facts, asking them as part of your CMAPP analysis will help you decide what is pertinent for your message. In fact, the majority of engineering communication is concerned with what can be observed, quantified, or verified—that is, with facts.

Facts and Critical Thinking

But just what do we mean by the word "fact"? On one level, we seem to know pretty clearly what we mean when we say that something is a fact: We mean that it's true; and we tend to understand "true" as being something that can be proved; and we like to think that we can prove something by testing it against evidence; and, finally, we usually look at evidence as something that is factual. But, note: We have (unintentionally, perhaps) succumbed to a circular argument: We have merely concluded, by a somewhat convoluted route, that "a fact . . . is . . . a fact." Thus, we have generated no *real* or even useful "information"—a definite problem for engineering communication.

Some centuries ago, several generations of people—less technologically advanced, but just as intelligent as anyone today—knew "for a fact" that the sun moved around the Earth. You can still observe their "evidence": Look to the East in the morning and notice the sun "rising," and look to the West in the evening and note the sun "setting." Obviously (they concluded), therefore, the sun circles the Earth. For a shorter time, but until not very many decades ago, it was to most of the world a "fact" that the atom could not be split. (The word "atom" itself derives from two ancient Greek terms meaning "not able to be cut or divided.") How is it, then, that the "facts" appear to have changed?

Consider another issue. On the evening of July 20, 1969, many people (authors Dave Ingre and Bob Basil among them) watched the live TV broadcast of Neil Armstrong setting foot on the moon, and shortly thereafter went outside, looked at the moon, and thought, "There's actually someone walking on that!" It was an amazing experience.

But, let's ask: Is it a "fact" that the moon landings actually took place? A search for corroboration might produce, for example:

- I've seen it on TV . . . and yet we all acknowledge that TV imagery can be faked.
- There are photographs . . . and yet we know that photos can be faked, too.
- I was taught about it in school . . . and yet most of us learned long ago that not everything we were taught in school is true.
- It's in the history books . . . and yet we also know that history writers have typically "rewritten" history to match their own or their society's prejudices. As a matter of fact, what is now often called historical revisionism or revisionist history has become a controversial and contentious topic among academics, politicians, special-interest groups, and even historians themselves. The issue is, in a sense, one of trying to define—or redefine for one's personal purposes—truth itself. For a few divergent points of view on the matter, and in indication of how difficult it is to distinguish objectivity from bias (which is something that, as an engineer, you'll need to do), have a look at the historical revisionism sites listed at the end of this chapter.
- I've seen moon rocks . . . and yet, the moon rock that author Dave Ingre personally saw in the National Museum of Science and Technology in Canada's capital of Ottawa in the late 1970s looked just like a piece of rock from his backyard. So the only "evidence" was the plaque beside it. And PlaqueMaker.com or Zazzle.com could likely provide something similar—at a much lower cost.

Perhaps the most convincing evidence for the reality of the moon landings derives from another "fact"—the Cold War. (See, for example, en.wikipedia.org/wiki/Cold_War.) At that time, the "space race" was but one of the more ostentatious manifestations of the ideological conflict between West and East, represented most prominently by the United States and the former Soviet Union, respectively. Neither side ever knowingly missed an opportunity to defame, disparage, or—if at all possible—ridicule the other. And yet, neither Bob nor Dave can remember or now find any media accusation by the Soviets that the Americans had falsified the moon landings. Remember that the Soviets had initiated the "space race" when they launched Sputnik in October of 1957, and taken a significant lead when, in April of 1961, their cosmonaut Major Yuri Gagarin became the first person to orbit the Earth. For many years, their technology continued to be the likely equal of ours. (The Soviets, don't forget, put up the world's first space station, Mir, in 1986.) And Soviet scientists, like those in other countries, tracked that first American moon landing, and the later ones.

Many people do, of course, acknowledge that in 1969 it would probably have been technologically and financially easier to fake the landing than to actually accomplish it. And, over the years, many "conspiracy theories" have arisen to "prove" that the landings never occurred. To our surprise, by the way, we've found over the years that a larger and larger percentage of the post-secondary students we see have decided that the "conspiracy theorists" are telling the truth. In that light, we'd really recommend that you examine the Conspiracy Theory sites listed at the end of this chapter.

But, here is something else. Is it, for example, a "fact" that $2 + 2 = 4$? It seems likely, no? However, note that we define 2 as $1 + 1$; we further define 4 as $1 + 1 + 1 + 1$; thus, $2 + 2$ *is by definition* 4. This "fact" is merely saying that if we *define* something as something else, the *something* must therefore *be* that *something else* . . . hardly the kind of truth to make much news.

Perhaps you've taken a bit of geometry. If so, you likely learned the "fact" that the shortest distance between two points is a straight line. Were you, though, taught the diameter of each of those points, or the thickness of that straight line? In Euclidean geometry (which is the "type" you were learning), the points don't have a diameter and the line has no thickness. They're "abstract concepts," and the "fact" involved depends in part on that idea.

Consider, though: What is the *real*—not *theoretical*—shortest distance between the two points of Seattle, Washington, and London, England? Obviously, you cannot fly a "straight line" between them: You'd be trying to fly through part of the planet. Thus, in this example of a kind of "non-Euclidean" geometry (and there are many), the "fact" is that the shortest distance between the two points is what we call a "great circle route."

Neither the relative movement of the Earth and the sun nor the structure of the atom has changed, of course; nor has the reality (or the conspiracy theorists' denial) of the Apollo moon landings. Our standard arithmetic continues to work for us, whether it is "true" merely by its own definition or by something more. Geometrical axioms—Euclidean and otherwise—still allow us to plan and to fly, and to make engineering students suffer. . . .

So, what *are* "facts"?

Frankly, all we can deduce is that when we use the term "fact"—or even "truth"—we really mean nothing more than *something we have chosen to accept and/or believe*. Although this realization may make some of us uncomfortable, it is practical and, perhaps more to the point, pragmatic. It permits us, for example, to no longer accept an Earth that is the center of our solar system, and to plot efficient airline routes. And yet, it in no way detracts from our ability to deal with everyday "facts" in the way we have always done. And, we'd suggest, it means that when we go online to look for 5WH "facts," for instance, we need to remain skeptical and we need to ask ourselves another set of 5WH questions: *Who* wrote this material? *What* is its function on this site? *When* was or is it valid? *Where* did the author find it? *Why* did the author make it available here and now? *How* critically should I receive it?

3–2b KISS

The acronym KISS translates as "Keep it simple, stupid." While its wisdom may apply to many things, view it as good advice for engineering communication. And remember: Simple does not equate to simplistic. Engineering communication often deals with complex, highly technical matters. The more simply you can express such ideas, the more effective they're likely to be. Consider that the beauty of one of the most complex ideas of our time derives from the simplicity of its formula: $E = MC^2$. (The author of this formula, Albert Einstein, elsewhere helpfully noted: "Everything should be made as simple as possible, but not simpler.") Here are some points to consider:

- Don't overcomplicate your CMAPP analysis. Look at each component, but avoid second-guessing yourself or allowing the answers to your CMAPP questions to become a first draft of the Great American Novel. In all your engineering communication, try to find the least complicated response to the "problem" that your communication is addressing, and try to find the simplest way of conveying that response.
- Avoid the temptation of using the vast variety of fonts or clipart images or slide themes offered by your software. Just because they're available to you doesn't mean they'll enhance your audience's understanding.
- Don't make the mistake of assuming that your message will seem more educated, formal, or sophisticated if you use complicated sentences and uncommon vocabulary. Normally, the opposite is true: Effectiveness usually increases when you keep your message as simple as possible. As an engineer, of course, you will need to consider your audience and your context to decide on the appropriate level of technicality—the degree to which you use specialized vocabulary.

3-2c ABC

The abbreviation ABC refers to three aspects of your message and your product: accuracy, brevity, and clarity.

Accuracy

When composing your message, ask yourself the following:

- Have I chosen the right facts for this situation? In other words, are all my facts pertinent to my context, my audience, and my purpose?
- Are all my data correct? (Have I checked?)

Remember that your audience "needs" your information. As you learned in Chapter 1, engineering communication is audience-driven. But your audience does not need data that will muddy the issue. Recall as well that your message represents you to your audience, and thus your reputation hangs on it. Imagine the consequences if your audience were to find an error in your information: Your message would lose credibility, as would you, the "messenger." Once lost, your credibility would be very hard to regain. The consequences of inaccuracy in engineering communication can be even more dramatic: Think of the potential impact of inaccurate stress specifications in documents sent to steel detailers, for example.

Brevity

A practical definition of brevity might be, "Say what you need to say, and then stop." If you include in your document or presentation material that is not relevant to your context, message, audience, and purpose, your audience may be confused, irritated, and/or bored—and all your efforts will have been wasted.

Clarity

If your audience cannot readily understand exactly what you mean, your message is ineffective, and could produce dangerous results. Clarity derives from the words and grammatical structures you use, from the organization of your information, from the logic and cohesion of your arguments, and from the way you present your message to your audience. Like accuracy, clarity is an absolutely necessary characteristic of good—and safe—engineering communication.

3-2d CFF

The abbreviation CFF refers to content, form, and format. In "The Total Package" (see Figure 3.1, below), a short article that appeared years ago in a publication aimed at small and home-based businesses, the author discusses each of these components. Perhaps surprisingly, the points he makes still seem valid, and they apply just as strongly to an engineering student writing a report as they did to the business audience he was addressing.

FIGURE 3.1: The Total Package

Everyone these days seems to want us to buy a "package deal." Whether it's a stereo (a "sound system"), a computer (an "integrated system"), or clothing (a "lifestyle"), it has become more and more difficult to purchase a single item. According to my own rather limited research, I'm always being asked to walk away with a combination of some kind.

To be honest, most of us want to sell a "package deal" as well. For some of us, it may be merchandise; for others, it's services. To compete, however, we have to be ready, willing, and able to offer a whole rather than a part. Coordinating the various pieces can be time-consuming and, occasionally, frustrating. But it is the way of the commercial world we live in, and it does have its undeniable value.

Making sure the package is complete and that all the tab As will fit into the appropriate slot Bs often requires additional expertise, which we sometimes have to learn from scratch. Since, for most of us, time is more than just money, that necessity can easily become a burden we'd rather not shoulder.

Each and every piece of writing you send out represents you to the people who pay your bills. Just as you'd never sell them a package of goods that don't fit together properly, you shouldn't send them a document that doesn't either. You should look at everything you write as being a set of integrated components: the content, the form, and the format.

Adapted from *Hats Off to English* by David Ingre, published by David Ingre Written Image Services, 1990. © David Ingre.

Content

This is something only you can decide on. But you should ask yourself a question each and every time: Am I writing the right thing to the right person at the right time? Unless your answer to each of these questions is an unqualified "yes," think again about what you want to do and why you're doing it.

That may sound like a profound statement of the obvious. However, most of us have a tendency to just sit down and start writing. A lot of planning (not a little—a lot) can make just as much of a long-term difference. Ask anyone who's later had to defend in court what he or she wrote in a hurry.

Form

The language you use when confronting your teenage son who has just crunched the fender on your new car is not the same as the language you use in a contract proposal. Similarly, constructing a logical argument isn't all-important when you're having a backyard beer with your neighbor—but it is when you're making a presentation to a potential client.

Form—the language you use—should vary according to the situation. This is important each time you use your company stationery. You have to make sure that the language you've chosen is right for your context, and that you've put things into an order that is both easy to follow and very convincing. Otherwise, your reader may react just the way your son did.

Format

A letter's just a letter, right? Not so. It's unlikely you'd be impressed if your lawyer or your biggest supplier sent you an invoice scrawled illegibly on an old piece of foolscap, forgot to date it, or left out your company name or address.

Your readers aren't going to be impressed with your company if you don't choose an accepted business letter format, either. There are a number of such standard formats. (The most common is called "block.") Which you choose is not at all as important as using the same one accurately and consistently. And it would be well worth your while to make that decision yourself, rather than leaving it to an assistant. After all, you're the one who should be most concerned with the image you're creating.

One of the consequences of the technology that surrounds us is that people's expectations have risen. We used to be fairly blind to things like poorly spaced text on a page, evidence of corrected letters or words, inconsistent or poor-quality print, and even the occasional typo. We knew how long it took to do it over, and so we were usually sympathetic.

We're more discriminating now. We notice that crossed-out address and the new one typed beside it. We can tell what type of printer you've used, and we make judgments about the quality of your stationery. That may not be fair, and it may not really have much to do with the quality of what you're selling. But it is reality. And if you ignore it, your bottom line will eventually suffer.

To sum up . . . Whatever your message and your audience, put together the best total package you can afford. Don't assume that people won't notice or that, if they do notice, they won't care. They'll do both.*

3-2e CAP

The acronym CAP represents the adjectives "concise," "accessible," and "precise." (The corresponding nouns would be concision, accessibility, and precision.) These elements relate to the way you construct your message and your product, and all play a role in ensuring that your "total package" is effective.

Concise

The meaning of "concise" could be summed up in the sentence "If you can say something in six words, don't use 26." Assume that your audience is busy and will want to gain the necessary information as quickly as possible. Assume, too, that concision is also a facet of the KISS principle.

Consider the following two notifications, each advising regular committee members of an upcoming meeting:

> All of the various individuals who regularly present themselves in attendance at the normal meeting of the committee usually referred to as the TSSC (the Technical Standards and Specifications Committee) are invited to the upcoming regular meeting of the Committee, to be held from 3:00 p.m. to 5:00 p.m., on Wednesday, March 15, 2016, in the customary meeting place, Room 215.

> The next TSSC meeting will be held March 15, 2016, from 3:00 p.m. to 5:00 p.m., in Room 215.

Which would you prefer to have to read during a hectic day at the office?

Accessible

Accessibility refers to the way the document has been structured to permit the audience to extract important information quickly and easily. It, too, reflects good application of the KISS principle. Again, assume that your audience has neither the time nor the inclination to read every word of your document. Rather, he or she will probably want—at least at first—to scan, noting the significant ideas and gaining a sense of the document's organization.

You make information accessible through clarity of language and presentation. Consider the documents in Figures 3.2 and 3.3. Both present information about Griffin Radisson, the CEO of Radisson Automobiles Inc. However, the information in Figure 3.2 is presented in the form of a standard paragraph (traditional prose format), while the information in Figure 3.3 appears in a format that an engineering communication audience would consider more accessible. Note how the arrangement of headings and lists in the latter figure takes the place of the more traditional topic, supporting, and concluding sentences in Figure 3.2. Notice, too, how the organization and the presentation of the information in Figure 3.3 allow you to grasp the essential facts at a glance.

---

*Adapted with permission from *Hats Off to English*, by David Ingre ©DIWIS 1990.

FIGURE 3.2: Traditional Prose Format

Griffin Radisson is the 56-year-old president and chief executive officer (CEO) of Radisson Automobiles Inc.

As a young man, Griffin left Dallas and hitchhiked to Vermont. With much energy but little enthusiasm, he completed an automotive mechanics diploma at Grandstone Technical Institute in Randolph, a small town about an hour's drive from Burlington. Then, having suddenly identified a goal, Griffin spent a tiring 16 hours driving nonstop to Ann Arbor with a friend. There, he registered at Ann Arbor University. While completing his studies over the next four years, he worked part-time and lived frugally. Upon receiving a Master of Business Administration degree, he accepted his father's gift of a first-class plane ticket and returned to his home in Dallas to work with Maurice in the family car business.

Griffin began as a lowly mechanic's assistant but soon obtained his mechanic's ticket. Shortly thereafter, his father promoted him to the sales floor, where he proved himself an eager and able salesman. Sometime later, Maurice had Griffin take over a junior position in the company's administrative office, so that he could become familiar with the business's financial operations. Over the next 10 years, he advanced into management positions. Eventually, Maurice decided to retire and persuaded the board of directors to appoint Griffin president and CEO. Since then, Griffin has devoted himself to Radisson Automobiles Inc., overseeing the creation of new dealerships in Atlanta, Boston, Chicago, Denver, Los Angeles, and Seattle.

Despite the rapid and successful expansion of what he continued to view as "his company," Griffin insisted on maintaining his head office in the Dallas neighborhood in which his father had set up the first dealership. The office address is 7500 Lemmon Ave, Dallas, TX 77802. The head office staff can be contacted by phone at (979) 555-0099, by fax at (979) 555-0091, by email at hq@rai.com, or through its website, www.rai.com. Though in the past he did so commonly, Griffin now rarely offers a client or an employee his home telephone number of (970) 555-7374 or his personal cell number of (970) 555-1122. More and more, this self-made multimillionaire enjoys the time he spends away from what he views as the constant struggle of the marketplace.

FIGURE 3.3: More Accessible Engineering Communication Format

Griffin Radisson:	President and CEO, Radisson Automobiles Inc. (RAI)
Age:	56
Home phone:	(970) 555-7374
Personal cell:	(970) 555-1122
Education:	Automotive mechanics diploma, Grandstone Technical Institute, Randolph, VT MBA, Ann Arbor University, Ann Arbor, MI
RAI positions:	Mechanic's Assistant (1998) Certified Mechanic (2000) Junior Sales Associate (2002) Sales Associate (2004) Administrator (2008) Partner (2010) President/CEO (2012–present)
RAI Head Office:	Address: 7500 Lemmon Ave, Dallas, TX 77802 Phone: (979) 555-0099 Fax: (979) 555-0091 Email: hq@rai.com Web: www.rai.com
RAI dealerships:	Dallas, Atlanta, Boston, Chicago, Denver, Los Angeles, and Seattle

Precise

Engineering communication requires precise language—wording that is explicit, clear-cut, and specific: the *right* word in the right place at the right time, given your audience, context, and purpose.

Distinguish, however, between "precise *versus* vague" and "precise *versus* incorrect." Figure 3.4 shows examples of the former; using vague phrasing can not only confuse your audience, it can endanger it. Figure 3.5 offers examples of the difference between two terms, both of which are *precise*, but only one of which is *correct* in the respective context.

FIGURE 3.4: Precise versus Vague

Precise	Vague
The transformer is an 8" cube.	The transformer is about the size of a small box.
Our Web order of April 13, 2015, was for five Husky brand, 3.2 running Hp, 60 Gallon Cast Iron Air Compressors, model VT6314.	We recently ordered some compressors.
Use the front thermostat to ensure that the motor's internal temperature does not rise above 55° Celsius.	Be careful not to let it run too hot.

FIGURE 3.5: Precise versus Incorrect

Precise	Incorrect
The bearing is spherical.	The bearing is round. (Round is two-dimensional; spherical is three-dimensional.)
Each cycle sees the piston head advance and return at a constant speed of 3 meters/second.	The piston head's velocity is 3 meters/second. (Velocity requires the specification of direction; thus, while the piston head's *forward* velocity might be 3 m/s, and its *backward* velocity might be the same, its velocity over a complete cycle is, in fact, 0.)
The impact suppressor spring is helical.	The impact suppressor spring is spiral. (A spiral curls outward on a single plane, involving length and width. Think of a flat coil spring in an old watch, for example. A helix involves three dimensions—it expands up as well as "out"; think of the front coil springs in a car, for instance.)

Sometimes, however, the audience and the context will mandate what is, in fact, *imprecise* and factually *incorrect* terminology. For example, you would not tell a realtor that you wanted a house with a *helical* staircase, even though, technically, a *spiral* staircase could not be constructed.

Further, you will often have to balance the need for precision against the need for brevity and concision. Consider the TSSC meeting notice discussed under *Concise*, earlier in this chapter. Through your CMAPP analysis, you have to judge exactly how much information your audience actually needs. For example, in your smartphone calendar for March 15, 2016, you might require nothing more than

TSSC: 3pm–5pm #D215

A reminder email to the clearly defined audience of SCPI members would probably need to be a bit more expansive:

Next TSSC meeting:

March 15, 2016

3:00 p.m. to 5:00 p.m.

Building D, room 215

RSVP local 4538

If you were sending a notification to someone who was not a member of the TSSC, you might include details such as the full name of the committee, the street address of building D, the complete phone number rather than just the local extension of 4538, and, perhaps, a list of TSSC members, as well as an agenda.

In each successive notification described above, the message would be more precise but increasingly less brief and concise.

As always, tailor your language to your audience, context, and purpose.

CASE STUDY

3 AAU Registration

SITUATION

Like most institutions, Ann Arbor University (AAU) requires online registration for courses. Students seem to agree with the university's External Marketing and Promotions Office that the system is "intuitive, user-friendly, accurate, and effective."

To access AAU's intranet, students must enter their student number and a password. (They are also required to choose and verify a new password prior to the beginning of each semester.) The password must be from 10–15 characters; it must contain at least one lowercase and one uppercase letter, at least one numeral, and at least one of the following special characters: &, %, #.

Once AAU students' overall academic standing has been approved, they receive specific dates and times indicating when they may go online to register for their courses for the upcoming semester. Since many classes fill quickly, students try to register as soon as the system allows them to. The system is able to show immediate acceptance or rejection for their choice of classes. Once their admissibility to their selected classes has been verified, students must pay online, using American Express, MasterCard, or Visa.

On May 18, 2015, Lester Mont, a 20-year-old AAU student, received approval to register as a sophomore in Civil Engineering for the following September. He was notified that he could log in to register as of 8:00 a.m. on Monday, June 15, 2015. Unfortunately, Lester's summer job as a Help Desk Assistant at Canberra Industries in Duluth obliged him to take an intensive training course that day; thus, he would be unable to log in to register at AAU until after 7:00 p.m., meaning that he might well find that the classes he wanted were already full. He would have been happy to ask his girlfriend,

Chris, to register for him; however, he knew that she would be on a canoeing trip in Colorado in mid-June. His parents, too, would be unable to help him, as they would be on their long-planned European river cruise.

He then thought of his maternal grandfather, with whom he had a warm relationship. Lester had spent a week of his previous Christmas vacation with Ernie and Alice Halburton in the small town of Chelsea, about 15 miles from Ann Arbor. At that time, he'd given them an iPad as a Christmas present, arranged for Verizon as their ISP, and set up Wi-Fi in their home. Overcoming their initial reluctance, Lester had taught them to use email and do some basic Web surfing, and—less successfully—had shown them how to use the Facebook account he'd opened for them. At the beginning of June, therefore, Lester phoned to ask Ernie to register for him. His grandfather agreed, but pointed out that he was still very much of a neophyte on the Web, and was concerned that he'd therefore be more of a hindrance than a help. Lester assured his grandfather that the procedure was quite simple, and that he would soon email him all the necessary information. It wasn't until Saturday, June 13, that Lester remembered to do that, however.

Figure 3.6 shows the content of Lester's email, and Figure 3.7 is the attachment he included.

Saturday, June 13

Hey, Granpa!

Sorry it took me so long to get this off to you. Things got hectic. Thanks very much for registering for me. Now that you're using your iPad, it'll be fun. Particularly since I'm only taking 4 regular courses this coming semester. I made arrangements last term when I took an extra one so I'd have an easier time this Sept when I'm going to be really busy anyway, what with the part-time job and all. I'm also hoping to have the time to get back out to Chelsea to see you and Grandma Alice, by the way. That'll be fun. Maybe we can go to the Country Fair again, like we did when I was a kid.

Anyway, here's the info you need to register for me. You can see that it's pretty simple. Even one of those robots I'm supposed to be learning to program here at Canberra could do it. Just go to www.aau.edu/register and follow the prompts. And don't forget . . . you're going to be me ☺ . . . My login is my student number, 200933987, and my PW is OuCan2dothis% (btw, that's a zero to start). By the way, the job is just great. I'm learning all kinds of cool stuff. I can't wait till I see you guys later on to tell you about it. Some of the things they're building out here are really awesome!

In case you get stuck, here's a URL for a YouTube tutorial that the college put up: http://www.youtube.com/watch?v=AAUrgt58l8=MM30Qrm30_48Lmn33. I'm attaching a jpeg screen cap of a clipping from the AAU calendar, just so you'll see how they list the courses when you go to register. Oh, and would you mind putting all this on your own VISA and I'll pay you back later in the summer. So you'd better have the card handy when you go online.

I'm going to need either one of the two Hydro sections, and either one of the survey-meth, plus the design and cmns courses. If you can, I'd prefer to get my classes over with earlier rather than later, but sometimes it's hard to get the sections you want. Oh, please try to get me Goldberg and Cornelian, they're both really good.

I'll give you a call when I get back. I'm going to Moosehead Lake tomorrow with some friends from here, but I'll be back at work dark and early on Monday for that training session that keeps me from registering myself. So I guess I'll phone you sometime Mon pm. Please don't forget that you need to get online right at 8 on Mon morning. If you leave it until later, a lot of the classes will likely be filled up and I won't get what I want. The whole thing should take you about half an hour I figure.

Thanks again, Granpa! Love you guys!

Les

FIGURE 3.6: Lester's Email to Ernie Halburton

FIGURE 3.7: Screen Capture from AAU Course Timetable

ENG 2100 Civil Engineering Design				
A339940	001	M W	08:30 10:00 F2100	Shen, A.
A339941	002	M W	11:30 13:00 F2239	Sayers, R.
A339942	003	T H	10:00 11:30 F2100	Parl, R.
A339950	004	W F	15:30 17:00 R2240	Singh, B
SRV 2200 Surveying Methodology				
A489901	001	T H	1000 1130 F2100	Goldberg, P.
A489903	002	M W	10:00 11:30 F2230	Parl, C.
A489905	003	W F	08:30 10:00 R2240	Singh, E.
HDD 2500 Hydraulics and Drainage				
C11100	001	T H	10:00 11:30 P1000	Harby, I.
C11103	002	W F	1530 17:00 P1005	Weischart, F.
ECC 2112 Engineering Communication				
M22001	001	W F	1130 13:00 A1100	Gatmar, E.
M22003	002	T H	08:30 10:00 A1115	Dijani, F.
M22005	003	T F	10:00 11:30 A1115	Hollings, A.
M22007	004	T F	14:00 15:30 A1100	Porseau, R.
ENG 2118 Research Methodology				
M32222	001	M W	08,30 10:00 A2110	Dijani, F.
M32233	002	T F	08,30 10:00 A2100	Colworth, W.
M32244	003	W F	14,00 15:30 A1100	Atwal, A.

ISSUES TO THINK ABOUT

Consider Lester's email in terms of the CMAPP attributes covered in this chapter.

1. Discuss the message's appropriateness for its audience.
2. Discuss the organization of the message.
3. Explain Lester's primary purpose.
4. Discuss the content of the message in terms of Lester's purpose.
5. Discuss the format's effectiveness in terms of the message and the audience.
6. Justify to what extent you think Lester's communication reflects the aspects of accuracy, brevity, and clarity.
7. Using examples, show to what extent you think his email reflects the attributes of concision, accessibility, and precision.
8. What improvements would you make to the email?
9. To what extent do you think the screen capture will help Lester's grandfather?
10. How will the screen capture's form and format make the message easier or more difficult for Lester's grandfather to understand and use?
11. How well does the screen capture reflect the information Lester provided about it in his email? Give examples.

You'll recall that you often have to strike a balance between precision and brevity, and that to do so, you have to think about your context, your audience, and your purpose. As you answer the following questions, put yourself in Ernie Halburton's position. Keep in mind what you know about his familiarity with online tasks.

12. How is Ernie likely to react to the level of accuracy, brevity, and concision expressed in the timetable information?
13. What do you think might present the greatest difficulties for Ernie? For example, to what extent will he understand what each of the columns represents?
14. What do you think that Ernie, as Lester's audience, actually needs and wants to know?
15. What do you think Lester should have done, on the assumption that he still needed to ask his grandfather to register for him?

EXERCISES

3.1 Give brief definitions of the following complementary CMAPP attributes:

- 5WH
- KISS
- accessibility
- accuracy
- brevity
- clarity
- concision
- precision

3.2 Assume that you have decided to purchase a new computer system. Inventing all the details you may require (e.g., about the system and the vendor), create one message for each of the following audience/context combinations:

- You are leaving yourself a note about purchase details so that you can later update your expense log (whether on paper or software such as Quicken®).
- You are writing or emailing a friend, indicating your delight (and slight apprehension) about the learning curve you will probably face.
- You are informing a computer-literate professor who has suggested that your computer-generated assignments "could be more professional."

3.3 Using what you have learned about engineering communication so far, create revised versions of each of the documents prepared by Lester in the case study.

3.4 Obtain an engineering document from a private- or public-sector source. Illustrating your points with examples from the document, briefly assess its effectiveness in terms of each of the following:

- 5WH
- KISS
- accessibility
- accuracy
- brevity
- clarity
- concision
- precision

USEFUL WEBSITES

Note: For ease of reading, we have removed the "http://www." from most of the URLs below. However, some browsers might still require that you insert it to precede some URLs so as not to receive an error message.

■ The 11 Rules of Writing, Grammar, and Punctuation site offers useful guidelines for writing more effective prose, including "Omit unnecessary words" (rule 11). Examples of each rule and links to related sites are provided.
Visit: junketstudies.com/joomla/index.php/11-rules-of-writing/the-rules

■ The Acronym Finder is a searchable database that you can use to look up more than 131,000 acronyms/abbreviations and their meanings.
Visit: acronymfinder.com

■ Conspiracy theory sites range from logical examinations to hateful fantasies. Here are a few that appear to take a reasoned approach.
Visit:

- "A Debunking of the Moon Hoax Theory" (Robert Braeunig), at braeunig.us/space/hoax.htm
- "Debunking the 9/11 Myths: Special Report" (Popular Mechanics), at http://www.popularmechanics.com/military/a49/1227842/
- "Did We Land on the Moon?" (Keith Mayes), at http://www.thekeyboard.org.uk/Did%20we%20land%20on%20the%20Moon.htm

■ Examining Famous "Conspiracies" (Robert Sheaffer), at http://www.debunker.com/conspiracy.html

■ The Faculty of Applied Science & Engineering of the University of Toronto in Canada offers an online handbook with tips for effective communication for engineering students.
Visit: http://www.engineering.utoronto.ca/Directory/students/ecp/handbook.htm

■ Historical revisionism sites are common, and range from the thoughtful and objective to the prejudiced and offensive.
Visit:

- Perspectives on History (American Historical Association), at https://www.historians.org/publications-and-directories/perspectives-on-history/september-2003/revisionist-historians
- Historical Revisionism—The Huffington Post, at huffingtonpost.com/tag/historical-revisionism/
- "How Revisionist History Works," at history.howstuffworks.com/history-vs-myth/revisionist-history.htm

■ Jack Lynch's "Resources for Writers and Writing Instructors" page offers a wealth of links to information on effective communication.
Visit: andromeda.rutgers.edu/~jlynch/Writing/links.html

CHAPTER 4

Research and Reference Works

When you communicate as an engineer, you'll want to make sure that what you write (or present) is supported by applicable research, that you document that research appropriately, and that the language you use is correct for your situation (see *Correct Language* in Chapter 1). This chapter looks at the kinds of tools you will want to use on a regular basis.

4–1 Research

Broadly speaking, research sources can be *primary* or *secondary*. In this context, the terms are not synonymous with "major" and "minor." Primary source information is unpublished, and you obtain it firsthand. Thus, primary sources normally include interviews, surveys, your own observations, photographs or maps that you have created, as well as printouts of your own company's financial data.

When you look up information that has already been prepared and published by someone else, however, you're using secondary sources. Typical examples include books, periodicals, directories, government publications and—more and more—websites and other electronic databases.

4–1a LIBRARIES

Today most students' first inclination when doing research is to turn to the Internet, which does indeed provide easy and rapid access to data. However, its very convenience often leads us to forget that university and college libraries have been accumulating information for literally hundreds of years. Their resources, furthermore, have been gathered, vetted, and organized by literate specialists—librarians who can provide personalized assistance and broad expertise. Recall, too, that today's libraries offer access to online journals, indexes, and databases.

Ignoring the enormous value of a library would be foolish and unprofessional, whether you are a first-year college student or an alumnus who now runs a large engineering firm. This cliché is only a slight exaggeration: *A librarian can be your best friend.*

4–1b ONLINE SEARCHES

You're likely already familiar with standard Internet search engines and directories, from the now ubiquitous Google to Yahoo! and Bing. There are, of course, many others. (See the Web links at the end of this chapter for more information.) As well, there are countless specialized databases—some free, some costly, some accessible to the public, some available only to members or clients. Note that as a student you likely have access to numerous technical and engineering-focused databases via your school's intranet system.

Using the Web as an effective research tool, however, is a skill with a growing learning curve. Its greatest assets are the seemingly limitless amount and variety of data now available, and the fact that it remains an arena largely unrestricted by any centralized controller. However, the Web does not readily distinguish authoritative fact from whimsy, fantasy, or invective.

The growing phenomenon of blogging has added to the difficulty of discerning highly biased opinion from substantiated, objective assertion. A personal blog is, in fact, simply a Web page; the contentions of its author can be presented as convincingly as those of any other Web poster, from the U.S. Department of Agriculture to Harvard University to the American Society of Civil Engineers.

Consequently, we offer three pieces of advice.

- Take advantage of the truly massive resources of the Internet.
- Always question the reliability of the source of what you find on the Web; typically, if the source is reputable, the information is more likely to be so, too.
- Remember that the Internet is *not* the repository of *all* the information in the world. Visit your university or local library. Befriend a librarian.

4–2 Documenting Sources

Whether you have used primary sources or identified secondary sources through the library or the Web, you have to document them. First, you need to show your audience that you are not pretending to claim ideas and/or words that belong to someone else as your own. And second, you need to give your audience ready access to your source material, should they wish to find out more about your topic. The first issue concerns avoiding plagiarism; the second relates to the use of style guides. They are applicable equally to your activities as a student and to your engineering career.

4-2a PLAGIARISM

Plagiarism refers to passing off another's work as your own. This may involve

- verbatim transcription of material you have found and inserted into your report or presentation, but without a proper source credit;
- a quote (or other data or visual) you have "modified" before including it as part of your own submission, but for which you have not provided due credit;
- material you submit that is not entirely your own composition: that is, material that has been written at least in part by someone else or that has been edited by someone else to the extent that your own language use and idiosyncrasies are no longer congruent;
- presentation of another's idea(s) as your own: that is, the attempt to have your reader believe that you are the originator of something that is not entirely your own creation;
- material that you submit that you had submitted previously, whether for another course, at another institution, or in other circumstances—unless you have received prior permission to do so and have indicated this to your reader.

4-2b ETHICS

Plagiarism—intentional or accidental—is unethical communication. Just as it will be your professional responsibility to know and follow relevant engineering codes, it is your obligation to ensure that you do not plagiarize. Plagiarism is theft, and that's the fact. In a college or university, it can lead to severe academic consequences; in the marketplace, it can lead to lawsuits.

4-2c STYLE GUIDES

Many organizations strive to achieve a consistent image in their documents. Their document style is likely to include specific criteria for everything from document visuals such as fonts and margins (to be discussed in detail in Chapter 7) to citing sources. No particular style guide is necessarily "better" than another. Each, however, sets out slightly different rules for presenting information.

In businesses, the compilation of such rules for a consistent image is usually known as a *house style*. Although any organization can develop its own house style, many companies and institutions have adopted well-established style guides, including the following:

APA

If you are producing documents for the social science disciplines (psychology, sociology, and so forth), you will probably need to follow the guidelines set out in *The Publication Manual of the American Psychological Association*, commonly referred to as the APA.

Chicago

Many North American organizations have modeled their house style on *The Chicago Manual of Style* (University of Chicago Press). For example, the house style of Cengage Learning, the publisher of this textbook, draws extensively on the Chicago style.

CSE

The style guide of the Council of Science Editors is often preferred by writers in biology, geology, chemistry, mathematics, medicine, and physics.

IEEE

Many technical organizations, particularly in the electrical and electronics fields, prefer to use the style guides of the Institute of Electrical and Electronics Engineers: the *IEEE Standards Style Manual* and the *IEEE Computer Systems Manual*.

MLA

If you are producing documents for humanities subject areas (literature, fine arts, philosophy, and so forth), you will most likely have to follow the style presented in the *MLA Handbook*, published by the Modern Language Association of America. The seventh and, at the time of writing, most recent edition of the handbook offers guidance in handling documents in electronic form.

Your choice of style guide will depend on your particular circumstances and be based, of course, upon your CMAPP analysis.

4–3 Language Use Tools

Every engineering communicator should make regular use of three language tools: a dictionary, a thesaurus, and a grammar and usage guide.

4–3a DICTIONARIES

A dictionary of a language is not the same as the language itself. Rather, it is a collection of its editors' opinions regarding the usage of the language. Occasionally, the editors are wrong. The eighteenth-century English writer, critic, and lexicographer Samuel Johnson was once asked why he had defined a term incorrectly; his response was, "Ignorance, pure ignorance."

A dictionary wears out, not because people use it too much, but because language changes. New words come into existence (think of "email," "Internet," and "app"), while established words take on new meanings (not too long ago, "gay" meant "lighthearted and carefree" and a "browser" was someone who skimmed a text). New words have been entering the language with increasing frequency—so much so that if your dictionary was last revised more than 10 years ago, it is already out of date.

Dictionaries reflect the language of a particular location. For example, most people consider the *Merriam-Webster Dictionary* to be the authority on American English, while the *Oxford English Dictionary* is regarded as authoritative for British English.

4–3b THESAURUSES

The word "thesaurus" derives from a Greek word meaning "treasure." Although its normal plural is "thesauruses," you will sometimes find the form "thesauri." A thesaurus is a book in which words are listed in categories according to their related meanings. Because no two editors see words in the same way, lists of synonyms will vary from thesaurus to thesaurus. Like a dictionary, a thesaurus can become outdated. For example, in James G. Fernald's *Standard Handbook of Synonyms, Antonyms, and Prepositions* (1947), "gay" is listed only as a synonym for "happy," with the explanation that "we speak of a gay party, gay laughter."

The function of a thesaurus is to offer options, normally in the form of synonyms and antonyms. Although synonyms have similar meanings, they are never exact substitutes; their use permits nuances that allow for finer distinctions of meaning and help maintain the richness of the language. Although some words seem to have exact opposites ("good" versus "bad," for example), the contrary meanings are not precise ("good" and "bad" are both highly subjective terms). Thus, antonyms allow for further nuances.

Recently published or revised thesauruses include *Roget's Thesaurus* (Random House, 2001), *The Merriam-Webster Collegiate Thesaurus* (Merriam-Webster, 2010), and *The Oxford American Desk Dictionary and Thesaurus* (Oxford University Press, 2010).

Merriam-Webster gives free access to both its dictionary and thesaurus online, at merriam-webster.com. You can also profitably visit Thesaurus.com.

4-3c GRAMMARS

Books offering comprehensive overviews of grammar, style, and usage abound. While some remain prescriptive/proscriptive ("do this but don't do that"), most good ones are more descriptive ("this is what people usually do"). Such grammars include sections on punctuation, capitalization, numbers, paragraphing, spelling, bias-free language, and documentation of references. (Note that Chapter 15 of this textbook is, in effect, a brief grammar compendium.) Which guide you use matters much less than whether you use one; at times, we all need reminders of standard practice. Competent ones include Salomone and McDonald's *Inside Writing: A Writer's Workbook (Form A)*, 8th edition (2014), Wadsworth, Cengage Learning, and Stageberg and Oaks' *An Introductory English Grammar*, 5th edition (1999), from Thomson.

You also can find a large number of Internet sites dealing with grammar and language use that offer both advice and practice. Purdue University's Online Writing Lab is one of the best (see its Web listing at the end of this chapter).

4-3d ELECTRONIC REFERENCES

Most of today's word processors offer the electronic equivalents of a dictionary (spell-checker), a thesaurus, and a grammar and usage guide (grammar checker).

Spell-Checker

A spell-checker is essentially a long list of words. As it checks your document, it compares every word it finds against that list. If it cannot find a match, it alerts you by flagging the offending word. While your spell-checker is very effective at finding typos, it pays no attention to problems related to meaning. The spell-checker used in writing this textbook, for example, is quite happy with the sentence "Stare Weigh Two Heaven is a peace of music thatch eye likes."

Thesaurus

An electronic thesaurus works by offering possible synonyms (and, often, antonyms) for a selected word. It is not uncommon for a word processor to feature a thesaurus that is as comprehensive and useful as a good print version. The one problem with an electronic thesaurus lies in its inconspicuousness.

If you don't "call it up," it will be of no assistance; and if you are unaware of distinctions of nuance (which often reflect cultural differences), you are unlikely to use it.

Grammar Checker

Grammar checkers attempt to match what you have written against a complicated series of programmed rules. Most allow you to choose among various sets of rules, perhaps labeled "business writing" or "casual writing." While grammar checkers can help you to avoid some careless errors and to become more familiar with some common grammatical patterns, they are not without drawbacks. Good writing is complex and subjective. Thus, depending on an electronic grammar checker could be seen as analogous to assuming a first-year college engineering text will allow you to understand the work of Buckminster Fuller, one of America's best known architects, designers, and inventors.

Grammar checkers also have a generally poor reputation for accuracy. They may ignore nonparallel structure, for example, or find "errors" in material considered highly literate. In reviewing the last two paragraphs of Patrick Henry's 1775 "Give Me Liberty or Give Me Death" speech, author Dave Ingre's grammar checker suggested replacing "Gentlemen" with "people"; it flagged as possibly incorrect each use of the passive; it objected to beginning a sentence with the word "But"; it identified what it called "non-standard question form"; it suggested changing "when we are totally disarmed" to "when we are very disarmed"; and it recommended changing most uses of "which" to "that." (Read about the speech at en.wikipedia.org/wiki/Give_me_liberty,_or_give_me_death!)

If you have an electronic grammar checker, by all means use it. Just don't expect it to be a reliable judge of the caliber of your engineering communication.

4-4 Professional Development: Staying Up to Date

As a student and, later, as a credentialed engineer, you will be expected to "stay on top" of what is happening in your area of expertise. How have recent newsworthy events or governmental regulations affected your industry, for instance? Your boss, colleagues, and clients will expect you to be able to answer that question, accurately and in depth, seven days a week. To keep up to date, we recommend that you use two types of online platforms: news alerts and news feeds.

4-4a NEWS ALERTS

News alerts scour the Internet for news, opinion, video, books, and discussions related to the search term(s) of your choice. These items are then sent automatically to an email address you have specified. In a sense, these alerts help the news find you, rather than the other way around, thus saving you time and making sure you are not missing something important.

Figure 4.1 illustrates the creation of an alert via Google News (google.com/alerts). You can also create alerts via Yahoo! and Bing, for example.

Figure 4.2 shows a list of engineering-related alerts that bring news stories, video, and blog posts into author Bob Basil's email in-box (Figure 4.3) daily.

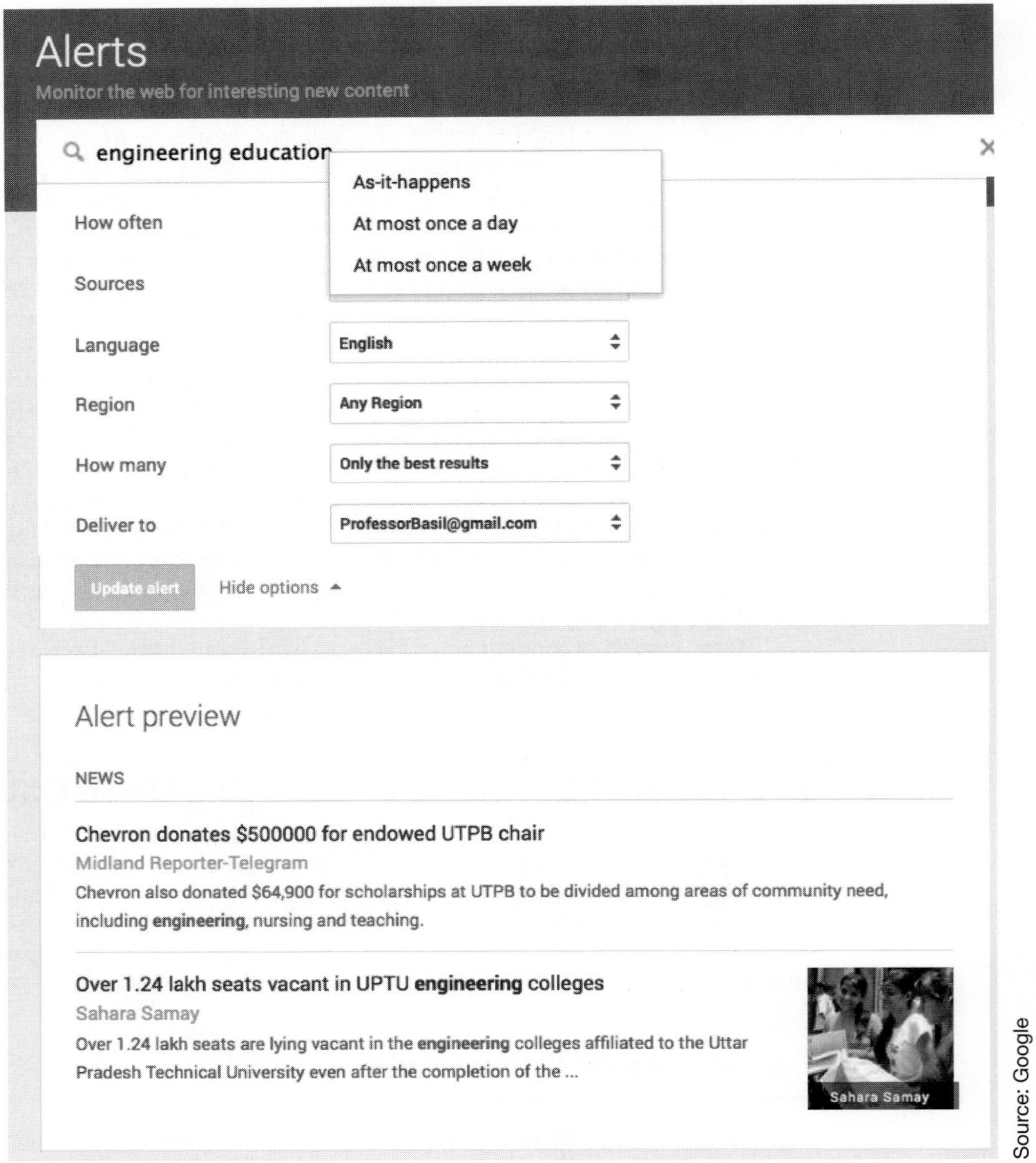

FIGURE 4.1: Creating a News Alert for "Engineering Education" via Google News

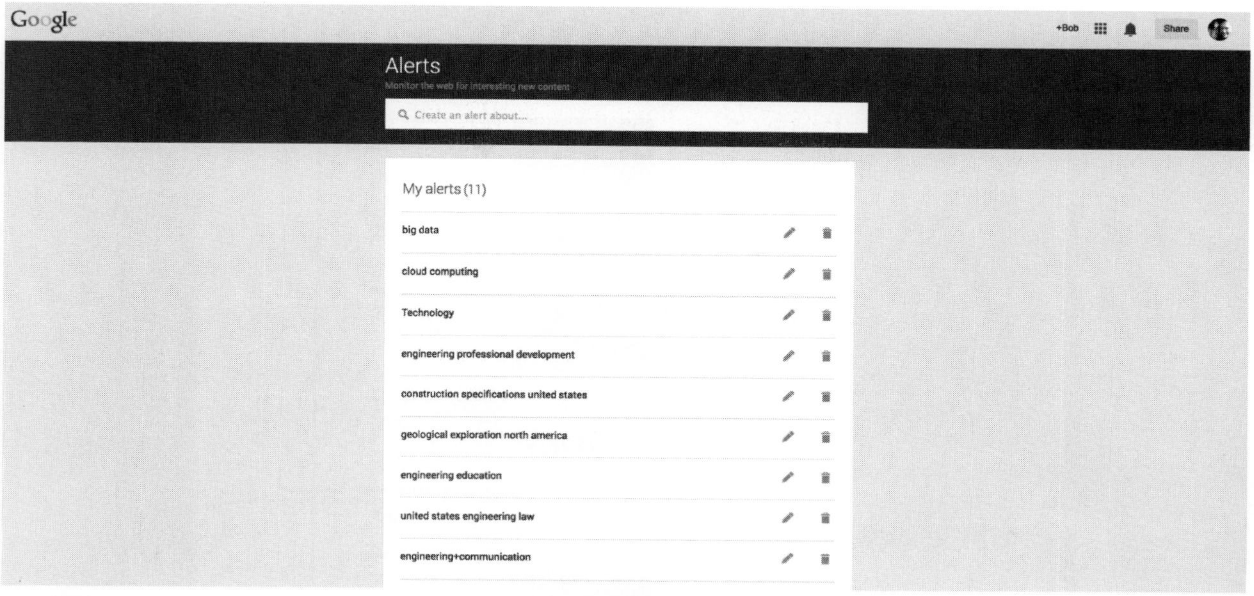

Source: Google

FIGURE 4.2: Collection of Engineering-Focused News Alerts Created via Google News Alerts

4-4b NEWS FEEDS

Another way of making the news come to you is to use an "aggregator" that collects articles from websites, forums, and blogs, for example, and organizes them onto a single page for you. Typically you can then access this page from any of your electronic devices such as tablets, smartphones, and computers. Popular personal aggregators include Feedly (feedly.com) and Digg (digg.com).

Figure 4.4 is an image of what author Bob Basil peruses every day. Note all of the engineering, digital-media, and economics feeds that appear on a single page. This saves the author a lot of time and energy visiting several dozen individual websites each day to see if anything new has been published.

We will go into more detail on the topic of digital and social media in Chapter 6.

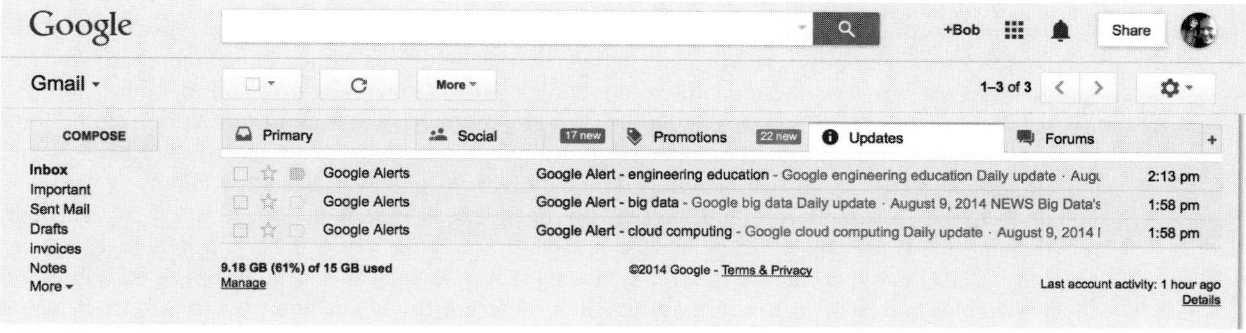

Source: Google Gmail

FIGURE 4.3: Engineering-Related Stories, Blog Posts, and Video in Bob Basil's Email Inbox

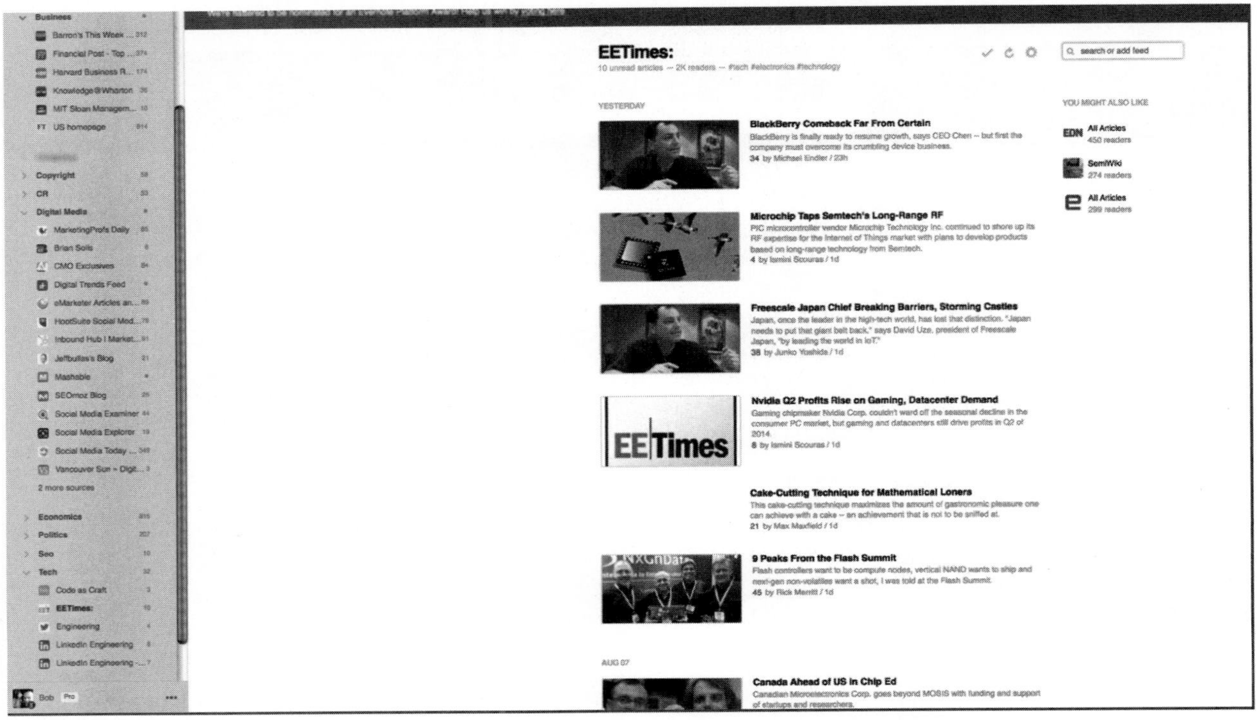

Source: Feedly

FIGURE 4.4: Feedly Aggregator Page

CASE STUDY

4 AEL Documentation

SITUATION

Melanie Smith is a recent civil engineering graduate of AAU, hired as a Junior Associate in AEL's Detroit office. Her boss, Dr. Mitchell Chung, the Detroit office's Senior Consultant, has asked her to prepare a report on the feasibility of building a curbside recycling program in a new suburban subdivision close to Chicago. The report is to be presented in one month to AEL's client, a joint government–private sector committee comprising representatives from Chicago's city government, Roundhouse Developers (who developed the subdivision), and EverGreen Recycling.

In her research, Smith goes to the City of Chicago's website at cityofchicago.org. Through the site, she finds cityofchicago.org/city/en/depts/streets/supp_info/recycling1/blue_cart_faqs.html, a FAQ page on blue-cart recycling. Recognizing the relevance of most of the questions and answers to her own work, she modifies the FAQ accordingly and includes the material in her report. From the City's Streets and Sanitation Department website at cityofchicago.org/city/en/depts/streets/supp_info/recycling1/recylingaz.html, she downloads a PDF copy of the *Chicago Recycling Guide: A–Z*. Again wishing to avoid reinventing the wheel, Smith incorporates several items from this file into her report.

At the end of her report, Smith provides a bibliography, listing the more than 50 sources that she consulted. Part of that bibliography is presented in Figure 4.5.

FIGURE 4.5: Extract from Smith's Bibliography

Suter, Jeffrey. *Citizen Participation in Recycling Waste*. New York: Desktop Press, 1999.

United States Department of Energy. *EREC Facts Sheets*. Energy Efficiency and Renewable Energy Network. http://www.eren.doe.gov/erec/factsheets/savenrgy.html.

———. Environmental Protection Agency. *EPA Environmental Fact Sheet: Municipal Solid Waste Generation, Recycling and Disposal in the United States: Facts and Figures for 2012*. www.epa.gov/osw/nonhaz/municipal/pubs/2012_msw_fs.pdf.

Burkley, Amy. *A Community Guide to Curbside Recycling*. Chicago: Burnside Press, 2001.

Auer, J. "The Green Room Debate: Genuine 'Green Rooms' Should Offer More Than Just a Marketable Amenity." *Lodging* (February 1994): 63–66.

Wagner, M. "Food Waste Recovery: Just Another Way to Reduce Waste." *Resource Recycling* (October 1995): 75–77.

Jenkins, R. *The Economics of Solid Waste Reduction*. Brookfield, Vermont: Edward Elgar Publishing, 1993.

American Plastics Council. *Stretch Wrap Recycling: A How-To Guide*. Washington DC, 1994 (22 pages).

Shrum, L. J., et al. "Using Marketing and Advertising Principles to Encourage Pro-Environmental Behaviors." (Chapter 11). *In Marketing and Consumer Research in the Public Interest*. Edited by Ronald Paul Hill. Thousand Oaks, CA: Sage Publications, Inc., 1996.

Staff of BioCycle, eds. *BioCycle Guide to Maximum Recycling*. Emmaus, Pennsylvania: The JG Press Inc., 1993.

Purcell, Arthur H. "Better Waste Management Strategies Are Needed to Avert a Garbage Crisis."

Helen Cothran, ed. *Garbage and Recycling: Opposing Viewpoints*. San Diego: Greenhaven, 2003: 20–27.

Environmental Protection Agency. *Cleaning Up Electronic Waste (E-Waste)*. http://www2.epa.gov/international-cooperation/cleaning-electronic-waste-e-waste.

ISSUES TO THINK ABOUT

1. After examining the FAQ and the PDF file mentioned in the Situation, discuss how Smith might have judged the reliability of those sources.
2. How do you view the legitimacy of her inclusion of the materials in her report? Explain whether she adhered to the standards of ethical communication. State what, if anything, you would have done differently.
3. Who is the principal audience for her report? What, if any, secondary audience do you identify?
4. Assuming Figure 4.5 is a representative sample of her bibliography, discuss to what extent it works in terms of her context, audience, purpose, and product.
5. What if any changes would you make to the bibliography, and why would you make them? (Once again, take the perspective of a CMAPP analysis.)

EXERCISES

4.1 What are three types of primary sources? When might you use each in creating engineering communication at college or university or at work?

4.2 Name at least five types of secondary sources. What might be the particular application of each to your studies or your work?

4.3 What concerns should you have when doing research on the Web? How might such sources impact your communication products as an engineer?

4.4 Suppose you are an engineer writing a feasibility report on renovating a client's computerized manufacturing process for tennis racket frames. Your research has uncovered a set of specifications relating to the manufacture of squash rackets in England. You take down the numbers, modify them for your own purposes, and create an effective chart that you include in your report. Discuss the ethics of your action and your alternatives.

4.5 Create a list of search items that you would use to create appropriate and helpful news alerts, and explain your selections.

4.6 Using Feedly or Digg (or another aggregator of your choice), create a list of resources (websites, blogs, and so forth) an engineer in your field might need to look at frequently, and explain your choices.

4.7 Dolores Lopez works for a manufacturing company outside of Troy, Michigan. The company produces cold-headed, hot-forged, screw-machined, stamped, and wire-formed fasteners for sale to wholesalers and retailers throughout the United States. The officers are interested in expanding internationally. They are aware that the demand for U.S.-made fasteners in foreign markets has grown considerably over the past several years. However, having sold their products only in the domestic market, they are unsure of the regulations involved in selling specialized metal and plastic products abroad. The company is particularly interested in selling its products in Russia, South Korea, and Saudi Arabia. They have asked Dolores to gather information and prepare a report for them regarding the advisability of expanding the business in these directions. Specifically, Dolores needs to determine (1) the technical standards required, (2) the specifications for the most common markets, (3) the packaging, labeling, and shipping requirements, and (4) any restrictions on the sale by either the United States or the importing country. Where and how should Dolores begin her search?

USEFUL WEBSITES

Note: For ease of reading, we have removed the "http://www." from the URLs below. However, some browsers might still require that you insert it to precede some URLs so as not to receive an error message.

- IFLA is the International Federation of Library Associations and Institutions. It maintains a portal page dedicated to style guides and related resources. It contains links to a broad variety of research- and documentation-oriented advice and assistance. Visit: ifla.org

- Infoplease is a portal with links to a variety of research sites and tools, from almanacs to dictionaries, and from U.S. history and government to major white page telephone listings. Its tagline is "All the knowledge you need."
Visit: infoplease.com

- Mashable.com is an American and British blog and perhaps the Internet's preeminent news site for developments in social/digital media, mobile media, and high-tech innovation. The link below is its "take" on the Feedly news aggregator.
Visit: mashable.com/category/feedly

- MIT's online learning module on academic integrity offers a comprehensive discussion of plagiarism and related issues.
Visit: mit.edu/aeroastro/academics/graddocs/academic_integrity.pdf

- Purdue University's Online Writing Lab, often referred to as "The OWL," is a highly regarded source for all kinds of advice on good writing—including grammar and language use.
Visit: owl.english.purdue.edu/owl

- RefDesk is a useful information portal. Billing itself as "The Single Best Source for Facts," it offers links to a very broad variety of information repositories, from current news to online dictionaries, and from "today's birthdays" to genealogy sources and maps.
Visit: refdesk.com

- Scambusters provides an "Urban Legends and Hoaxes Resources Center" that is a highly useful resource for determining the reliability of many claims published on the Internet.
Visit: scambusters.org/legends.html

- Snopes's urban legends references pages are, despite the many ads that clutter them, an excellent resource for checking the validity of the warnings and claims that people often receive by email.
Visit: snopes.com/snopes.asp

- The Institute of Electrical and Electronics Engineers provides a PDF version of its widely used *IEEE Standards Style Manual*.
Visit: ieee.org/documents/style_manual.pdf

- UWM, the University of Wisconsin-Madison, provides its Internet Citation Guides that offer specific advice on citing electronic sources in research papers and bibliographies when adhering to diverse style guides.
Visit: researchguides.library.wisc.edu/citing

- "Web Searching Tips" is part of Search Engine Watch. It provides excellent tips on using Web search engines more effectively, as well as news stories and commentaries on search-engine-related matters.
Visit: searchenginewatch.com/facts/index.php

CHAPTER 5

From Data to Information

This chapter deals with outlining—using the CMAPP approach to convert data into useful information. This outlining process has particular relevance to formal reports and long presentations, but its underlying principle applies to *all* effective communication. In order for your audience to grasp your meaning, your thoughts must be organized and your understanding of what you want to convey must be absolutely clear. You've probably heard the old saw that "engineers build things." How can you build anything sturdy—including communication—if you haven't planned and organized first?

5–1 Data versus Information

While the words "data" and "information" are often used interchangeably, you *can* look at them differently. Note, by the way, that *data* is actually the plural of *datum*—"a fact or proposition used to draw a conclusion or make a decision" (*Online American Heritage Dictionary*, 2014). The use of "datum" has become relatively rare, and many writers now use data with either a singular verb ("the data is available") or a plural verb ("the data are all in"). Data are almost always empirical; that is, they can be objectively verified and/or measured. Though data on a given topic might be plentiful, its lack of organization makes it difficult to interpret. For example, as an engineer, you could readily obtain a multitude of data on the tensile strength of various types of steel. For that data to be useful, however, you must analyze it and cull from it what you need to build a safe structure. You convert what you and your audience need into useful *information*.

5-2 Process and Structure

Both primary and secondary sources can yield enormous amounts of data. The challenge for you as a professional is translating that data into information you can use to produce a CMAPP product such as a report. To accomplish this, many engineers apply their professional approach to their communication: Before they begin the actual writing, they collect their data, organize that data into a formal multilevel outline, manipulate that outline, and then finalize it.

A formal multilevel outline takes the form of a series of headings of different levels, called *level heads*. Two level-head numbering systems are shown in Figure 5.1.

FIGURE 5.1: Alphanumeric and Decimal Numbering Systems

Alphanumeric System	Decimal System
I. Level-1 Head	**1. Level-1 Head**
A. Level-2 Head	1.1 Level-2 Head
B. Level-2 Head	1.2 Level-2 Head
1. Level-3 Head	1.2.1 Level-3 Head
2. Level-3 Head	1.2.2 Level-3 Head
II. Level-1 Head	**2. Level-1 Head**
A. Level-2 Head	2.1 Level-2 Head
B. Level-2 Head	2.2 Level-2 Head
1. Level-3 Head	2.2.1 Level-3 Head
a) Level-4 Head	2.2.1.1 Level-4 Head
b) Level-4 Head	2.2.1.2 Level-4 Head
2. Level-3 Head	2.2.2 Level-3 Head
C. Level-2 Head	2.3 Level-2 Head
III. Level-1 Head	**3. Level-1 Head**

5-2a A NOTE ON PROCESS

For generations, people created multilevel outlines using pen (or, even better, pencil) and paper. Since working with an outline involves making many changes, the process often involved rewriting "clean copies." Admittedly, some people still prefer to work with this "old-school" method. Most, however, now use software to create and work with a multilevel outline. Although you can buy specifically designed applications, standard word processors such as Word (from the Microsoft Office suite) and Writer (from the open-source LibreOffice suite) include an "outline" function. If your word-processing package includes such an option, we think it would be well worth your while to learn to use it.

5-3 Collecting Data

Imagine that you needed to report on your day's activities. Think of the list of activities presented in Figure 5.2 as the data you have quickly compiled. (Note that it will not yet be in a form that you would want to show to anyone else—your professor, for example.) But how you translate that data into useful information in the form of a report will depend on the answers to the CMAPP questions you devise.

FIGURE 5.2: Day's Activities

1) Woke: 7:00 a.m.
2) I fed my dog Prince
3) Did homework: 1.5 hours
4) Breakfast
5) Lunch was in the cafeteria
6) I ate supper at home as usual
7) New jacket looks great
8) Drove car to campus
9) Attended BusMgmt 305
10) Comm. 302 class
11) Bit late for Digital Circuits 313
12) Engineering Applications 315L: interesting class today!
13) TV: one hour only
14) Went to bed
15) Carlos told me about his new car
16) From Comm. Instructor—info re next assignment
17) Spoke to BusMgmt instructor about term paper
18) Continued Chapter 3 of Engineering Applications project
19) Read part of latest issue of Time magazine
20) Internet search on History of Electronics
21) Bought new Engineering Applications textbook
22) Researched Digital Circuits project in library
23) Taped TV program on History Channel
24) Saw results of last Engineering Applications midterm: awful!
25) What the 3 new library books are about
26) Composed part of Comm. assignment on computer
27) Spoke to Jan on phone
28) Asked Dad for raise in allowance: unsuccessful
29) Walked Prince: 1/2 hour
30) Finished Chapter 2 of Engineering Applications project
31) The cover article in Time is going to be useful for Engineering Applications project
32) Discovered how my Digital Circuits term paper also relates to my Engineering Applications project
33) Outline for Engineering Applications term project needed adjusting
34) Got cash from ATM
35) Had coffee with Les
36) Downloaded Google search results

5-4 Organizing Data into an Outline

Assume that your CMAPP analysis reveals that your Engineering Applications instructor has asked you for an update on your project. You decide that your product will be a memo and that you will concentrate on three main topics:

- the Engineering Applications project itself
- other courses that have a bearing on these applications
- related activities

You conclude that the items in Figure 5.2 numbered 9, 10, 12, 16, 17, 18, 19, 20, 21, 22, 24, 26, 30, 31, 32, 33, and 36 might be relevant. The others will not translate into useful information, and so you ignore them. Your next step is to organize under appropriate headings the 17 data elements you have kept, along with the introductory information your instructor will expect to see in the memo. This stage is often complicated by the fact that some data items may appear to fit under more than one topic, so you have to make composition decisions as you work. Completion of these tasks will result in a first draft of a formal multilevel outline similar to the one shown in Figure 5.3.

FIGURE 5.3: Initial Draft of a Multilevel Outline

I. Why I'm Writing a Memo
  A Instructor asked for it
II. Engineering Applications Project Itself
  A Finished Chapter 2 of EngApp. project
    1 Continued Chapter 3 of EngApp. project
    2 Outline for EngApp. project needs adjusting
  B Bought new EngApp. textbook
  C EngApp. 315: interesting today
    1 Saw results of EngApp. midterm: awful!
  D Discovered that DigCir. term paper relates to my EngApps. project
III. Other Courses That Have a Bearing
  A Comm. 302 class
  B Comm. instructor—info re next assignment
  C Was a bit late for DigCir. 313
  D Attended BusMgmt 305
  E Composed part of Comm. assignment on computer
IV. Related Activities
  A Spoke to BusMgmt instructor about term paper
  B Researched DigCir. project in library
  C What the 3 new library books are about
  D Did homework: 1.5 hours
    1 Internet search on History of Electronics
    2 Read part of latest issue of Time magazine
      a) The cover article in Time is going to be useful for EngApps. project
  E Downloaded Google search results
V. General Conclusion
  A Ask instructor to give me feedback

5–5 Refining the Outline

Now you have to refine your outline. This stage in the process, of translating data into useful information, is a painstaking one, particularly if you are working with an outline for a longer, more complex document. As we noted previously, most sophisticated word-processing software has an outlining function that you should learn how to use: It will greatly simplify this refining task, because the software automatically handles renumbering and constantly shows a clean copy. If you do decide to develop your outline by hand on paper, you might want to recopy it from time to time so that you can "see through" and analyze all the corrections and modifications.

After you have reviewed your initial outline carefully and perhaps made changes (adding or deleting points, changing the position of items, and so forth), you will finalize the exercise by going through the outline three more times: You will apply three principles—*subordination*, *division*, and *parallelism*. This process will require you to examine your outline in detail, and to think carefully and precisely about every line it contains.

5-5a SUBORDINATION

The principle of subordination is an exercise in logic. It says that every item that appears under a particular level head must *logically be a part* of the subject matter of that level head. Conversely, the item must not deal with a different issue, and must not be of equivalent or greater importance or scope.

If you look again at Figure 5.3, you will notice that the level-2 head numbered III. B ("Comm. instructor—info re next assignment") is not really part of the idea of its level-1 head (numbered III., "Other Courses That Have a Bearing"). Probably, III. B should go under the Level-1 head numbered IV ("Related Activities"). You may find other anomalies of this kind, but note that there isn't necessarily any single right answer. Your CMAPP analysis will allow you to determine what is appropriate in your circumstance. When you find an entry that violates the principle of subordination, you may decide to change the sequence of items, to create another, separate level head, or to change the wording of a level head so that it reflects what you really mean.

5-5b DIVISION

The principle of division is, in a sense, an exercise in arithmetic. The principle states that you cannot subdivide the content of any level head into fewer than two parts. For example, if you have a I. A, you must have at least a I. B. Similarly, if you have a VI. A1, you must have at least a VI. A2. If you look at item I. A in Figure 5.3, you will notice that the first level-1 head has been divided into a single level-2 head, violating the principle of division (and logic, too, for that matter). Similar problems exist with items II. C1, IV. D2 a), and V. A: There is no II. C2, IV. D2 b), or V. B.

Once again, how you rectify the problem depends on the results of your CMAPP analysis. You might decide to remove an item, to add an item (taking care to follow the principle of subordination), or to make the item itself a higher level head. What you do depends on what you decide you really mean.

5-5c PARALLELISM

The principle of parallelism requires that all level-1 heads exhibit the same grammatical structure, that all level-2 heads exhibit the same grammatical structure, and so forth. Note that level-1 heads may be grammatically different from level-2 heads, which may be grammatically different from level-3 heads, and so on. If you look at Figure 5.3, you can see that there is no such uniformity of grammatical structure. The level-1 heads numbered I. and III., for example, are subordinate clauses, while the level-1 heads numbered II., IV., and V. are noun phrases. Items III. A and III. B are noun phrases, while III. C, III. D, and III. E are elliptical clauses. (Incidentally, if you are unsure about the meaning of these grammatical terms, you are likely not alone! As a serious engineering student, however, you will not want to remain ignorant of something that could affect your later career. You will certainly want to look them up in one of the grammar sources mentioned in the previous chapter.)

The solution to this problem is to reword items until each of the same-level heads exhibits the same grammatical structure. When applying the principle of parallelism, you may well find that the easiest grammatical structure to work with is a noun or noun phrase. The level-1 heads in Figure 5.3, for example, could become the following:

I. Rationale for memo
II. Engineering Applications Project
III. Related coursework
IV. Complementary activities
V. Conclusion

Note, incidentally, that the principle of parallelism applies not only to formal multilevel outlines, but also to bulleted and numbered lists in a document: Every item in a list should have the same grammatical structure. Ensuring they do increases the CMAPP complementary attributes of clarity and accessibility, and thus makes your message more effective.

Figure 5.4 offers a possible completed outline (remember, there is no single, correct version).

FIGURE 5.4: Completed Outline

I. Rationale for memo
II. Engineering Applications Project (EngApp 325)
    A. Outline (completed)
    B. Chapter 2 (completed)
    C. Chapter 3 (partially completed)
    D. Digital Circuits course Term Paper (mentioned)
III. Related coursework
    A. Engineering Applications 315
        1. Issues related to project
        2. New textbook
    B. Digital Circuits 313
    C. Business Management 305
        1. Course relevance
        2. Conversation with instructor
    D. Communications 302
        1. General relevance to Engineering Applications project
        2. Instructor's comments re assignment
IV. Complementary activities
    A. Library research re Digital Circuits 313
    B. Relevance to Engineering Applications project
    C. Usefulness of three books withdrawn from library
    D. Homework
        1. Internet search
        2. *Time* magazine cover article
        3. Complementary sources
V. Conclusion
    A. Summary
    B. Request for feedback

5–6 The Information-Evaluation Distinction

Imagine that you are a mechanical engineer working on the design of an industrial ventilation system for a manufacturing facility in Minneapolis. One of the sets of guidelines to which you will adhere includes the requirement that "Air foil or backward inclined fan wheels must be used on fan systems requiring more than 2.0" water column (w.c.) of total fan static pressure or 1 inch w.c. of external static pressure." (Design Parameters, Minnesota State Architect's Office, mn.gov/admin/images/RECS-CS-iaq_guide.pdf).

Now imagine that the MN state government had written, instead, "Everyone prefers to see really up-to-date fan wheels that strive to push lots and lots of air away from the center, and we totally want to see them when we're afraid of really high total fan static pressure or pretty high external static pressure." As a professional, what would you have thought of the communication skills that led to the latter version?

If you examine the two, however, you'll notice several things.

- The first is specific and precise; the second is ambiguous and vague.
- The first uses a higher level of technicality than the second; thus, it better suits its intended audience, purpose, and context.
- In the first version, the word "must" shows obligation, which is, in a sense, a judgment. And a judgment, according to the *American Heritage Dictionary* (2014, ahdictionary.com), is "the formation of an opinion after consideration or deliberation." Any such indication of responsibility, in fact, derives ultimately from an opinion; it is, therefore, what we can call an *evaluative* statement. The remainder of the first version, however, is *informative*. It is empirical, factual, and quantitative.
- The second version, though, is predominantly *evaluative*. It uses terms such as "prefers," "really up-to-date," "strive," "lots and lots," "totally want," "afraid," "really high," and "pretty high." All these words and phrases show opinion. They mean different things to different people in different circumstances; they convey points of view, and they cannot be quantified or, for that matter, objectively verified.

Apart from the caliber of the message, the principal difference between the "real" and the "imaginary" versions is that the former is, fundamentally, *informative*, while the latter is, overall, *evaluative*. And for engineering communication, that distinction is crucial. As an engineer, you *must* (notice the evaluative term) be able to differentiate between fact and opinion, between information and evaluation.

Note that *information* is neither "better" nor "worse" than *evaluation*, but it is different. For an engineer, both are necessary. But you have to make sure that you understand the difference, and that you word your message in such a way that your audience receives the one that's appropriate.

When you are trying to convey *information*, you need to word your message carefully, restricting your terms to what is verifiable, quantifiable, factual. If you've included terms that show judgment, opinion, assessment, analysis, and so forth, you're giving your audience an *evaluation*. Doing so unintentionally would be sloppy communication for an engineer; doing so intentionally would also be unethical communication.

5–6a SIMPLE EXAMPLES

Here are a few examples of the difference between *informative* and *evaluative* statements. Note that the first (evaluative) sentence of each pair cannot be the equivalent of the second (informative) one: Subjective expression is not empirical information.

1. I strongly suggest that the problem be reported to the authorities in charge. (The words "strongly" and "suggest" are evaluative.)
   Company policy dictates that all infractions be reported promptly to the Security Office.
2. Reserved parking always has empty spaces. (The word "always" is evaluative.)
   The building manager's report indicated that, on average, 10 percent of parking spots were unoccupied before noon during the work week.
3. That is an appalling percentage. (The word "appalling" is highly subjective.)
   The increase amounted to 54 percent over one month.
4. The college really should make registration more convenient for students. (The word "should" gives an opinion; "convenient", too, is subjective.)
   Twenty of the 25 students polled reported that they found the registration procedure confusing.
5. Having to fill out a five-page form is outrageous. (The term "outrageous" is highly subjective.)
   Jones' opinion is that a five-page registration form is outrageous. (This sentence is informative because it factually relates someone else's opinion.)

CASE STUDY

5 GTI Student Union Facilities Leasing Report

SITUATION

The Grandstone Technical Institute Student Union (GTISU) has for the last three years leased space from the Institute: a large student common room and three small adjoining rooms for offices. Its lease with GTI includes utilities (electricity and heat), but GTISU deals independently with the phone company for telephone, fax, and Internet connections.

On August 4, Heather Hong, the Executive Secretary of GTISU, drops by the campus and finds in her mailbox a formal notice from GTI. It requires the Student Union to immediately vacate the facilities it has been renting.

Hong realizes that Vlad Radescu, the GTISU President, is on vacation in Florida, and that she will have to deal with the issue herself. Angry, she dashes off an email (shown as Figure 5.5) to her friend Sally Johal, who is also on the GTISU Executive. Then, after drinking a coffee and calming down, she places a call to the local law firm of Chung Gomes McNamara, and explains the situation to Franco Gomes, GTISU's legal counsel.

Gomes, aware that the relationship between the Institute and the Student Union has for the last year been rocky, and thinking that things have finally come to a head, asks her to prepare a report for him.

FIGURE 5.5: Hong's Email to Johal

Hi, Sal!

I guess I'm just venting here. Got a notice from Admin telling us to vacate right now. And Vlad is off on another of his Florida jaunts, of course! What on EARTH do they think they're doing? Term starts again in a few weeks! And now they've given us just a couple of weeks to get out. Can you believe it? And we've been here for years and years now, too! I don't know . . . the Institute keeps telling us that they need the space for more classes, and that there was never any guarantee that we'd get to renew the lease, and that they've been trying to do something else with the rooms for ages, and that our reps just wouldn't get together with them—well, like why should they? Admin's been totally unfair with the rent for ages, anyway. So what's to talk about? Anyhow, I think I'll try to get a petition together. We'll keep talking with them, I suppose, for all the good that'll do. But in the meantime, we can scout other places, too, just in case we do have to find somewhere else. And all this just a month before term starts, too! I'll keep you informed.

Well, hope your summer was better than mine. I guess I'll see you when you get back into town in a couple of weeks or so. By the way . . . how was Vegas? I loved it last year. See you! Oh, and by the way, it seems that in a pinch we'd be able to rent a really big room from Apex Holdings just down the block. Pat Chen—you remember Pat . . .—has an uncle who's a big-shot there, and can get it for us pretty cheap. I know . . . it is farther away. But it might work!

See ya!

Conscientiously, Hong undertakes the task by gathering her data and then working to convert it into information that will be useful to Gomes. Thus, she creates the brief multilevel outline that appears in Figure 5.6.

Data > information allows realistic interpretation.

Introductory information takes CMAPP into account, allows audience to place situation in context.

Precise details are crux of most technical communication.

Graphic *might* be useful; but would it *enhance*? Would it be superfluous?

Balanced approach: Audience will likely need/want *both-sides* arguments.

Note: Use of *apparent* required for accuracy: "Information" derives from "secondhand data."

Options listed because CMAPP analysis reveals reader will *need to know*.

Recommendations and **Deadline** point out writer's *purpose*, and specify *action requested* and *time frame* required.

Logical grouping of varied items (throughout) assists reader in following argument; also suggests later use of bulleted or numbered lists in final document.

I. Situation background
  A. Current Student Union (SU)/Institute (GTI) agreement
    1. Implementation date
    2. Length of agreement
    3. General terms of lease
    4. Foreseen SU Executive petition
  B. Premises
    1. Location
    2. Size
    3. Function
  C. Essentials of "disagreement"
    1. SA perspective
      a. Value of SU location to GTI for attracting registrants
      b. High rent
      c. Categorical approach of GTI
    2. Apparent GTI perspective
      a. Several tardy rent payments
      b. Urgent space requirement for additional classrooms
      c. Lack of "responsible attitude" on part of SU representatives
      d. Lack of time: agreements with contractors
II. SU's options
  A. Continued negotiation with GTI
    1. Advantage: possibility of improved lease
    2. Disadvantage: possibility of wasted time/effort
  B. Relocation to Apex Holdings premises
    1. Advantage: resolution of problem with GTI and good lease terms
    2. Disadvantage: distance from campus and lower student interest
  C. Abandonment of SU facilities—no real advantage
III. SU's Recommendations
  A. Attempt at reconciliation/renegotiation with GTI; *then*
  B. Relocation to Apex Holdings premises
IV. Decision deadline: within 2 weeks
  A. Must notify all parties by August 20.

FIGURE 5.6: Heather Hong's Outline for Gomes

The outline is accompanied by marginalia (notes in the margin) to assist you in analyzing her efforts.

ISSUES TO THINK ABOUT

1. If you were Gomes, what explanations could you see for the action of GTI's administration?
2. To what extent would you expect to see them reflected in the report you want from Hong?
3. What differences do you notice between the information in Hong's email to Johal and the information in the outline she has prepared for her report to Gomes?
4. How would you explain and/or justify them?
5. When you construct your own multilevel outlines, it is very unlikely that you would create the kind of marginalia that accompany Figure 5.6. If you examine them, however, of what type of analysis do they remind you?
6. Can you think of several questions that Hong might have been asking herself as she created her multilevel outline in this case? What are they?

EXERCISES

5.1 Describe the differences between data and information.

5.2 Provide an example of a situation in which data are translated into information.

5.3 List some of the benefits of using a multilevel outline in the creation of engineering communication documents.

5.4 Define and briefly describe the principles of subordination, division, and parallelism, as they apply to a formal multilevel outline.

5.5 In what other situations should you apply the principle of parallelism?

5.6 Refer to Hong's multilevel outline (Figure 5.6) in answering the following questions:

1. Who is Hong's primary audience?
2. Might a secondary audience emerge at some point? If so, who?
3. What does the primary audience already know?
4. What does that audience likely need and/or want to know?
5. What is Hong's primary purpose here?
6. What might her secondary purpose be?

5.7 Based on the information you have, do you see any omissions in Hong's outline?

5.8 Are there elements of Hong's outline that do not comply with the principles of subordination, division, and parallelism? If so, what are they, and how would you change them?

5.9 Examine a piece of engineering writing such as a report or a lengthy letter or memo. Based on your examination, create the multilevel outline that the author might have used while developing the communication.

5.10 Explain how the following sentences are evaluative rather than informative. Then, create an analogous informative sentence for each one. (If you need to refresh your memory about this, review section 5-6a Simple Examples.)

1. The bookstore's prices are totally unreasonable.
2. Parking is expensive for students.
3. This type of added stress should not be put on students.
4. I am sure it will greatly increase our chances of reaching viable solutions.
5. The labeling of genetically modified foods should be mandatory because the public deserves to know.

USEFUL WEBSITES

Note: For ease of reading, we have removed the "http://www." from the URLs below. However, some browsers might still require that you insert it to precede some URLs so as not to receive an error message.

- Diffen, a Seattle-based wiki, offers a "Data versus Information" page that provides a slightly different perspective on this chapter's topic.
Visit: diffen.com/difference/Data_vs_Information

- The "Guide to Grammar and Style" page, created by Jack Lynch, Professor in the English Department of the Newark campus of Rutgers University, offers a variety of information on writing, and points to other related URLs.
Visit: andromeda.rutgers.edu/~jlynch/Writing

- The Capital Community Technical College in Hartford, Connecticut, offers this "Guide to Grammar and Writing," with links organized by sentence, paragraph, and essay level.
Visit: grammar.ccc.commnet.edu/grammar/

- Joe's Technology Blog offers a page titled, "Mastering Word 2007 . . . 2013 Outline Structure and Headings," which can facilitate creating and working with an outline.
Visit: joes-tech-blog.blogspot.ca/2013/04/mastering-word-2007-2013-chapter.html

- MakeUseOf.com's "How to Create Outlines & Organize Document in Word 2007" refers to a now dated version of the software; but its clear instructions also work well with Word 2010.
Visit: makeuseof.com/tag/create-outlines-organize-document-ms-word-2007

- The Microsoft Office Support site's "Create an Outline" page provides instructions for using MS Word 2013 to create a multilevel outline.
Visit: office.microsoft.com/en-ca/word-help/create-an-outline-HA102922596.aspx

- The Online Writing Lab of Purdue University is an excellent source for all writers. Its "Developing an Outline" page offers a slightly different approach to ours when it comes to transforming data to information. (From OWL's home page, use its navigation bar to go to *General Writing / The Writing Process / Developing an Outline*.)
Visit: owl.english.purdue.edu/owl/resource/544/1

- Penn State University's College of Engineering offers excellent advice on creating various types of communication, both written and oral.
Visit: writing.engr.psu.edu

CHAPTER 6

Digital/Social Media

This chapter is intended to help engineering students (and engineers) intelligently navigate the current worlds of digital and social media. We focus on the key platforms that engineers use to find and deliver information, to network, and to promote their businesses and industry.

For the purposes of this chapter we have made a distinction between social and digital media.

Social media comprises platforms that promote interactivity and "community building" among users. At the moment, these platforms include blogs, Facebook, Google+, Twitter, LinkedIn, YouTube, image-sharing sites like Flickr and Pinterest, and open-source servers and collaboration services like Github and Mendeley. *Digital media*, strictly speaking, is a larger category. It includes "social media," along with applications that are not interactive. Chief among these latter would be smartphone and tablet mobile apps, for which communication is more generally one-way, from app to user.

6–1 Social Media

Let's start with the main interactive platforms.

Over the last few years, social media has been trumpeted as a universal "change agent" that will transform all human activity. That's not the case, alas. It is unlikely that Facebook or LinkedIn will change how geologists drill for diamonds, for instance, or how manufacturers produce textiles . . . or even the hardware via which we access those applications. What social media *has* done and likely will continue to do is affect how people communicate with one another. Thus, social media

- is about listening as much as conveying a message.
- builds communities that don't need to be centralized.
- is a conversation rather than a one-way speech.
- allows professionals and the organizations for which they work to listen to what clients (and potential clients) say about their products, personnel, and services.

Back in early 2013 the British engineering technology company Calgavin published a survey showing that engineers had already started using social media in their profession in large numbers (calgavin.com/2013/01/engineering-social-media-survey/). What were these engineers using social media for?

- Reading news
- Finding products and new suppliers
- Learning about engineering solutions
- Reading product/company recommendations
- Viewing white papers

The most popular social media platforms were Facebook (67 percent), LinkedIn (43 percent), Google+ (33 percent), followed by Twitter (22 percent). We suspect these percentages might have skewed high (the survey was, after all, conducted online), but we also believe numbers of social media users among engineers have increased since the survey was published.

6–1a FACEBOOK

There is a very good chance that you are already using Facebook for personal use to keep in touch with friends and family members, and for sharing stories, photographs, and videos. Before we move on to professional uses of Facebook, we do want to stress that your current and future employers and colleagues can easily access your *personal* profiles. Assume that everything you post will be seen by all of these people—and, probably, others. Be careful to cultivate a professional image at all times: no profanity, no stupid or embarrassing photos, and so on—ever.

Right now Facebook is probably the most efficient, least costly way for professional organizations to reach their desired demographic, their customers, and their potential customers. Below are examples of a construction company using its Facebook pages in a professional, intelligent manner. Note in Figures 6.1, 6.2, 6.3, below, how Hitachi Construction uses its Facebook page (facebook.com/HitachiConstruction [Note: You may need a Facebook account to access some of these and other Facebook links]) to communicate with customers, "show off" its construction equipment and its current projects, and maintain up-to-date contact information. As of this writing, the page had more than 100,000 "likes."

FIGURE 6.1: Hitachi Construction's Facebook Home Page

Source: Hitachi Construction

6-1b GOOGLE+

Google+ is also very often used by engineers and engineering companies. Because it is a Google product, Google+ pages often appear "high up" in the Google search engine, which helps companies gain exposure and potentially reach new customers.

A helpful, intuitive feature of Google+ is called "Circles." The Circles feature enables people to organize their contacts. In the very useful e-book, *The Definitive Guide to Social Media for Engineers* (Pinnacle Marketing Communications), author Lewis Tonkinson notes that engineering companies can use the Circles feature of Google+ as a Customer Relationship Management (CRM) platform to maintain relationships with contractors and other project stakeholders.

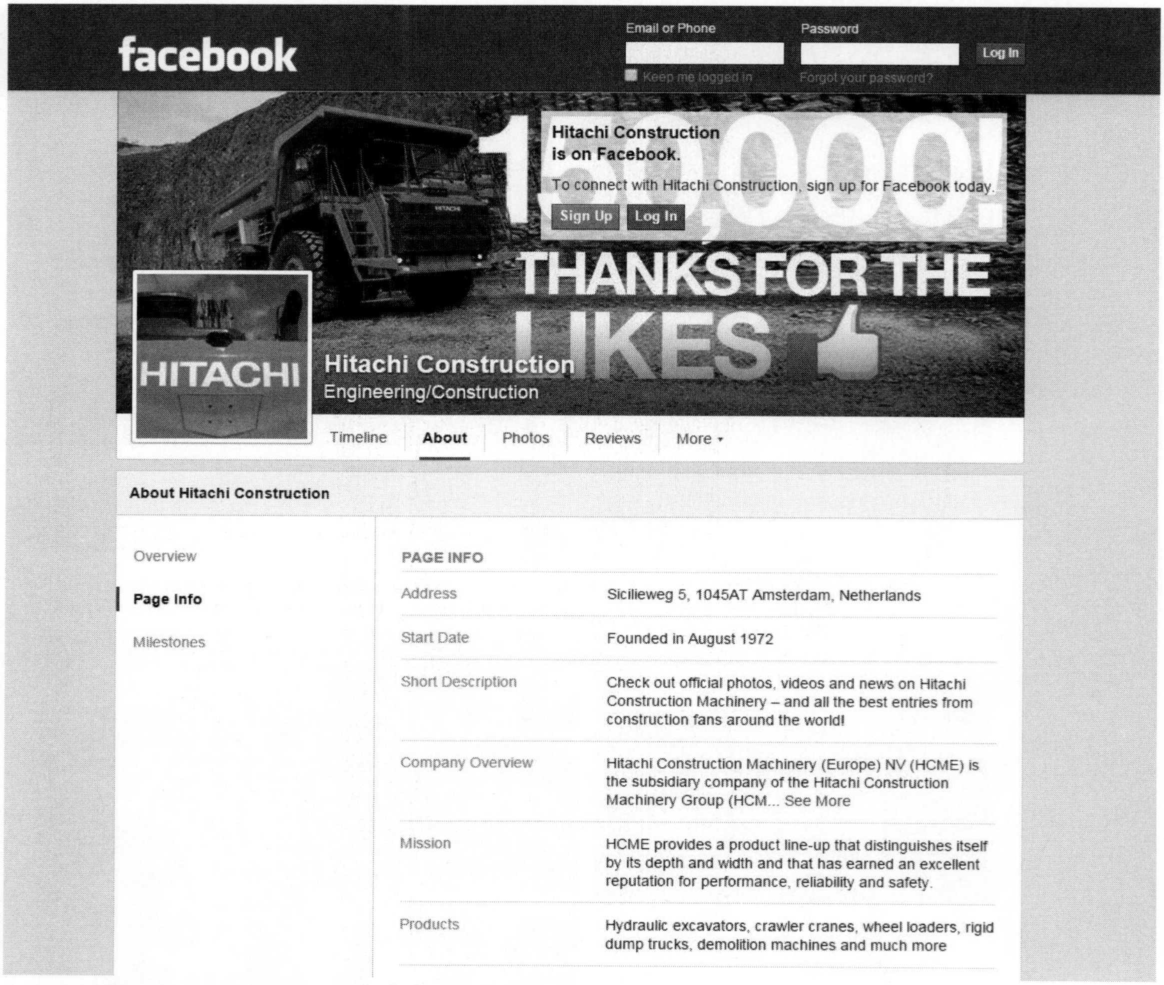

Source: Hitachi Construction

FIGURE 6.2: Hitachi Construction's Facebook Contact Page

6–1c BLOGS

"Blog" is short for "Web log," an unwieldy phrase that yielded that funny-sounding word. Essentially blogs are websites that are devoted to a single topic or a narrow range of topics. Engineering blogs give industry professionals, as well as students, opportunities to ask and to answer questions, to network, to post company news, and to present highlights of recent projects.

Image-sharing, for instance, is widely used on engineering blogs to depict specifications and to provide photographs of work engineers have done. "How-to" instructions can be accompanied by pictures to show people how to perform specific engineering tasks. Engineering companies can also use blogs to build a positive public image for their brands, showing, for example, sustainable technologies and materials they use in their engineering process.

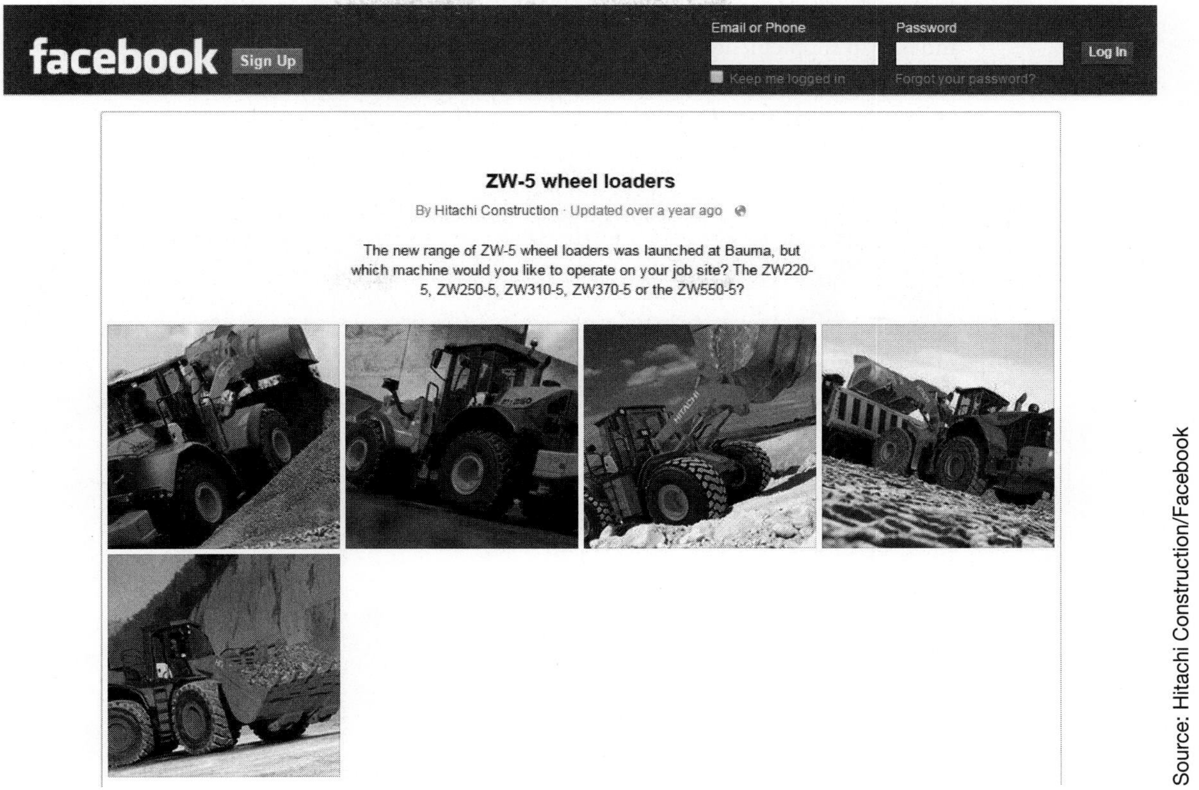

FIGURE 6.3: One of Hitachi Construction's Facebook Photo Gallery Pages

The Definitive Guide to Social Media for Engineers (pinnacle-marketing.com/blog/2013/01/18/the-definitive-guide-to-social-media-for-engineers) recommends two especially helpful blogs for engineers:

- *Engineering Exchange* (engineeringexchange.com): "A social network that allows users to connect with other engineers in a similar position, location or industry to share pictures, videos, blogs and forums."
- *Engineering.com*: "An intersection of technology and media, delivering content across social media, video, internet destinations and mobile devices. Videos, job listings, calculators, articles and an extensive engineering library are among some of the useful free tools."

6–1d TWITTER

We discussed news feeds and news alerts in Chapter 4. Twitter is the next best way of keeping up to date in your field. This microblogging platform gives engineers the ability to network with colleagues in their field, to build their professional profiles, to plan and promote events, to conduct research, to advertise jobs, and to share news—all in 140 or fewer characters per tweet.

The two aforementioned blogs each have very active and popular Twitter feeds (see Figures 6.4 and 6.5) to supplement their professional communications:

Source: Engineering EX/Twitter

FIGURE 6.4: Engineering EX Twitter Feed

As of this writing, Engineering EX's feed has more than 14,000 followers, and Engineering .com's has more than 32,000. We recommend you get on Twitter and start following feeds, too. One neat feature of Twitter is that you can become a prolific contributor without having to create your own original posts—by using Twitter's "retweet" feature. Your Twitter feed can become an "aggregator"—a collection of "greatest hits," as it were—presenting important, intriguing, engineering-related tweets composed by other experts and posted to the timelines of your own followers.

6–1e LINKEDIN AND OTHER GROUPS

There are now more than 300 million members of LinkedIn, which correctly calls itself the world's largest online business network. In Chapter 14, "Seeking Employment," we will show you have to create a superb LinkedIn profile that will help you in your search for professional positions. But, here, we want to mention how helpful this online platform is in terms of professional development and networking.

Source: Engineering.com/Twitter

FIGURE 6.5: Engineering.com Twitter Feed

MachineDesign.com, a site you will likely want to visit throughout your engineering studies and career, currently estimates that there are about 25,000 groups on LinkedIn that are focused on some aspect of engineering. (There are almost two million groups in total.) A very small sample is shown in Figure 6.6.

These groups act as forums through which you can get engineering experts from around the world to answer your questions and to provide feedback on your projects. These groups also offer you the opportunity to demonstrate your own expertise, providing educated observations to others in your field, thereby increasing your own professional stature.

There are, of course, many other groups and forums online where engineers can share discussions and upload news. One popular forum is CR4 (cr4.globalspec.com—see Figure 6.7 on page 81), which calls itself "the Engineer's place for news and discussion." Anybody can read the forums, but you need to register with the site in order to post yourself.

6–1f YOUTUBE

YouTube is the world's second most popular search engine, after Google (the company that has owned YouTube since 2006). Engineers and engineering companies use YouTube videos to describe their services and products and to show off their projects, technology, and worksites.

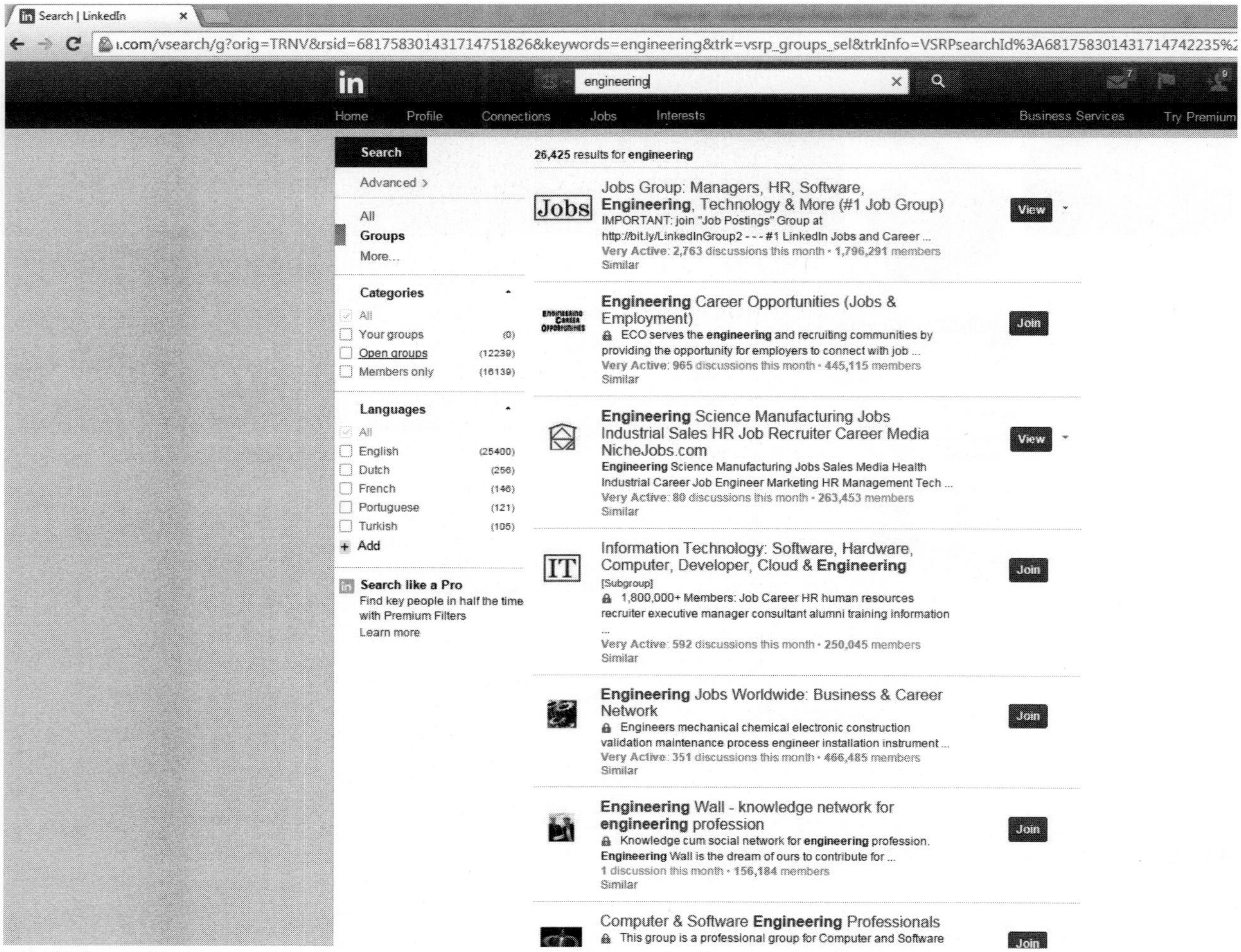

Source: LinkedIn

FIGURE 6.6: LinkedIn Engineering-Focused Groups

Especially helpful to engineering students, however, are the many superb teaching channels on YouTube that focus on engineering. A 2014 survey by Engineering.com listed these as the top five:

- Crash Course (youtube.com/user/crashcourse)
- Minute Physics (youtube.com/user/minutephysics)
- Veritasium (youtube.com/user/1veritasium)
- Engineer Guy (youtube.com/user/engineerguyvideo)
- Khan Academy (youtube.com/user/khanacademy)

Figure 6.8 on page 82 is a screenshot of Hank Green, of Crash Course, explaining how "network solids" work.

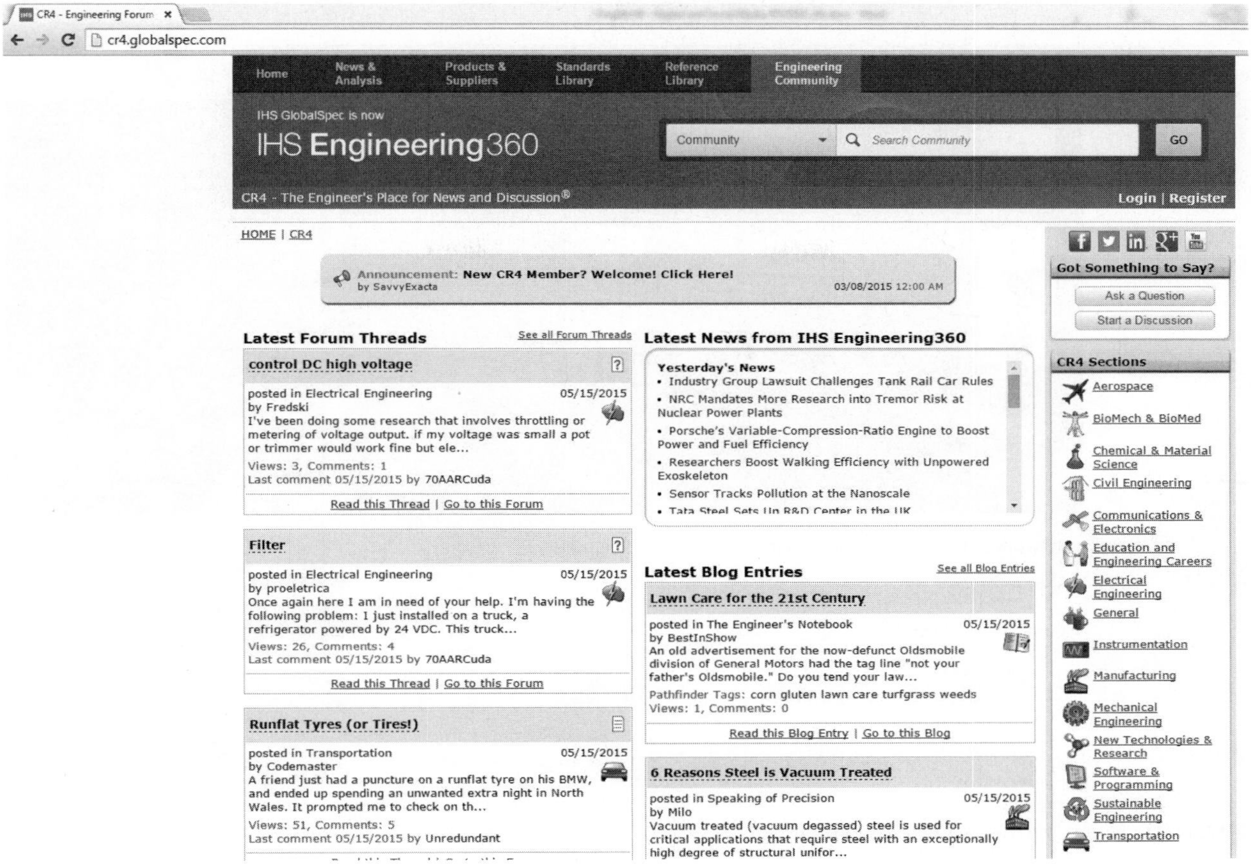

Source: CR4 - Engineering Forum/IHS Engineering360

FIGURE 6.7: CR4 Engineering Forum

6–2 Image-Sharing Platforms

The three most popular image-sharing platforms are Pinterest, Flickr, and Instagram. These platforms will allow you to show clients and colleagues what you have designed and built in your engineering career. Figure 6.9 on page 83 shows a Pinterest page devoted to illustrating aspects of engineering with concrete.

6–3 Creating a Social Media "Hub"

Many of you will want to use a number of the social media platforms described in this chapter. You might want to create a LinkedIn page, for example, to network with professionals in your field and show off your strong resume, Flickr to showcase photographs of your recent work, and Facebook to help you coordinate events. You will need to make sure that all of your platforms are *congruent*; that is, you will want to make sure you convey the same message and "look" across the various platforms you are using.

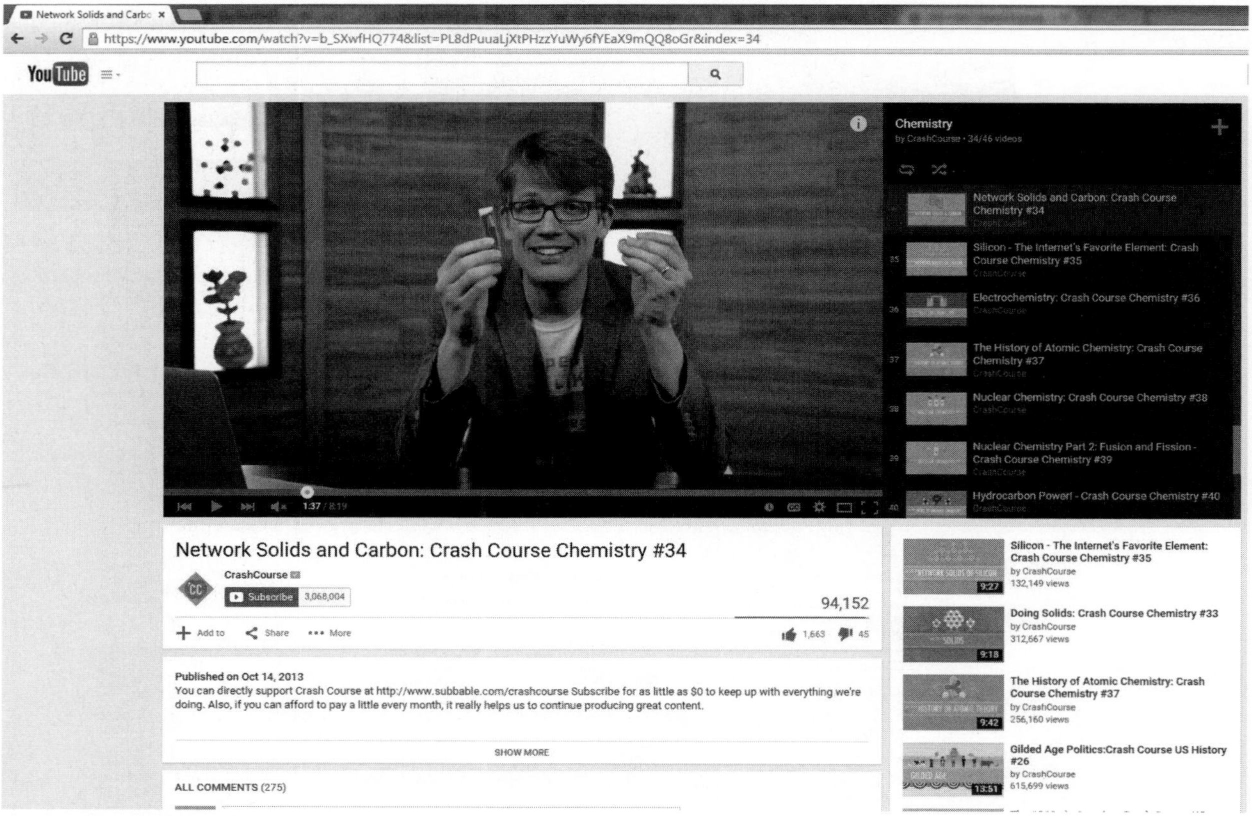

Source: YouTube

FIGURE 6.8: YouTube Instructional Video

You will always want to make it easy for your colleagues and associates to navigate across your platforms. A smart way to accomplish this is by creating a simple website that acts like a "hub" connecting all of your social media. Elliance, Inc., a digital marketing company, calls this "doughnut marketing," an example of which is depicted in the infographic in Figure 6.10 on page 84.

Figure 6.11 on page 84 is an example of a website, created by author Bob Basil, that links together a number of social media platforms: LinkedIn, Flickr, Twitter, a communications blog, an iPhone blog, and Facebook. The website also links to other online projects created by the author and some of his associates.

Creating a professional, personal website is neither difficult nor expensive. The home page of basil.ca, shown in Figure 6.11 on page 84, was created in an afternoon using the free Wordpress.com platform. For other projects Basil has used Google's Blogger.com platform. Both Wordpress and Blogger provide templates that you can modify according to your own requirements. (Find out more about these and other website/hub-creation platforms in the "Useful Websites" section at the end of this chapter.)

Note this important point: You will need to keep all of your social media platforms up to date. If you won't be able to do that, don't put them up. Nothing says "I don't care that you have wasted your time visiting me" more than a Facebook or LinkedIn page that hasn't added new (and interesting) material in several months. As we wrote earlier, professional communication is about fostering and maintaining relations with your colleagues and clients; it is about "not screwing up," too.

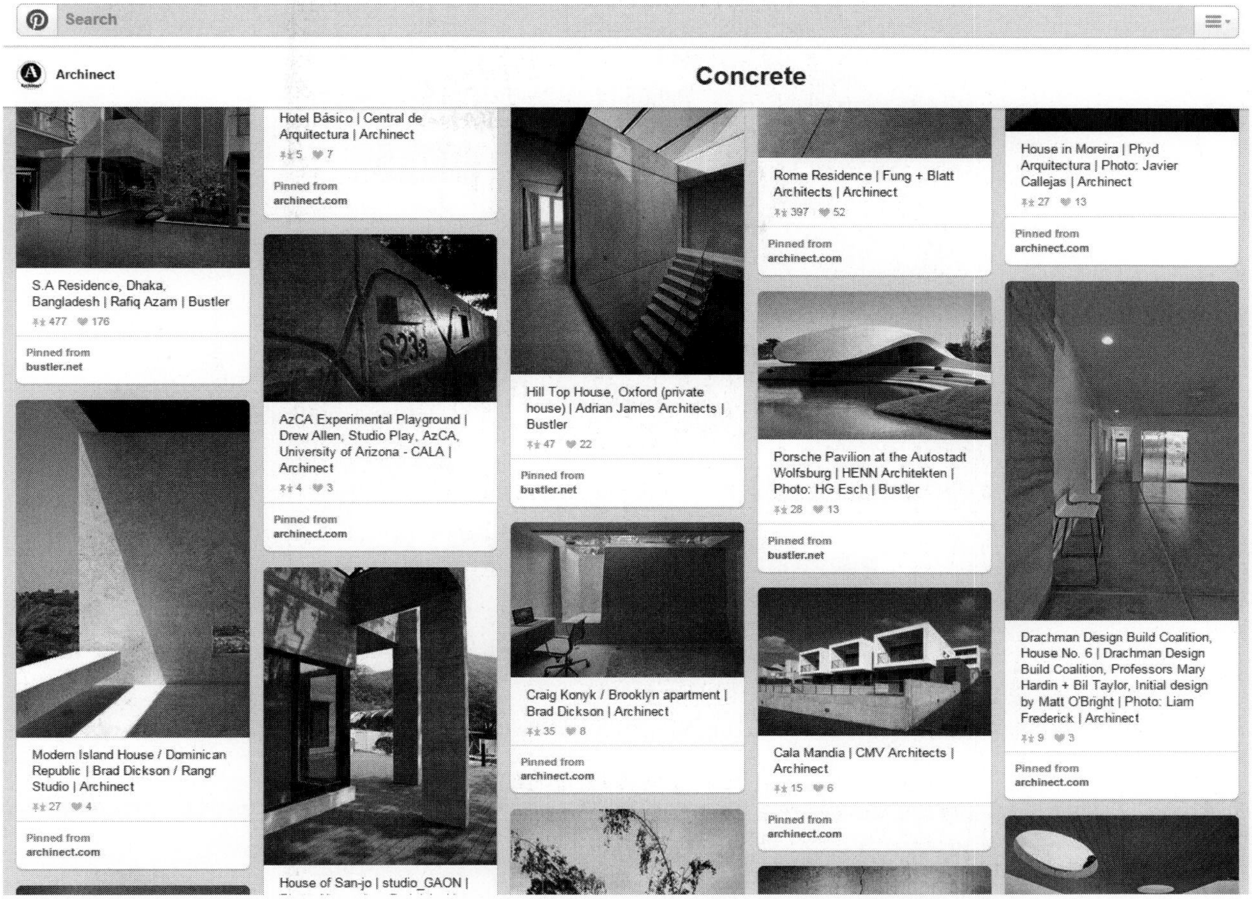

Source: Pinterest

FIGURE 6.9: Pinterest Page—Engineering with Concrete

6–4 Open-Source Collaboration

Open-source sites allow technical folk to cooperate on technical projects in an organized fashion. Two sites that engineering students in particular will find helpful are the following:

- **Github**: Computer and software engineers should check out Github (github.com/about). As of late 2014, more than seven million people were collaborating on almost 15 million software projects via this robust platform.
- **Mendeley**: You will find that Mendeley (mendeley.com) can make engineering team projects go remarkably smoothly. This platform helps you conduct, organize, share, and annotate research; create and organize your citations; and track and respond to feedback.

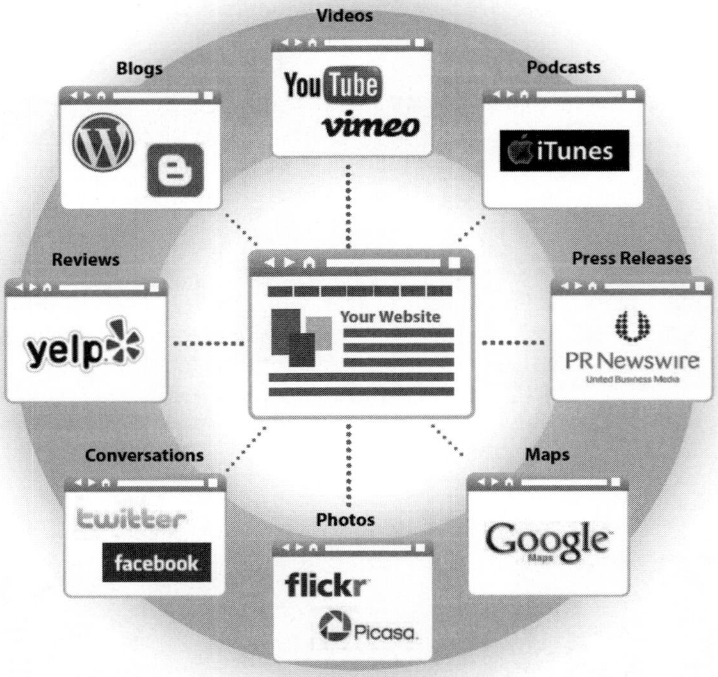

Source: Elliance, Inc. http://www.elliance.com/

FIGURE 6.10: Connecting Your Social Media via a Website Hub

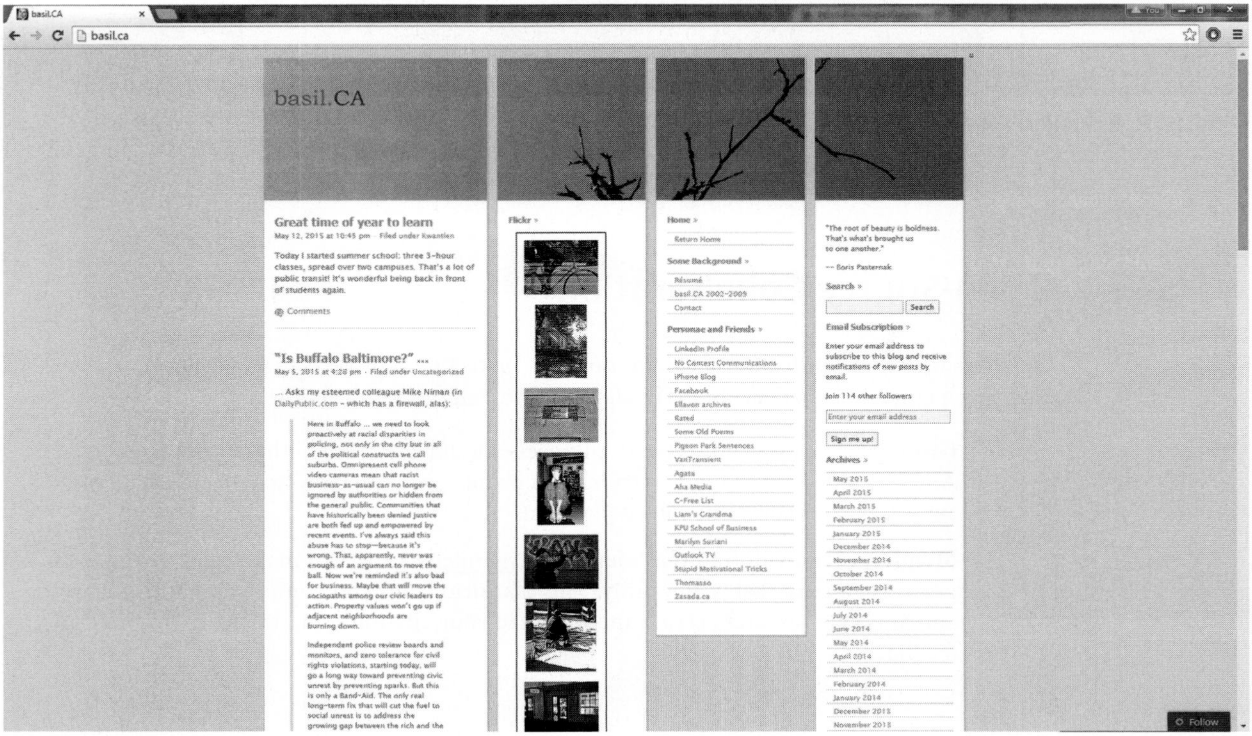

Source: Basil.ca

FIGURE 6.11: Screen Capture of Basil.CA

6–5 Smartphone and Tablet Mobile Apps

Applications for smartphones and tablets are proliferating at a rate that astonishes even seasoned digital media professionals. To keep up requires almost daily study. Below, however, are three apps that we believe you will find useful in your current studies:

- **Engineering Pro**: Mashable.com describes this app as a "hub for more than 650 need-to-know formulas from all sectors of engineering, including chemical, civil, electrical, environmental, hydrology and mechanical. Plus, you can save your favorite formulas for quick access."
- **FingerCad**: Apple's iTunes describes this app as the "first CAD application for technical drawing with fingers assisted by computer. . . . [You] can draw houses, bridges, floor plans, mechanical components, geometrical figures, anything that can be designed with a common PC CAD. Useful for engineers in construction site, architects, surveyors, designers, students, etc."
- **Teambox** is a "handy-dandy" app that allows professionals as well as students to collaborate on school and workplace projects "on the go" and stay "on top of things." With your smartphone you can track a team or organization's tasks, conversations, to-do lists, and meeting times/places in a very easy-to-follow format. You can also share files.

We recommend you take our lead and visit appcrawlr.com ("The App Discovery Engine") often, to download these apps if you like and to find out what new app might make your life easier (and more fun).

CASE STUDY

6 AAU Student Association Function: Before, During, and After

SITUATION

In Chapter 2's Case Study 2a, you read about the difficulty the Ann Arbor University Student Association (AAUSA) had when it attempted to rent Fuller Hall from the University for a Thanksgiving weekend concert.

Let's follow up on this story.

Jack Lee, the AAUSA Functions Coordinator, is elated when the room booking is finally confirmed. The concert, headlined by the bands Hodgepodge and Really Me, would indeed be going on.

Jack still has a lot of work to do to organize the promotion of the event. He decides to complement standard advertising—posters and flyers—with posts to various social media platforms. Here is his plan, put together in a rush (and without a CMAPP analysis) because the concert is only a day away:

- Local promoters from the company House Specialties will update the AAUSA Facebook page and Twitter feed.
- Because AAUSA President Dorothy Palliser is still out of town, Jack has asked first-year student Kevin Lopez, Palliser's assistant, to post an announcement regarding the concert on the University website's "What's Coming Up" page.
- Since Jack does not want to share the passwords for AAUSA's LinkedIn and Google+ accounts with anybody, he will post announcements about the concert on these platforms himself.

Jack wants the concert event to be lavishly documented, in particular by photographs for AAUSA's Flickr feed and by videos for its YouTube channel. His roommate is Sarah Blair, a talented, responsible artist who has a high-end DSLR (digital single-lens reflex) camera that takes HD video, so he hires her to take care of this assignment.

ISSUES TO THINK ABOUT

1. What do you think of Jack's social media plan in terms of event promotion?
2. How might a CMAPP analysis have changed his plan?
3. In this chapter we mentioned the importance of maintaining *congruence* of message and imagery across social media platforms. What challenges does Jack face in maintaining such congruence?
4. How might he intelligently address these challenges?
5. The principal intent of Jack's promotional messages is informational, to get the word out about the concert. Might these messages have additional purposes? Explain your reasoning.
6. What would be the primary and possibly secondary purposes (again, in terms of the CMAPP model) of the YouTube and Flickr content that are to be recorded at the concert? To achieve these purposes, what should this content convey?
7. Has Jack neglected certain social media platforms in his planning of the promotion and coverage of the concert event? Explain and justify your response.

EXERCISES

6.1 Using examples, distinguish "social media" from "digital media" as they relate to this chapter's contents.

6.2 How do you think social media in a workplace with which you are familiar governs or affects communications among colleagues and management? Be specific in describing the social media platforms in use.

6.3 Pick an engineering company involved in the sector you are studying. Then answer the following questions.

- What social media platforms does this company use?
- How does the company present itself?
- Are the company's "messages" congruent from platform to platform?
- Are they all up to date?

6.4 Look at another engineering company in the same sector, and compare its platforms with the company chosen in 6.3. List some "best practices"—that is, techniques that almost always seem to succeed—for social media in your industrial sector.

6.5 In this chapter we noted that it is important *not* to establish a social media platform unless you keep it up to date. At this point in your academic career, which social media platforms, if any, do you believe you both need and can keep up to date. Explain.

USEFUL WEBSITES

Note: For ease of reading, we have removed the "http://www." from most of the URLs below. However, some browsers might still require that you insert it to precede some URLs so as not to receive an error message.

■ Brian Solis is one of the world's foremost theorists and best writers on the intersection of technology, communications media, and the workplace. His website and Twitter feed are "must reading" for those who see workplace and business models being transformed by digitization.
Visit: BrianSolis.com and Twitter.com/briansolis

■ CMO.com (*CMO* stands for "chief marketing officer") focuses on social and digital media for marketing purposes, but its materials are quite helpful for anyone in a profession that uses this media. Check out its superb collection of slide shows and info graphics.
Visit: CMO.com

■ Lewis Tonkinson, the author of "The Definitive Guide to Social Media for Engineers," mentioned in this chapter, is also the author of numerous other worthwhile items relating to technical and engineering communication.
Visit: pinnacle-marketing.com/blog/author/lewistonkinson

■ Mashable is the premier online site for news in high-end technology as well as social and mobile media. It has a wonderful smartphone app and a very lively Twitter feed.
Visit: Mashable.com and Twitter.com/mashable

■ The Wayback Machine is an Internet archive that allows you to see what current URLs looked like in the past, going back as far as 1996. (URL stands for *Universal Resource Locator*, by the way.) This can be very useful when you are doing research on a company's history. The archive also demonstrates, albeit inadvertently, that websites are no longer always companies' primary online presence: Platforms like Facebook, LinkedIn, and Twitter, not to mention smartphone apps, have taken a large share of influence away from websites.
Visit: archive.org/web

■ Wikihow.com has admirably lucid explanations that can help you quickly set up useful websites using the Wordpress and Blogger platforms.
Visit: wikihow.com/Create-a-Wordpress-Blog and wikihow.com/Start-a-Blog-on-Blogger

CHAPTER 7

Visual Elements in Written and Oral Communication

For many people, the term "visuals" can include graphs, charts, photographs, sketches, and the like, whether in a document or a presentation, as well as typeface features such as bold, italics, and underline. CMAPP terminology, however, makes the following distinction:

- *Document visuals* include typographical and related features affecting the appearance of a document
- *Visuals* comprise illustrations of specific ideas within a document or presentation

7-1 Document Visuals

To make effective use of document visuals, you should be familiar with relevant terminology and have a working knowledge of the various features offered by your word-processing software. Although you may use a graphics program such as InDesign® or PhotoShop®, a business presentation program such as PowerPoint®, a desktop publishing program such as Publisher®, a web-page authoring program such as Adobe Dreamweaver®, or specialized engineering applications such as AutoCAD®, you will likely create most of your documents with a word processor such as Word. While different applications may use different techniques to accomplish the tasks, all sophisticated word processors now offer substantial control with regard to document visuals.

7–1a BODY TEXT AND LEVEL HEADS

The term "body text" refers to the text that makes up your document's paragraphs and bulleted or numbered lists. Body text can be thought of as the "main" text of your document. If you prefer, you could think of it as any text that is not a heading or subheading of some kind.

The term "level heads" refers to titles, subtitles, headings, and subheadings of all kinds. This term is useful in that it permits a ready numerical reference. In this chapter, for example, the heading "Chapter 7, Visual Elements in Written and Oral Communication" is a level-1 head, the subheading "Document Visuals" is a level-2 head, and the subheading "Body Text and Level Heads" is a level-3 head.

7–1b HEADERS AND FOOTERS

A *header*, also known as a *running head*, is text that appears at the top of every page. Its content will depend on the nature of the document, but common elements include the author's name, the document's title (sometimes abbreviated), and/or the page number. By convention, headers generally do not appear on the first page of a document. If the document is printed on both sides of the pages, left- and right-hand pages may feature different headers.

Usually having the same kinds of elements as headers, but appearing at the bottom of every page, are *footers*, also known as *running feet*.

7–1c WHITE SPACE

White space refers to the portions of a printed page that are blank. Examples of white space include margins, and line and paragraph spacing.

Margins

Some word processors measure all four margins from the top, bottom, left, and right edges of the page. Others measure the bottom margin from the top of the footer and the top margin from the bottom of the header. In principle, however, margins should normally keep body text from 1" to 1.5" from all edges of the page.

Line and Paragraph Spacing

"Line spacing," or "line space," refers to the space between lines of text within a paragraph. Your word processor will apply a default (preset) line spacing automatically each time the text wraps to the following line as you type. Currently, for example, Microsoft Word defaults to a line spacing of 1.15 times line height (explained later in this section). You can adjust the line spacing.

Paragraph spacing is the space between the last line of one paragraph and the first line of the next. On a typewriter, it was common to hit the return key twice between paragraphs. Many novice word processor users do the same. More effective, however, is to set paragraph spacing so that it conforms to what you generally find in printed books: greater than a single line space but less than two line spaces. A common practice is to set paragraph spacing at 1.25 to 1.5 times the line space. How you do this, of course, depends on the particular software you are using. (Note that Microsoft Word currently defaults to paragraph spacing of 10 points more than line height. We will explain this in detail later.)

7–1d JUSTIFICATION

Justification refers to the way in which lines of text relate to the left and right margins of the page. Examples (and descriptions) of left, center, right, and full justification appear in Figure 7.1.

FIGURE 7.1: Line Justification Examples

Left Justification

When you write in longhand on loose-leaf paper, you begin each line at the left margin and conclude it when you run out of room at the end of the line. This results in *left justification*, also known as *flush left*. (An exception would be an indented first line of a paragraph, of course.) Since different lines will have slightly different lengths, left justification is also known as *ragged right*. This paragraph, like those of many documents, is left justified.

Center Justification

Sometimes lines of text, particularly titles and subtitles, appear equidistant from the left and right margins. This arrangement, not surprisingly, is known as *center justification* or *center-justified* text. These lines are center justified.

Right Justification

To create particular visual effects, lines of text (usually level heads) may be aligned with the right margin but not with the left. This results in *right justification*, also known as *flush right* or *right-justified* text. Since there is no left alignment, this arrangement of text is also known as *ragged left*. This paragraph is an example of right-justified text. Right justification is generally not suitable for body text.

Full Justification

In most books (including this one), body text is set so that it aligns with both the left and right margins. This is known as *full justification* or *fully justified* text. (Note that Microsoft Word refers to it, rather, as "justified text.") The problem with this format is that when you use proportional-space fonts (discussed a bit later in this chapter), different combinations of letters form words of different lengths that, in turn, require different line lengths. Ensuring that each line fits precisely between the margins, while ensuring as well that the apparent space between the words remains constant, is painstaking work. The shorter the line of text and the larger the font, the greater the problem. Most newspapers use full justification, setting text in fairly narrow columns. This can produce undesirable effects such as those apparent in Figure 7.2, comprising a repetition of this paragraph's text, but fully justified in a three-column format, using a slightly larger font size. The "white currents" you see running down the columns are called *rivers*, and are, in fact, a kind of optical illusion. In most cases, your word-processed documents are likely to look better if you use left justification.

FIGURE 7.2: Three-Column Fully Justified Text

In most books (including this one), body text is set so that it aligns with both the left and right margins. This is known as *full justification* or *fully justified* text. (Note that Microsoft Word refers to it, rather, as "justified text.") The problem with this format is that when you use proportional-space fonts (discussed a bit later in this chapter), different combinations of letters form words of different lengths that, in turn, require different line lengths. Ensuring that each line fits precisely between the margins, while ensuring as well that the apparent space between the words remains constant, is painstaking work. The shorter the line of text and the larger the font, the greater the problem. Most newspapers use full justification, setting text in fairly narrow columns. This can produce undesirable effects such as those apparent in Figure 7.2, comprising a repetition of this paragraph's text, but fully justified in a three-column format, using a slightly larger font size. The "white currents" you see running down the columns are called *rivers*, and are, in fact, a kind of optical illusion. In most cases, your word-processed documents are likely to look better if you use left justification.

7-1e FONTS

While some typographers and professional printers still distinguish between the terms "font" and "typeface," most word processors do not. (You'll find differing views on this subject at, for example, mindgruve.com/blog/advertising/typeface-vs-font-what-is-the-difference, http://www.fastcodesign.com/3028971/whats-the-difference-between-a-font-and-a-typeface, and graphicdesign.stackexchange.com/questions/12717.) For most engineering communication, it is sufficient to think of a font (or a typeface) as a particular style of letters, including their size and shape. Figure 7.3 illustrates some common font attributes.

FIGURE 7.3: Common Font Attributes

This line of text is in bold.

This line of text is in italic.

This line of text is in bold italic.

This line of text is underscored (or underlined).

THIS LINE IS IN UPPERCASE.

THIS LINE IS IN SMALL CAPS.

There are literally tens of thousands of different fonts, and typographers are creating new ones every day.

Fonts can be either monospace or proportional space. Different letters take up different widths. The letter "i," for example, takes up a fraction of the width of the letter "m." In monospace fonts, however, the same amount of horizontal space is accorded to every letter. Such fonts were necessary for typewriters, since the carriage had to move a set distance to the left, regardless of which type key was pressed. Thus, in a monospace font, the distance between the double "i" in "skiing" would be slightly greater than that between the "s" and the "k." (Here is "skiing" in a monospace font: `skiing`.)

Most word-processed and published documents use proportional-space fonts. For each of these fonts, width across the line is determined by the respective widths of the letters.

Figure 7.4 shows the same lines of text in both a monospace font (in this case, Courier New, a common word-processor font that closely approximates an "old-fashioned" monospace font) and a proportional-space font (in this case, Times New Roman).

FIGURE 7.4: Monospace and Proportional Space

```
The quick brown fox jumps over the lazy dog. The quick brown
fox jumps over the lazy dog. The quick brown fox jumps over
the lazy dog. The quick brown fox jumps over the lazy dog.
```

The quick brown fox jumps over the lazy dog. The quick brown fox jumps over the lazy dog. The quick brown fox jumps over the lazy dog. The quick brown fox jumps over the lazy dog.

Monospace fonts are measured in characters per inch (cpi): the number of characters—including letters, numbers, punctuation, and spaces—that would be printed within one inch horizontally on the page. This means, of course, that the higher the cpi number, the smaller the font.

This kind of measurement would obviously not work for proportional-space fonts, since the number of characters that would occupy a "horizontal inch" would depend on the characters involved and their respective width. Thus, proportional-space font size is measured not by width but by *line height*, and is usually counted in *points* (pt), of which there are 72 to an inch. Consequently, as seen in Figure 7.5, the length of the same line of words in different proportional-space fonts may be quite different. Note that they are all 10pt text.

FIGURE 7.5: Font Samples

Arial, 10pt	The quick brown fox jumps over the lazy dog.
Bookman, 10pt	The quick brown fox jumps over the lazy dog.
Brush Script, 10pt	*The quick brown fox jumps over the lazy dog.*
Calibri, 10pt	The quick brown fox jumps over the lazy dog.
Cambria, 10pt	The quick brown fox jumps over the lazy dog.
Comic Sans, 10pt	The quick brown fox jumps over the lazy dog.
COPPER PLATE GOTHIC, 10PT	THE QUICK BROWN FOX JUMPS OVER THE LAZY DOG.
Garamond, 10pt	The quick brown fox jumps over the lazy dog.
Old English Text, 10pt	**The quick brown fox jumps over the lazy dog.**
STENCIL, 10PT	**THE QUICK BROWN FOX JUMPS OVER THE LAZY DOG.**
Times New Roman, 10pt	The quick brown fox jumps over the lazy dog.

Fonts fall into three general categories: *serif*, *sans-serif*, and *decorative*. Bookman, Cambria, Garamond, and Times New Roman (all in Figure 7.5), for example, are serif fonts. If you look closely at the 26 pt samples in the left-hand column of Figure 7.6, you will see that the letters have tiny "tails" (the meaning of the old Arabic word from which "serif" derives) at their extremities. Serifs were reputedly designed by typographers long ago to help the eye move more easily from one letter to the next. In almost all books (including this one), body text is set in a serif font.

Fonts that do not have "tails" are called *sans-serif* (from the French, meaning "without serif"). Examples of sans-serif fonts in Figure 7.5 are Arial and Calibri, also seen in the right-hand column of Figure 7.6. Sans-serif fonts are commonly used for level heads.

FIGURE 7.6: Serif versus Sans-Serif Fonts

Bookman	Arial
Cambria	Calibri
Garamond	
Times New Roman	

Although some technical communication writers do set their body text in a sans-serif font (maintaining that it "looks cleaner"), most typographers believe that sans-serif is harder to read. Note, however, that on screens, from smartphones to tablets to computer monitors to data-projections onto large screens, body text set in a sans-serif font seems to be easier to read. This might explain the following development. Until a few years ago, Microsoft Word (and most of its competitors) defaulted to a serif font (Times New Roman) for body text and a sans-serif font (Arial) for headings. The last few iterations of Word, however, no longer adhere to that practice. As mentioned at www.forbes.com/sites/quora/2013/12/18/why-did-microsoft-change-the-default-font-to-calibri/, Microsoft, it would seem, is counting on reading becoming more common on screen than on paper. The defaults for

Microsoft Word 2010 are a (Microsoft) sans-serif font called Calibri for body text, and a (Microsoft) serif font called Cambria for headings. For Word 2013, Microsoft continues to default to Calibri for body text, but defaults to Calibri Light for headings.

The Brush Script, Comic Sans, Copper Plate Gothic, Old English, and Stencil samples in Figure 7.5 are examples of *decorative* fonts. Such fonts, which may be serif, sans-serif, or a combination, are designed to create a special visual impression. When used for body text, decorative fonts tend to be difficult to read. Imagine trying to read page after page of text set in the decorative Brush Script font shown in Figure 7.7.

FIGURE 7.7: Decorative Font Text

On the Insert tab, the galleries include items that are designed to coordinate with the overall look of your document. You can use these galleries to insert tables, headers, footers, lists, cover pages, and other document building blocks. When you create pictures, charts, or diagrams, they also coordinate with your current document look. You can easily change the formatting of selected text in the document text by choosing a look for the selected text from the Quick Styles gallery on the Home tab. You can also format text directly by using the other controls on the Home tab. Most controls offer a choice of using the look from the current theme or using a format that you specify directly. To change the overall look of your document, choose new Theme elements on the Page Layout tab. To change the looks available in the Quick Style gallery, use the Change Current Quick Style Set command. Both the Themes gallery and the Quick Styles gallery provide reset commands so that you can always restore the look of your document to the original contained in your current template.

Source: Microsoft Support, How to insert sample text into a document in Word, https://support.microsoft.com/en-us/kb/212251

7-1f OTHER ATTRIBUTES

Document visuals may also include features such as those shown in Figure 7.8.

FIGURE 7.8: Additional Document Visual Features

Rules

Vertical and horizontal rules are simply straight lines that are intended to help the audience recognize divisions on a page. For example, you may recall the vertical rules between the columns in Figure 7.2. And below the line you're reading now is a double horizontal rule.

Boxes

Text may be enclosed in a blank or a shaded box. For example:

This center-justified line appears in a blank box.

A shaded box (with a "shadow") encloses this center-justified line.

Reverse Text

Reverse (also called *inverse*) text involves placing white letters on a black background. (Other combinations are possible, of course, if you are printing in color.) While reverse text can be effective in certain circumstances (e.g., in a heading), it should be used sparingly because it is harder to read than the same text set black on white.

This is a sample of reverse text.

Lists

Engineering communication makes frequent use of both bulleted and numbered lists. The two types of lists tend to serve slightly different purposes. A numbered list suggests a particular sequence or hierarchy for a particular reason—order of importance, for example, or the required order of a series of steps. Bulleted items don't imply sequencing or hierarchy; however, you should note that it would be more difficult to locate the ninth of fifteen items in a bulleted list than in a numbered one.

Which list should you use? As always, your choice should derive from your analysis of context, message, audience, purpose, and product.

7-2 Visuals in Documents and Presentations

7-2a INFORMATION AND IMPRESSION

You may recall from Chapter 1 the observation that audiences respond not only to the objective or factual meaning of words (denotation) but also to their emotional effect (connotation). Think again of author Dave Ingre's story in Chapter 2 about his beloved dog. Analogously, visuals carry two types of meaning as well. In CMAPP terms, when your audience sees your visual, it receives not only *information* from it—the facts that you are representing—but also a *visual impression* that evokes a particular response. These two meanings may be quite different. Imagine a photo of someone accused of embezzling funds from a company. The photo provides *information* that allows you to recognize the person's face. Now imagine that the photo had been taken from an unflattering angle, that the lighting had created stark shadows on the person's face, and that the subject had been scowling at the camera operator. The *visual impression* you receive will likely be one of the subject's guilt. When you use visuals, ethics and professionalism require you to remain aware of both types of meaning.

7-2b CONSIDERATIONS

Here are some points to consider when choosing or constructing your visuals.

Relevance

A visual should reflect or advance each of the CMAPP elements: context, message, audience, purpose, and product. In addition to being relevant, your visual should create interest. Your personal delight with a particular visual will be of no consequence if your audience looks at the same visual and immediately thinks, "Why on earth is that here?"

Simplification and Emphasis

Remember the adage "a picture is worth a thousand words"? Visuals can help increase your audience's understanding of your message by clarifying concepts or data. A visual that confuses an audience is one that has been poorly chosen and/or constructed. Just as a properly chosen visual can be effective in emphasizing a particular point in your message, a visual that lacks focus will tend to distract an audience.

Enhancement

The visual should enhance your message, not overwhelm it. For example, those PowerPoint slides you so painstakingly constructed will not have been a success if your audience walks away from your presentation thinking, "I have no idea who the speaker was or what the topic was, but those sure were fantastic pictures!"

Size

Your visual must be large enough for its details to be visible to your entire audience. For example, a car engine diagram that has been photographically reduced to 1" × 1" in your document will be of little utility. Similarly, no one in an audience of 30 engineers would benefit if you held up a 5" × 4" photograph of soil erosion at a development site.

Legibility

If its text is not readily legible to your audience, your visual will not contribute to your message; rather, it will generate impatience and frustration. While often a function of size, legibility also depends on clarity of reproduction. A poor-quality photocopy or projected image can severely compromise an otherwise effective visual. Remember, too, that even the high ppi (pixels per inch) resolution on smartphone and tablet cameras may not provide sufficient legibility for detailed visuals. And, as noted earlier, some fonts will be difficult to decipher.

Color

Consider whether color is necessary. An illustration of wiring in an electrical system will probably require the use of color—for safety reasons at least—while a graph comparing yearly sales figures may not. You have to take into account how you are planning to reproduce or display your visual. And, if in the case of a document to be distributed in large numbers, you may need to consider the significantly higher cost of color reproduction.

Level of detail

Consider the appearance of Figure 7.9, a photo of the controls of a gas fireplace insert. What kind of audience in what context and for what purpose would all the labels and arrows be necessary? Do they enhance understanding of the visual? Do they add to its clarity? When deciding how much detail to include in your visual, think about all CMAPP complementary attributes.

FIGURE 7.9: Gas Fireplace Controls

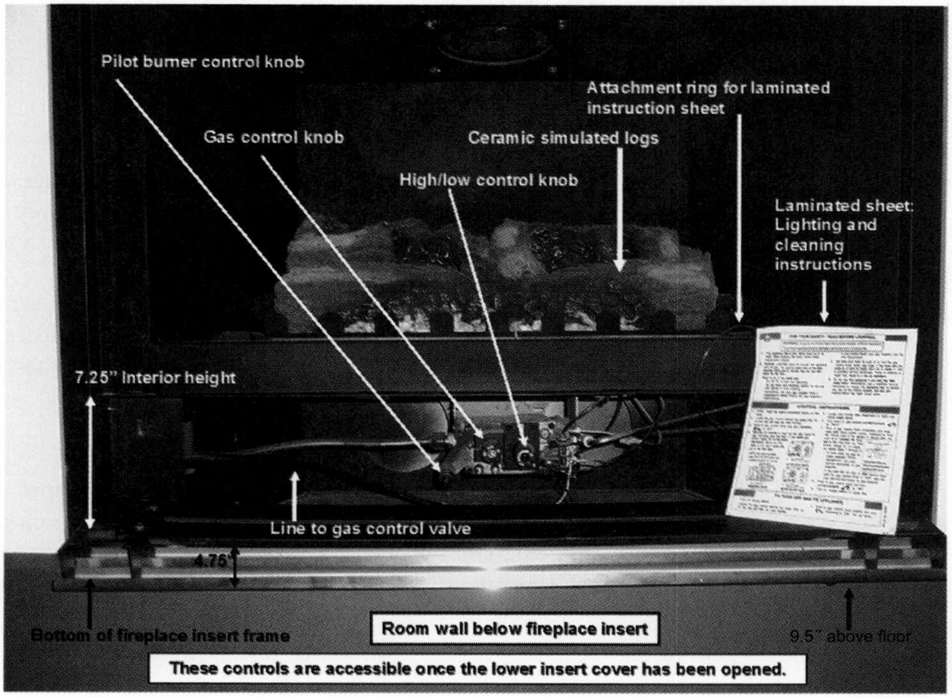

David Ingre

Scale

If you were involved in preliminary planning for East–West Interstate highway improvements across the country, the map in Figure 7.10 would be an appropriate scale. If, however, you wanted to show transportation improvement in the Juneau area, you would start with a larger scale map like the one shown in Figure 7.11. Applying a CMAPP analysis to the selection and construction of your visual will help you determine an appropriate scale.

FIGURE 7.10: Map of the United States

7-2c CHARTS AND TABLES

Charts (also called graphs) are commonly created from the data generated by a spreadsheet or database program. In both documents and presentations, graphs can convey concise visual representation of data. *Pie charts*, *vertical bar charts* (often called *column charts*), *horizontal bar charts*, and *line charts* are the most common types used in engineering communication.

A table consists of data arranged in columns and rows—an arrangement just like that of a spreadsheet. Like data in charts, data in a table may derive from a spreadsheet or database, but may also be compiled directly into its final form. Tables are more likely than charts to display text as well as numbers.

Pie Charts

Whatever the information provided by a pie chart, the visual impression created is that of parts of a whole. Each segment stands for a different category, and the data displayed in the segments add up to 100 percent. The pie chart in Figure 7.12 is an attempt to represent the numbers of registrants in GTI's Aeronautical and Agricultural Engineering departments over an eight-year period.

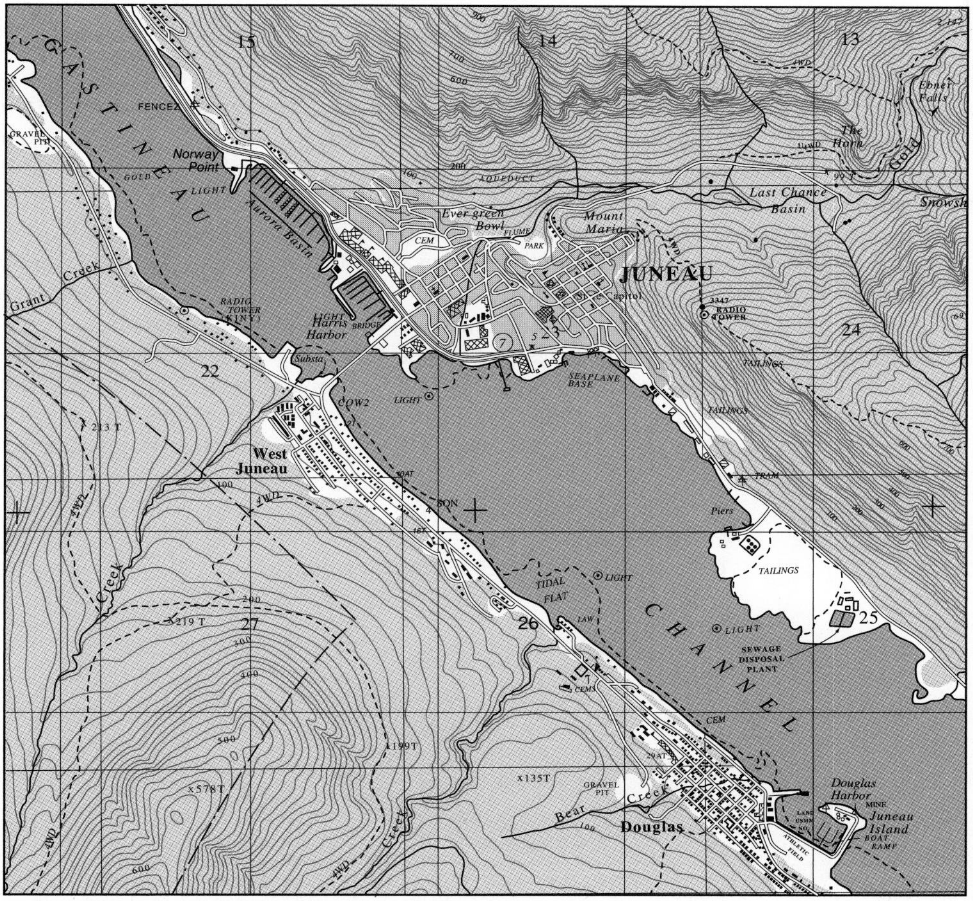

Perry-Castañeda Library Map Collection, University of Texas. Courtesy of the University of Texas Libraries, The University of Texas at Austin.

FIGURE 7.11: Topographic Map of Juneau, Alaska

FIGURE 7.12:
Ineffective Pie Chart

GTI Agricultural/Aeronautical Registration: 2009–2016

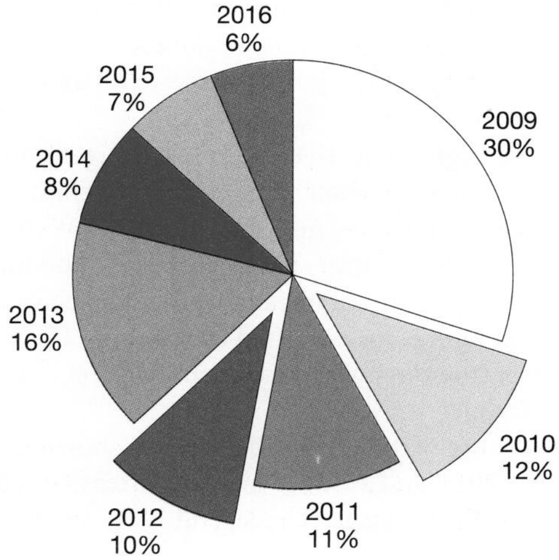

But a pie chart is not an appropriate choice here. First, it is unlikely that we want the audience to think of the eight years as parts of a whole. Second, the eight segments, several of similar size, create a visual impression of "a couple of bigger ones and some smaller ones." Although percentage labels (the visual's *information*) accompany the segments, precise differences are not visually apparent (part of the *visual impression*). Finally, since the purpose of "exploding" a segment is to focus your audience's attention on it, the two exploded segments (2010 and 2012) merely confuse.

A better use of the pie chart is shown in Figure 7.13. Here the visual's *information*—the registration percentages for the four program areas that make up GTI's 2015 Co-op Programs—lend themselves to representation in a pie chart. The 3-D perspective adds interest and enhances intelligibility. The use of only four segments avoids the confusing visual impression of Figure 7.12. The single exploded segment leaves no doubt as to which segment the audience should focus on. And finally, the use of a legend allows for less clutter around the pie. Note that the visual's effectiveness depends also on the choice of fills or patterns that allow an audience to easily distinguish one segment from another. Here, again, you may well need to balance the advantages and disadvantages of using color.

FIGURE 7.13: Better Use of Pie Chart

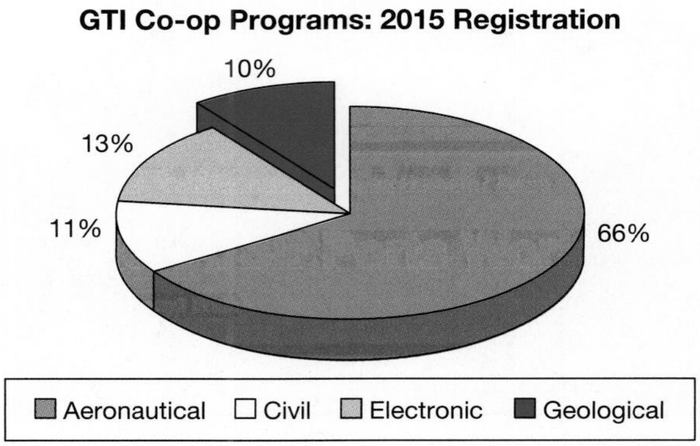

Column Charts

The visual impression created by a column chart is that of discrete quantities. The audience expects to see sets of data (often called *data categories* or *data series*) ranging along the *x*-axis (the horizontal axis), and quantities increasing up the *y*-axis (the vertical axis). Negative amounts are often shown descending below the *x*-axis.

Column charts are highly effective in representing comparisons and contrasts. The visual impression they create makes them better vehicles than pie charts for representing precise differences. As well, column charts can be effective in displaying multiple comparisons and contrasts.

A common misuse of the column chart is to include too many sets of data. Figure 7.14 exemplifies the clutter and confusion that typically result. The chart attempts to compare and contrast eight years of registration figures for seven GTI programs. Thus, the *information*—eight series (the years) of seven categories (the programs)—generates 56 columns, resulting in the visual impression of a bad view of a city skyline.

A far more effective use of the vertical bar chart is shown in Figure 7.15. Here the restriction of data to three series (2013, 2014, and 2015) of two categories (Agricultural and Aeronautical) enhances, condenses, and simplifies the message. The legend is far more legible than the one in Figure 7.14.

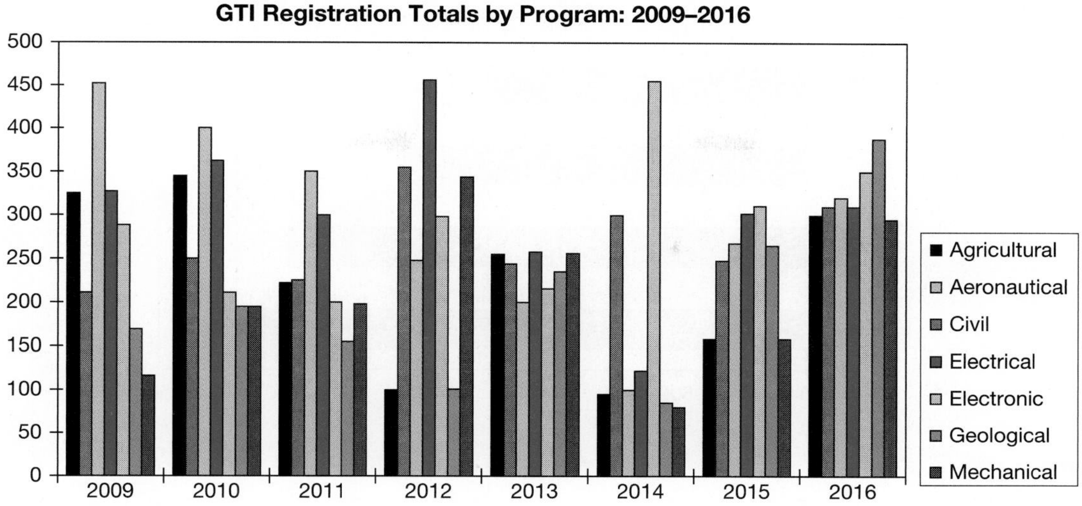

FIGURE 7.14: Ineffective Column Chart

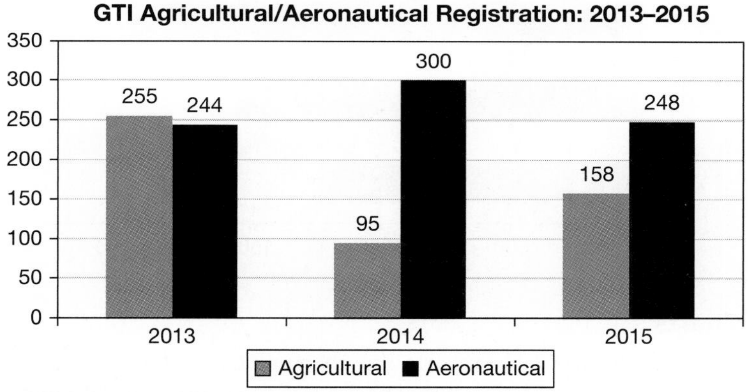

FIGURE 7.15: A Better-Constructed Column Chart

Finally, the horizontal gridlines allow for a more accurate visual impression of the differences in quantities. Be cautious, however, because gridlines can lead to clutter.

Horizontal Bar Charts

The horizontal bar chart looks like a vertical bar chart turned 90 degrees clockwise. It, too, is an effective vehicle for presenting one or more comparisons or contrasts (although, as mentioned, your audience's unconscious expectation is to see quantities increasing up the *y*-axis, rather than along the *x*-axis).

A horizontal bar chart's visual impression is conditioned by the overall shape of the chart. By way of example, compare Figures 7.16 and 7.17, which are derived from identical data. The *y*-axis in Figure 7.16 shows the earliest date at the top; Figure 7.17 places it at the bottom. One could make a reasonable case for either order. However, note how Figure 7.16 creates a visual impression of stability, while that of Figure 7.17 suggests a situation that is top-heavy (i.e., inherently unbalanced or unstable).

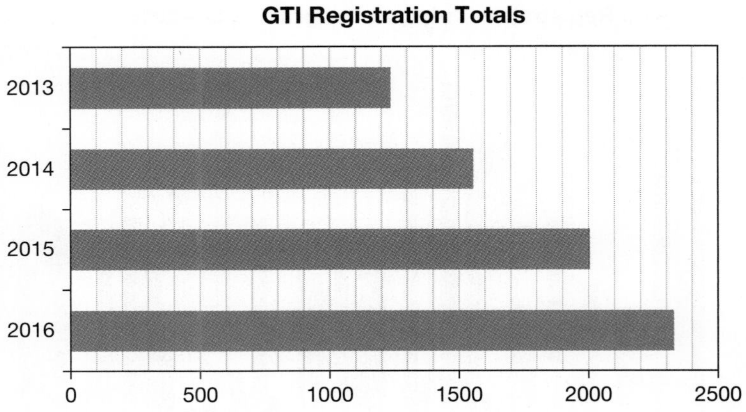

FIGURE 7.16: GTI Registration Totals—Version A

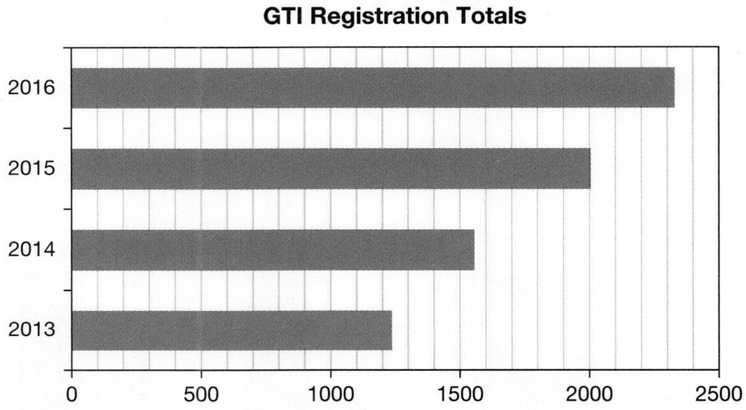

FIGURE 7.17: GTI Registration Totals—Version B

Line Charts

Like vertical and horizontal bar charts, line charts plot data along *x*- and *y*-axes. Accustomed to reading from left to right in English, we expect a line to progress in the same direction. And the visual impression of a line chart is normally that of progression *over time* along the *x*-axis. Thus, a line chart is particularly effective in showing chronological change. As Figure 7.18 shows, you can plot more than one data series in a line chart by including more than one line.

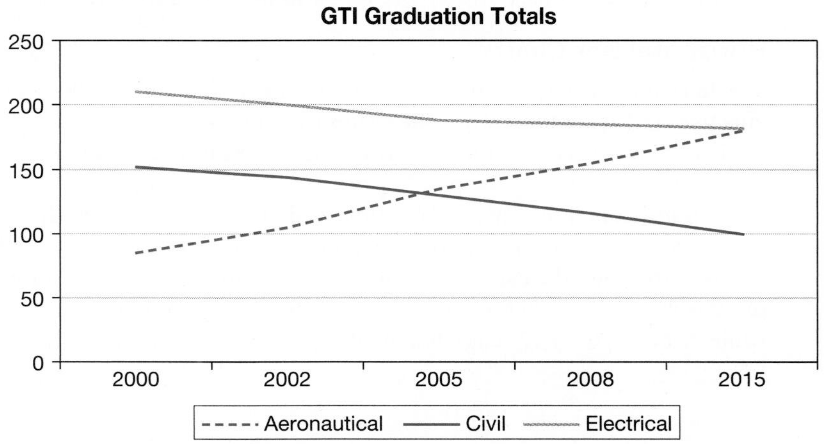

FIGURE 7.18: Multiple-Line Chart

Note another aspect of Figure 7.18's visual impression. The shape of the lines as they progress from left to right creates an image of partial convergence: Graduation numbers for the Aeronautical and Electrical programs are growing more similar, although those for the Civil appear to show a constant decline.

You can also increase the precision of the visual impression by using markers on the data points and including horizontal and vertical gridlines. The disadvantage, however, is that the chart may become too cluttered, as in Figure 7.19.

FIGURE 7.19: Multiple-Line Chart with Markers and Gridlines

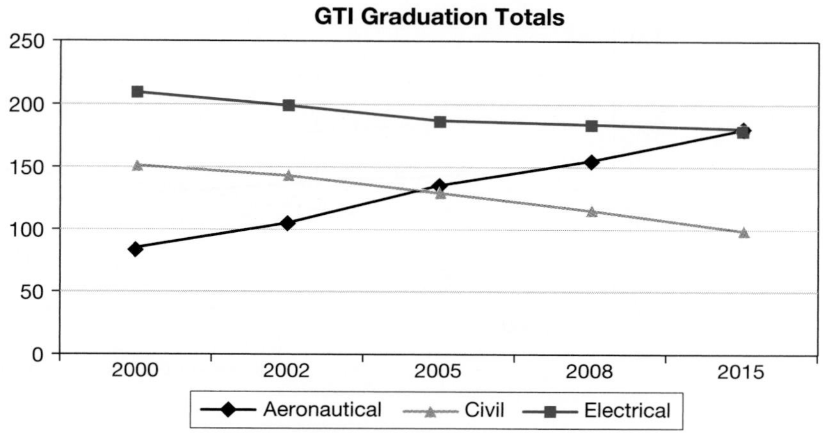

Certain aspects of line charts can make them deceptive. Note that the span of years between the x-axis points in Figure 7.19 changes, even though the physical distance separating them on the chart itself does not. (Software such as Microsoft Excel® defaults to such equidistant data points. You have to expend careful effort to make the chart's x-axis reflect the reality of the time spans.)

Note also that there are graduation numbers for each of the indicated years, but not for the intervening ones. However, in Figure 7.20, which uses the same data, the line chart's visual impression might result in an audience extrapolating the numbers indicated by arrows A, B, and C. Such extrapolations would be little more than assumptions—quite possibly inaccurate ones, since we in fact had no data for those particular years.

FIGURE 7.20: Assumed Data Points

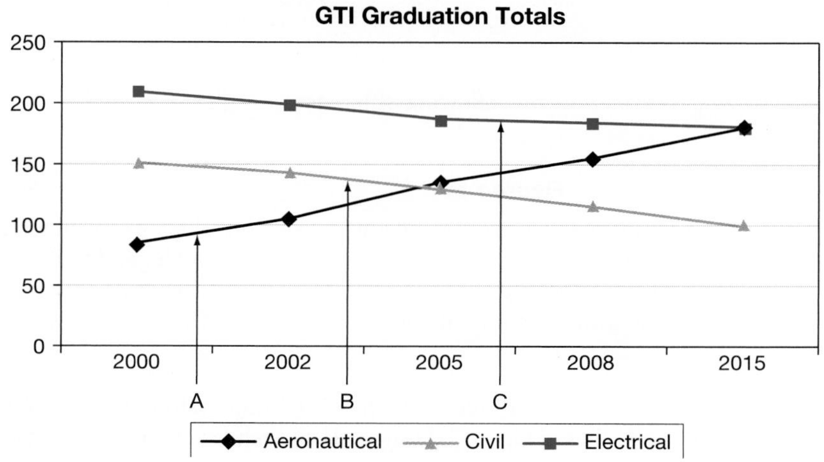

Another potentially misleading aspect of line charts has to do with the relative lengths of the *x*- and *y*-axes, the minimum and maximum points on the *y*-axis, and the size of the *y*-axis increments. Consider Figures 7.21 and 7.22. Both charts represent the same data.

FIGURE 7.21: "Flattened" Line Chart

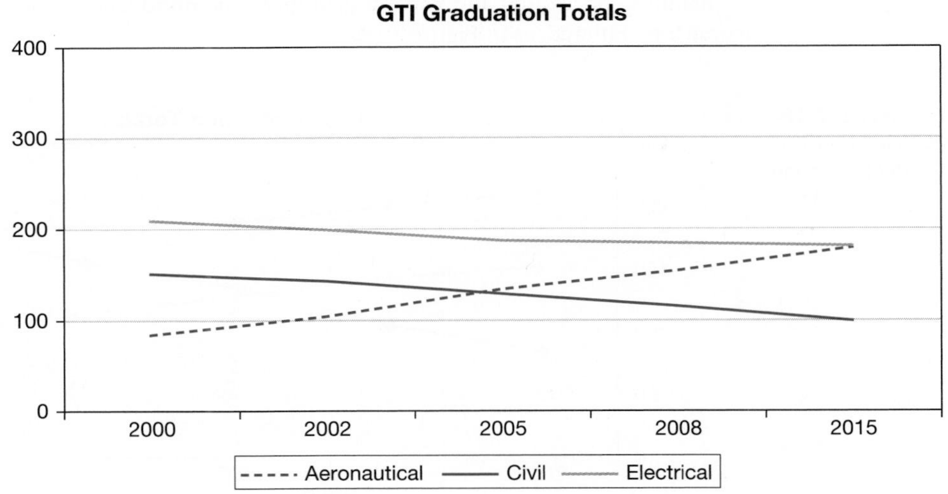

FIGURE 7.22: "Emphatic" Line Chart

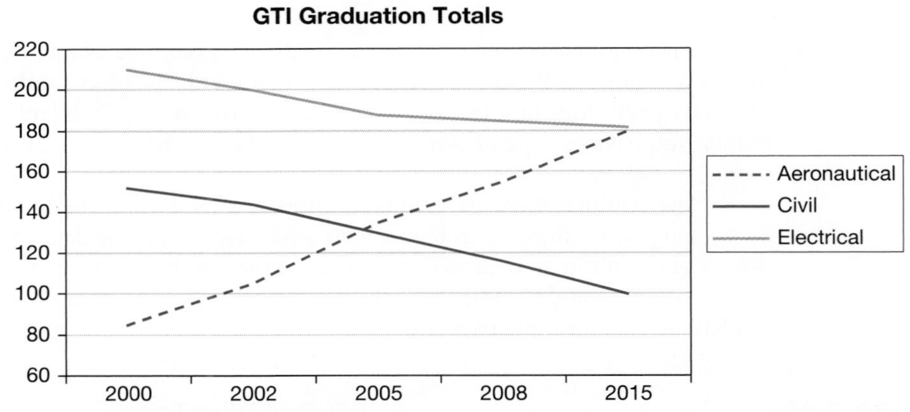

However, note the following differences.

Figure 7.21	Figure 7.22
1. The *x*-axis is made longer by having the legend below it.	1. The *x*-axis is made shorter by having the legend to its right.
2. The *y*-axis data points number from 0 to 400.	2. The *y*-axis data points number from 60 to 220.
3. The *y*-axis shows increments of 100.	3. The *y*-axis shows increments of 20.

Because of these differences, the visual impression of Figure 7.21 conveys "slow but steady growth," while that of Figure 7.22 suggests a much faster and more radical change.

When constructing a line chart, you must choose *something* for the axis length, the minimum and maximum points and the increment value. Does doing so mean that you are deceiving your audience? Recall our discussion of ethics in Chapter 1. If you deliberately mislead your audience, you are probably being unethical. If, on the other hand, you are merely making effective use of visuals to present your message in conformity with your purpose—and if that purpose is not deceptive—you are probably on solid ethical ground. In any event, when you construct line charts or interpret those of others, remember that the visual impression may strongly condition the interpretation of the information.

Tables

A surfeit of rows and columns of data in a table can produce what we might call "visual information overload." The table shown in Figure 7.23 is relatively uncomplicated visually, giving an impression of interrelated, if not readily understandable, numbers. Faced with a table in a document, a motivated audience could carefully examine every cell, discerning relationships, trends, and so forth. But if you use a table as a projected visual, your audience can do little with it but follow your lead.

FIGURE 7.23: RAI Sales to Service Event Ratio, 2006–2016

	Autos Sold/Leased	Repair/Service Events
2006	1676	15424
2007	1745	15377
2008	1756	14326
2009	1748	14114
2010	1751	13598
2011	1763	13112
2012	1798	12123
2013	1895	11112
2014	2024	10119
2015	2135	10168
2016	2368	9945

In talking to an audience about the numbers in Figure 7.23, for example, you could leave two very different impressions. On the one hand, you could say:

> In 2006, almost 1700 sold or leased vehicles accounted for the unacceptably high figure of more than 15,000 repair or service incidents. In 2016, a full decade later, the sales-to-service ratio showed moderate improvement, but the almost 10,000 repair or service events for the 2000-plus vehicles sold or leased were still unacceptably high.

On the other hand, you could say:

> In 2006, we were forced to deal with approximately 10 repair and service incidents for each and every vehicle we sold or leased. Within 10 short years, we effected a dramatic improvement. In 2016, there were only about four repair and service incidents for every vehicle leased or sold.

Thus, you could use the same table to leave either a negative impression of ongoing difficulty or a positive impression of successful change. Use the results of your CMAPP analysis to guide your treatment of tables, and never ignore ethical questions.

7-3 Clip Art and Other Web Visuals

7-3a CLIP ART

The term "clip art" refers to electronic artwork, including photos, sketches, drawings, and line art, that you can import and use as visuals for documents or presentation materials. Thousands of color or black-and-white clip art images are available through software packages and websites. Figures 7.24 and 7.25 are examples of public domain clip art available on the Internet.

Remember that a large number of electronic formats exist, both *vector* and *raster*[1], from *.bmp to the current Internet transfer standard, *.jpg. Not all are compatible with every computer application, and not all clip art can be effectively printed or projected. Further, you should consider the ethical—and legal—issue of copyright. You may generally use public domain material such as the photo of the Golden Gate Bridge in Figure 7.24 and the drawing of the old sextant in Figure 7.25 without seeking copyright permission. However, the majority of images on the Web are, in fact, proprietary and covered by copyright. Thus, you need to be careful. If you insert any copyrighted material into your course assignments, for example, you must provide an appropriate source credit. (How you present that attribution will depend on which style guide you have been prescribed. See Chapter 4 for examples.) If you use copyrighted material elsewhere, you will be breaking the law if you do not adhere to whatever licensing requirements are specified by the copyright owner. Note, also, that software licenses are often ambiguous about what you may do with the clip art included with applications such as word processors or presentation graphics programs.

FIGURE 7.24: Public Domain Photo of Golden Gate Bridge[2]

USGS photograph by Tom Reiss

7-3b OTHER WEB VISUALS

As you know, on the Web you can find a great variety of other material that might be included under the rubric of visuals. Video clips such as MPEG or RM files are available in the millions, as are a variety of sound files, from AVI to MIDI. (The related issue of downloaded music files, MP3 or other,

[1] For a brief explanation of these terms, see the *Useful Websites* section at the end of this chapter.

[2] Photo source page: http://www.usgs.gov/blogs/features/files/2014/01/20071026-TER-52.jpg

FIGURE 7.25: Public Domain Clip Art of Sextant[3]

is currently the subject of much controversy . . . and litigation.) And, of course, YouTube provides a seemingly infinite selection of videos. Ready-made graphs and charts are abundant, as are templates for pictures, spreadsheets, documents, business and technical forms, and so on, along with a multitude of software applications and utilities. Some are presented as being in the public domain, some are offered without copyright restriction, and some request simple attribution. Much is made available without a clear indication of your copyright responsibilities. Fundamentally, the fact remains: If you are going to use something that you have not created entirely on your own, it is your professional and ethical obligation to find out what you should do . . . and to act accordingly.

CASE STUDY

7 RAI Report on Participation in the GTI Internship Program

SITUATION

In 2014, Radisson Automobiles participated in Grandstone Technical Institute's Internship Program. Several students in GTI's Automotive Mechanics and Electronics programs secured work placements with Radisson dealerships. Upon graduating in the early summer of that year, most of these students applied for positions with Radisson and a number were hired. The following year, as a result of union-related and financial issues, Radisson declined to participate in GTI's Internship Program. Late that summer, however, Angelos Methoulios, GTI's Internship Program Coordinator, wrote to Lorna Hildebrande, RAI Boston's vice president, requesting (a) that Radisson Boston report on its assessment of its 2014 Internship Program participation, and (b) that the company consider participating again in the January 2016 session of GTI's program.

Hildebrande asked Carole Schusterman, Radisson Boston's general manager, to take care of the matter. Schusterman consulted company records and sought additional input from both Larry

[3] Clip art source page: http://www.wpclipart.com/tools/viewers/Sextant_BW.png.html

Cornetski, the service manager, and Emily Cardinal, the personnel manager. Having concluded that the company's previous participation in the program had been worthwhile, and that it would be in Radisson Boston's interest to take part again, Schusterman created for Methoulios the report that appears in Figure 7.26.

Radisson Automobiles of Boston
11245 Commonwealth Avenue Boston, MA 02145
Tel: (877) 555-4235 Fax: (877) 555-9097
Email: radiss.boston@radissoncars.com

Radisson
Automobiles Inc.
Where all the cars are stars

Your file: IP20-RA1
Our file: CC150915.7

September 9, 2015

Mr. Angelo Methoulios
Coordinator, Internship Program
Grandstone Technical Institute
400-11th Street
Randolph, VT 05060-4600

Radisson participation in Grandstone Internship Program

Dear Mr. Methoulios:

Thank you for your letter of August 16, 2015, to our Vice President, Ms. Lorna Hildebrande. She has asked me to reply.

Your letter made two requests:

- That we give you our views on our earlier 2014 participation in your Internship Program;
- That we consider participating again for January, 2016.

First, here are our overall reactions.

Observations

1. As a whole, Radisson Boston staff regard the Internship experience as having been positive for the company ... ☺;
2. Recorded feedback from the students who worked with us suggests that in general, they, too, found the experience productive and valuable ... ☺;
3. After their graduation, several students applied for full-time work with our company; three are now employees of Radisson Vancouver ... 👍;
4. Although we foresee little opportunity for later full-time hiring, we would be pleased to participate in the Internship Program for January, 2016.

Let me now deal with each of your queries in detail.

FIGURE 7.26: RAI Boston's Report to GTI

2014 Participation

In June of 2014, 14 students applied for Internship positions with Radisson Boston. As a result of the subsequent interview process, and in the light of our own projected needs, we accepted eight for placement for the period of January 3, 2014, through March 3, 2014.

The students were assigned a variety of tasks—none, I should stress, being artificial or "make-work". Their progress was monitored by assigned Radisson personnel, who then provided our management with written feedback. The following table summarizes the information collected.

Student	GTI Area	RAI Work Sector	Comments From Staff
Peter Bennett	Automotive 	Repair shop floor	Knowledgeable; conscientious; enthusiastic; learned quickly; good inter-personal skills; good communication skills; required little supervision.Hired as apprentice mechanic, May, 2014.
Penni Melele	Electronics 	Administration	Excellent understanding of MIS electronics; broad operating system background; worked very quickly; thorough and meticulous; some difficulty accepting criticism; average interpersonal but excellent written communication skills.
Francine Orti	Automotive 	Repair shop floor	Good independent worker; got along very well with mechanics and with customers; eager to learn; enthusiastic worker; intent on finding solution to problems; excellent mechanical diagnostic sense; good interpersonal and communication skills Hired as apprentice mechanic, May, 2014.
Jason Parakh	Automotive 	Parts dept.	Competent worker but rather unenthusiastic; argued at times with supervisor and with mechanics; seemed very keen on coffee/lunch breaks; often unwilling to accept personal responsibility for errors, problems, etc.

Student	GTI Area	RAI Work Sector	Comments From Staff
Shiraz Ram	Electronics 	Parts dept.	Eager worker; quick learner; enthusiastic; punctilious; dealt extremely well with customers; always calm under pressure; sent thank-you notes to his supervisor; stayed in touch with company until graduation. Hired into Parts Dept., May, 2014.
Dana Rowan	Automotive 	Repair shop floor	Good worker; good practical and diagnostic skills; good communicator; likable personality; good sense of responsibility; did not seek employment here.
Ahmed Sala	Automotive 	Repair shop floor	Excellent worker; liked and respected by supervisor and mechanics; well above average expertise—had worked as mechanic for three years before moving to Massachusetts; did not apply for employment here, despite supervisor's encouragement to do so.
Katie Wing	Electronics 	Repair shop floor	Excellent theoretical knowledge; very conscientious; excellent interpersonal skills; well liked by all; made interesting comments: glad she had chance to work here—learned she prefers programming; did not apply for position.

Potential 2016 Participation

After discussions with appropriate personnel, I am pleased to indicate that Radisson Boston will have five Internship openings for the spring of 2016:

- two Automotive Mechanics students in the repair bays
- one Electronics student in the automotive service area
- one Electronics student in our expanding Parts Department
- one Plumbing & Welding student in our Facilities Management area

Openings will be from January 4, 2016 through April 30, 2016. Your lead-hand contact with Radisson Boston will be our Personnel Manager, Ms. Emily Cardinal, who can be reached by phone at (877) 555-4235, by fax at (877) 555-9097, or by email at cardinale@radissoncars.com.

Conclusions

We are pleased that you kept Radisson Boston in mind when planning your Internship Program for 2016. It is our conviction that partnerships of this kind between post-secondary educational institutions and established businesses are of enormous value. They provide capable students with the practical experience that allows them to enter the workforce better prepared, and they confer on businesses such as ours the opportunity to get to know the people who often become the key to our own continued success. Believing programs such as yours to be essential to the growth of the local and state economy, Radisson Automobiles remains committed to them.

Thank you again for your interest in Radisson Automobiles. Ms. Cardinal looks forward to receiving your call over the next couple of weeks.

Yours sincerely,

Carole Schusterman

Carole Schusterman, General Manager
c.c. Lorna Hildebrande
L. Cornetski
E. Cardinal

ISSUES TO THINK ABOUT

1. What was the context that generated this communication product?
2. Why do you think that Hildebrande chose to have Schusterman deal with the matter instead of dealing with it herself?
3. From your answer to question 2, what can you infer about the relationship between Hildebrande and Schusterman?
4. How would you summarize Schusterman's primary message? Can you see a secondary message? If so, what do you think it is?
5. What do you think of Schusterman's choice of product? Would you have chosen a different one? If so, why?
6. What do you think of the way she formatted the information on past participation? Why do you think she might have chosen that presentation? What else might she have done?
7. What do you think of her use of clip art, emoticons, and document visuals such as boxes?
8. What can you say about her use of fonts? What might that say about her as a manager? What fonts would you have chosen?
9. What do you think of her choice of line and paragraph justification? Do you see any difficulties?
10. The primary audience is Methoulios. Do you see a secondary audience? If so, who do you think Schusterman would have had in mind, and why?
11. How do you think Methoulios will react to the report, and what do you think will be the reasons for his reaction?

EXERCISES

7.1 Define the following terms:

1. body text
2. level head
3. white space
4. header
5. footer

7.2 Describe the differences between monospace and proportional-space fonts.

7.3 Describe the main characteristics of serif, sans-serif, and decorative fonts.

7.4 Suggest the most common uses in standard technical communications documents for each of the three font categories.

7.5 How might you change your answer to 7.4 if "the most common uses" was in reference to text projected onto a screen?

7.6 What are the different purposes served by bulleted and numbered lists?

7.7 Briefly explain the CMAPP terms "information" and "impression" as they relate to the message carried by visuals.

7.8 Briefly explain why each of the following are important points to consider when choosing or constructing visuals:

- relevance
- simplification
- enhancement
- size
- legibility
- color

7.9 On a Web page or in a document, find an example on the Internet of a visual that you believe includes too much detail, and justify your decision.

7.10 On a Web page or in a document, find an example of a visual whose scale you believe to be too large, and justify your decision.

7.11 Figure 7.27 shows data obtained from a company called Gala Motors, concerning that firm's national auto sales over the past 20 years. Think of two different ways of presenting that data visually. For each of your suggestions, produce a brief CMAPP analysis. Then produce a draft of (or describe) the visual you would use for each analysis.

FIGURE 7.27: Gala Motors Sales Figures

In 1995, 9% of Gala Motors' vehicle sales across the United States came from its innovative Traveler line of sport utility vehicles. The Trend line of sport sedans accounted for 19% of sales, while the Security line of passenger vehicles accounted for the remaining 72%. In 2001, the Traveler line accounted for 14%, the Trend line for 23%, and the Security line for 63%. In 2005, the figures were 20% for Traveler, 21% for Trend, and 59% for Security. In 2010, the figures were 27% (Traveler), 25% (Trend), and 48% (Security). In 2013, 42% of sales were Traveler's, 39% were Trend's, and only 19% were Security's. Gala predicts that by 2017, fully 55% of its sales will be attributable to Traveler models, 30% to Trend vehicles, and 15% to the Security line.

Gala sold 7,000 vehicles in 1995, 11,400 in 2001, and 22,900 in 2005. In 2010, sales totaled 35,400, rising to 48,400 in 2013. A total of 57,000 vehicle sales is predicted for 2017.

7.12 Create a table that arranges the information in the paragraph below into rows and columns. Include a title for the table as well as labels at the top of each column. Write a sentence below the table explaining to your audience what key insight you intended the table to convey.

> The Allenbury MicroTech Human Resources Department submits to top management the following information about five new employees hired during the month of October 2015: Sally Anderson was hired as an IT specialist for the Chicago office at a salary of $40,000. Andre Boulanger was hired as an account representative in the New Orleans office at a salary of $47,000. Diana Cabrerra was hired as an auditor in the Boston office at a salary of $44,500. Akeo Matsumi was hired as a programmer in the Chicago office at a salary of $55,000. Sara Beth Wilson was hired as a lawyer in the Atlanta office at a salary of $71,000.

USEFUL WEBSITES

Note: For ease of reading, we have removed the "http://www." from the URLs below. However, some browsers might still require that you insert it to precede some URLs so as not to receive an error message.

■ Daniel Will-Harris, in "Choosing & Using Type" offers an excellent discussion on the effective use of fonts.
Visit: will-harris.com/use-type.htm

■ MotionElements' blog offers a brief explanation of common video file formats.
Visit: motionelements.com/blog/?s=video+formats

■ Reasoft offers a brief explanation of commonly used graphic file formats.
Visit: reasoft.com/tutorials/formats/formats_1.html

■ SUNY at Albany Libraries provides an online Interactive Media Center that offers a tutorial on Basic Web Page Layout and Design.
Visit: titus.ulib.albany.edu/imc/webdesign/index.html

■ Wikipedia offers links to a number of design-related articles, many of which are useful for creating more effective visuals.
Visit: en.wikipedia.org/wiki/Category:Design

■ Wikipedia also provides links to a host of public domain—as well as copyrighted—clip art and other visuals.
Visit: en.wikipedia.org/wiki/Public_domain_image_resources

CHAPTER 8

Reports, Summaries, and Abstracts

This chapter describes and classifies most kinds of workplace-focused reports, both formal and informal. We leave until Chapter 11, however, our discussion of proposals and business plans, report types very common in the engineering workplace. These latter reports have very specific persuasive purposes and warrant a separate examination.

8–1 Reports Overview

Regardless of your career, your private life will require you to submit an income tax return to the IRS, and you'll probably sign up for auto, home, and/or medical insurance. And in whichever branch of engineering you practice, you may well have to generate feasibility reports, progress reports, environmental impact reports, laboratory reports, engineering analysis reports, and/or research reports.

Few people get up in the morning exclaiming, "I want to write a report today! Now, what should I write about?" You do not produce a report because you feel like writing one; you produce a report because someone has asked for it. Thus, a report is an audience-driven document—a document whose content and style is determined almost entirely by its audience. As such, it requires you to focus particular attention on your audience and your context.

So what is a report? Every report—whether written or oral—can be defined as an organized set of information created in response to an expressed need. The term, therefore, could apply to something as simple—and whimsical—as the lottery ticket you filled out at the mall ("You might win a new car!"), to something as complex—and significant—as the 1986 Presidential Commission's Accident Report on the Space Shuttle Challenger disaster. In both cases, data (which, you recall, can be used as a singular or a plural noun) is organized into useful information because a particular audience needed the result. (Remember Chapter 5's discussion of the difference between "data" and "useful information.")

Written reports can be grouped into two broad categories:

- *informal reports*, also called *short reports*; and
- *formal reports*, also called *long reports*.

But the terms "informal" and "formal" can be misleading; although still in common use, they are vestiges of now outdated distinctions. In fact, some *informal* reports are written at a higher level of discourse (formality of language) than some *formal* reports. As with any other piece of engineering communication, your CMAPP analysis should allow you to identify the appropriate level of discourse. Furthermore, the terms "short" and "long" are no longer valid discriminators: Some *short reports* are longer than some short *long reports*. Today, the distinction between short and long reports is more one of structure and conventions.

8–1a RATIONALE

The needs met by written reports are virtually limitless. For example, a report can

- keep others informed of developing situations (e.g., consecutive reports on load-bearing tolerances of structural models under examination);
- maintain a permanent record of events (e.g., an engineer's detailed description of a series of experiments);
- reduce errors stemming from faulty recollections or confused interpretations (e.g., witness statements concerning a building collapse);
- facilitate planning and decision making (e.g., a comparison of the Linux® and Windows® computer operating systems);
- fulfill legal requirements (e.g., a corporation's annual report or an income tax return);
- fulfill administrative requirements (e.g., a company's pay and benefits records, or a medical expense claim form).

The university transcript in Figure 8.1 is an example of a report that fulfills administrative requirements.

8-1b HIERARCHY

If you are the originator of a report within an organization, one of your primary CMAPP considerations should be that organization's hierarchy. In general, you can consider your report as being either *lateral* or *vertical*.

Ann Arbor University

Seeking Excellence
In the Great Lakes State

Transcript of Student Academic Record

CONFIDENTIAL COPY

Not valid if removed from sealed envelope before delivery to requesting institution

Name: Rosalind Rebecca Greene
Date of Birth: 03/10/1997
Date of Issue: 08/17/2016
Envelope Issued To: Rosalind Rebecca Greene
44702 Leslie Lane
Plymouth, MI 48170

Student Number: 1123835988
Current Program: General Studies

R.M.Concorde
Signature of Registrar

Course	Title	Crd	Grd	GPA	Course	Title	Crd	Grd	GPA
Fall Session 2015					**Fall Session 2015**				
CMNS 220	Bus & Tech Comm	3.00	B+	3.33	ANTH 120	Anthro Survey	3.00	A–	3.67
CMNS 235	Report Writing	3.00	B+	3.33	PHIL 150	Survey	3.00	A–	3.67
ENGR 210	Engineering Ethics	3.00	B	3.00	SOCI 220	Historical Overview	3.00	B	3.00
ENGR 312	History of Civil Eng	3.00	B	3.00	SPAN 100	Introduction	3.00	A–	3.67
GEOE 210	Intro Geolog Eng	3.00	C+	2.33	TCWR 200	Technical Reports	3.00	B	3.00
		Semester:		**3.00**			**Semester:**		**3.40**
Spring Session 2016					**Spring Session 2016**				
MISY 214	System Design	3.00	C+	2.33	ENGL 352	Short Story	3.00	A	4.00
MISY 225	System Development	3.00	C+	2.33	ENGL 355	Drama Survey	3.00	A	4.00
MISY 229	System Evaluation	3.00	C	2.00	MISY 300	LAN Connections	3.00	C+	2.33
STAT 210	Intermed Statistics	3.00	C+	2.33	MIISY 320	Advanced Design	3.00	C+	2.33
TECH 310	Engineering Writing	3.00	C+	2.33	MISY 330	Advanced Development	3.00	B-	2.67
		Semester:		**2.26**			**Semester:**		**3.07**
Summer Session 2016					**Summer Session 2016**				
COMP 200	Basic Hardware Design	3.00	A–	3.67	MISY 350	Intro to OS Languages	3.00	B	3.00
COMP 210	Hardware Troubleshooting	3.00	A	4.00	MISY 355	Intro to System Tech	3.00	B	3.00
COMP 215	Intro to LAN/WAN	3.00	A	4.00	FRNC 200	Language & Literature	3.00	A–	3.67
COMP 218	System Recognition	3.00	B+	3.33	PHIL 200	Ethics	3.00	A	4.00
PHYS 110	Intro to Physics 2	3.00	A	3.67	TCWR 300	Manuals	3.00	A	4.00
		Semester:		**3.73**			**Semester:**		**3.53**
		YEAR:		**3.00**			**YEAR:**		**3.33**

FIGURE 8.1: Ann Arbor University Student Grades Transcript

Lateral

Suppose that you are the manager of the steel-detailing group in an engineering firm and have been asked to produce a report on your staff's progress on a particular job. If your primary audience is the manager of the structural drafting section, you are writing for an audience at approximately the same level in the organization as yours, and your report is therefore *lateral*. Your audience, another manager, will have concerns similar to your own with regard to day-to-day activities, level of detail, and so on. These factors, part of the *context*, will influence not only your purpose, but also your message (including the report's organization and the language you use).

Vertical

If the principal audience for your report is the vice president of the design division, someone above you in the organizational hierarchy, your report will be *vertical*. The VP will likely be interested less in the minutiae of your staff's accomplishments than in the big picture (e.g., the effects on the bottom line). In fact, the VP may very well not have the technical background and expertise to appreciate the finer points of detailing. By tailoring your report to the results of your CMAPP analysis, you will create a product quite different from the lateral report.

8–1c LEVEL OF TECHNICALITY

Because the level of expertise of your report's primary audience will dictate the language of your report, the level of technicality is not necessarily dependent on whether the report is lateral or vertical. For example, your primary audience in the first example above could be the personnel manager, whose knowledge of steel detailing is negligible. Although your report would still be lateral, you would try to avoid technical jargon. Conversely, if the primary audience for your vertical report were a vice president who is a practicing structural engineer, you would likely support some of your points by using the appropriate technical terminology.

But consider the following situation. Your supervisor, the drafting director, has requested your report and is thus your primary audience. Your supervisor can be considered a technical audience, but you've been told your report will also go to the vice president of public relations, a lay (nontechnical) audience. Common practice would be to write your report at a high level of technicality for your primary audience (your supervisor) and to append to your report a brief, nontechnical document (often in the form of an executive summary, discussed later in this chapter) for your secondary audience, the VP.

8–2 Informal/Short Reports

Informal reports share a number of common characteristics having to do with focus, format, length, and content.

8–2a SINGLE FOCUS

An informal report normally addresses a single issue. Note that "single" does not mean "simple." For example, Phillip Osterhuis, RAI's General Manager in Atlanta, might have submitted to his VP, Melinda Shaw in Dallas, a seven-page report with the *Subject* line "Decline in New Car Sales, Atlanta Dealership, 2005–2010." His report would certainly have included sales statistics and would inevitably have referred to the 2008 recession and perhaps to Global's reduced supply of vehicles. But it would probably also have focused on issues such as concerns about deficient Minotaur components and customer complaints following a Minotaur recall notice. Because the report is a short report with a single focus, however, Osterhuis would not have included a detailed analysis of the dealership's advertising budget, or his conversation with Shaw regarding the RAI reorganization.

8-2b VARIED PRODUCT FORMAT

Short reports can take the form of a variety of CMAPP products, including the following:

Letter

Noam Avigdor, an AEL Detroit senior associate specializing in civil engineering, might use a letter as the format for a report on seismic upgrading to the Plumbing and Electrical Building at GTI that he is preparing for Cal Sacho, GTI's senior facilities manager.

Memo

Before creating his report to Melinda Shaw, Phillip Osterhuis asked his manager of new car sales, Caroline Pritchard, to give her biweekly updates on both Minotaur sales and customer comments. As well, he asked Dean Wong, the service manager, to report every two weeks on Minotaur service incidents. Because both reports were written communications *within* an organization, Pritchard and Wong appropriately used the memo format—even though they *transmitted* them to him in electronic form.

Note: Letter and memo document conventions will be described in detail in Chapter 12.

Report

Whether *internal* or *external* communication (as defined in Chapter 12), an informal report may be a document titled "Report on . . ." For example, Avigdor might have chosen as his product a document headed "Report on Seismic Upgrading Requirements for the Plumbing and Electrical Building at Grandstone Technical Institute," which he would have sent to Sacho with a short *transmittal letter* (simply a brief, standard business letter that introduces and sometimes also presents highlights of the report). Analogously, Osterhuis might have emailed to Shaw a document entitled "Decline in New Car Sales, Atlanta Dealership, 2005–2010," introduced by a *transmittal memo*.

Prepared Form

Many informal reports are created on prepared forms. Examples might include the last speeding ticket you received on your way to campus, the transcript shown in Figure 8.1, an online survey you completed, or even your Facebook or LinkedIn profile. Such prepared forms proliferate in our society. Whatever their intended use, they allow information to be recorded in a standard format.

When completing a prepared form, you will probably be required to follow the organization pattern of the form itself, whether it is a student loan application, a marketing survey report, or the online application form at https://netforum.nspe.org/eweb/DynamicPage.aspx?site=MemberProf&WebCode=NSPENMJW to join the National Society of Professional Engineers.

8-2c LENGTH

Short reports normally do not exceed a few pages, although some may actually be longer than some long reports, as we noted earlier. Consider the following examples:

- When you receive a parking ticket, the official who fills out the form is, in effect, completing a short report: information with a single focus recorded on a prepared form. Such a report typically runs to no more than half a page.
- The income tax return you prepare for the IRS deals with a single issue: your income tax situation in a particular year. Likely running but a few pages, your tax return is also a short report.
- AEL submits an income tax return as well. But the complexities of the firm's operations are such that the return runs more than 75 pages. Nonetheless, because AEL's return (just

like yours) has a single focus and uses standardized forms, it qualifies as a short or informal report despite its length.

8-2d CONTENT TYPE

The content of an informal report may consist of the following:

- *Text only:* For example, several paragraphs, perhaps with one or more bulleted lists, or a prepared form such as your online university registration.
- *Text and visuals:* For example, Osterhuis's report to Shaw, which consists of tables of sales figures integrated into several pages of text.
- *Visuals only:* For example, the table of sales figures, shown in Figure 8.2, submitted to Phillip Osterhuis by Caroline Pritchard.

8-2e CLASSIFICATION

A traditional way of looking at informal reports is to classify them according to the main function they serve. The most common types include the following.

Incident Report

An incident report documents what happened in a particular situation. It also assumes that the incident is not likely to be repeated. Examples of incident reports would include a safety officer's report on an accident in a large industrial operation, a report on the installation of a new telephone exchange, or the transcript you saw in Figure 8.1.

Sales Report

A sales report presents sales figures for a particular business. Sales reports may consist entirely or almost entirely of visuals—usually tables or graphs—such as the sales data in Figure 8.2 prepared by Caroline Pritchard for Phillip Osterhuis.

FIGURE 8.2:
RAI Sales Report

	Jan. 2 to Jan. 15	Jan. 16 to Jan. 31	Feb. 1 to Feb. 14	Feb. 15 to Feb. 28	Mar. 1 to Mar. 15	Mar. 16 to Mar. 31	Totals
Chevrolet Malibu	12	10	15	13	14	9	73
Chevrolet Sonic	18	22	15	35	22	32	144
Chevrolet Equinox	15	18	22	15	19	20	109
Global Minotaur	49	55	56	58	21	15	254
Global Whirlwind	53	65	34	48	22	18	240
Global Fargo	22	25	28	25	29	32	161
Totals	**169**	**195**	**170**	**194**	**127**	**126**	**981**

Periodic Report

A periodic report appears at regular intervals of time and focuses on the same principal issue. Examples might include the annual *U.S. News and World Report*'s ranking of engineering schools, as well as the *Congressional Record.*

Progress Report

A progress report indicates the extent to which something has been completed. Examples might include the state of your investment portfolio, the progress of a construction project, or your midterm marks.

Trip Report

A trip report might include a site inspection by an engineer to determine the condition of a building site, a field trip description by a mining engineer seeking appropriate sites for test holes, or a professional's report on a recently attended conference.

Test Report

Test reports are often completed on prepared forms or according to strict format and wording guidelines. Such reports might include a research biochemist's report on the testing of certain pharmaceuticals, or a software engineer's report on debugging and recompilation of operating system subroutines.

It is important to remember that this classification system often results in considerable overlap. For example, a biweekly report on car sales would be both a sales report and a periodic report, while a series of regularly scheduled reports covering the various stages of developing a new antidepressant drug would at the same time be test reports, progress reports, and periodic reports. Therefore, unless you are following your organization's house style requirements, we recommend you conduct and apply the results of a CMAPP analysis, and thus generate a short report appropriate to your audience's needs.

8-2f CATEGORIES

Your CMAPP context and purpose will lead you to construct a report that falls into one of two general categories: *informative* (or *content*) *reports* or *analytical* (or *evaluative*) *reports.*

Informative Reports

Leaving your opinion out of a report is no easy task. Suppose that a recent fee increase at your school has left you and your fellow students incensed. But suppose also that you are supposed to write an *informative* report on the issue. If you refer in your report to "the unreasonable and unjustified hardship imposed on the unsuspecting student body by this untimely and disproportionate fee increase," you are not adhering to the rules for writing informative reports, because the connotation of your words expresses your own opinion. Rather, you might have stated, "Approximately 35 percent of the students interviewed believed that the $275 per term fee increase will have an adverse effect." Such wording is objective and verifiable and thus informative.

You need to organize the body of your report clearly and logically. Common organization patterns include the following:

- chronological (e.g., a student council meeting)
- spatial or geographical (e.g., the building of a new campus)
- topical (e.g., transportation options around your school)
- order of importance—from highest to lowest or vice versa (e.g., the impact of turning recreation space into new classrooms)

Analytical Reports

Variously called *analytical, evaluative,* or *persuasive reports,* these require you to present the "facts" and comment on them. You voice your opinion through your analysis and, commonly, your recommendations. Suppose that you are asked to suggest ways for your student council to respond to an announcement of raised fees. You might create an analytical report that would examine the issue, analyze its ramifications, offer alternatives to the student council, and, probably, indicate which option you prefer.

Normally, you will organize an analytical report in one of two ways. Very common in business settings, the *deductive* or *direct pattern* attempts to provide as quickly and clearly as possible the recommendations that the audience has requested. The following sequence of elements is typical:

1. *Problem* or *introduction*: Very briefly sets out the reason for the report
2. *Recommendations*: Precisely and concisely states the recommendations resulting from the analysis
3. *Background* or *facts*: Presents relevant details concerning the issues
4. *Discussion* or *solution(s)*: Discusses the issues, develops the arguments, analyzes the aspects of the problem, and demonstrates evidence for the report's assertions

In the example described above, you would likely have organized your report so that the student council could quickly note the topic of your report and immediately examine your recommendations. Reading the rest of your report, and thus determining how you reached those recommendations, would be optional.

Note the CMAPP implications: Your audience is already aware of the subject, and, having requested your recommendations, will respond positively to receiving them, even if their content raises concerns.

If your audience is unfamiliar with the issues or is likely to react negatively to receiving recommendations, you should probably choose the *inductive* or *indirect pattern*. It follows the steps you often use when you are thinking something through: You prepare the audience for your recommendations by first explaining your route. The inductive pattern incorporates the same elements as the deductive pattern, but the sequence is as follows:

1. problem or introduction
2. background or facts
3. discussion or solution(s)
4. recommendations

The success of your analytical report will depend in large part on how well you organize the background and discussion segments. Suppose, for example, that Ann Arbor University's senate has asked you to make recommendations on the issue of raising fees. You could organize the background and discussion segments by *option* or by *issue*, as illustrated in Figure 8.3.

The first column shows organization by option. For the first option—raising fees—you would look at each issue:

- the effect on students, since they will have to pay the fees
- the effect on AAU Administration, since the institution's funding will also be affected
- the potential effect on the education provided at the institution

FIGURE 8.3: Report Organization by Option or by Issue

By Option	By Issue
Raise Fees Impact on Students Impact on AAU Administration Impact on Education	Impact on Students Raise Fees Maintain Fees Lower Fees
Maintain Fees Impact on Students Impact on AAU Administration Impact on Education	Impact on AAU Administration Raise Fees Maintain Fees Lower Fees
Lower Fees Impact on Students Impact on AAU Administration Impact on Education	Impact on Education Raise Fees Maintain Fees Lower Fees

You would repeat the process for the second and third options: maintaining fees and lowering fees. By looking at the same issues for each of the options, you are in effect comparing apples with apples, not with oranges.

The second column shows organization by issue. The three issues are the impact of fees on students, AAU Administration, and education. For each of the three issues, you would "compare apples with apples" by examining the same options: raising, maintaining, or lowering fees.

How do you decide which organization pattern to use? Conduct a CMAPP analysis. In our scenario, your audience is the AAU Senate members. Because that audience would likely want to focus on financial implications, you would probably select the organization pattern that places emphasis on raising, maintaining, or lowering fees: organization by option.

But suppose that your primary audience are members of the student council who want to stimulate public opposition to higher fees by focusing on the negative impact on students and (if they could make the argument) on AAU itself, and on education as well. Such an audience might well prefer a report organized by issue.

8–3 Formal Reports

Formal, or long, reports need not run hundreds of pages, although some—the annual reports of immense businesses such as Boeing or Microsoft, for example—might well do so. Generally, the report's length is dictated by its content and structure. As Figure 8.4 shows, the body of a formal report mirrors the three-part structure of other technical communications.

Formal reports are distinguished from informal reports by their characteristics and their conventions.

8–3a CHARACTERISTICS

Formal reports differ from informal reports in the following ways.

Multiple Focus

The long/formal report deals with the interrelationships of a number of issues, while the short or informal report focuses on a single issue.

Formal/Long Report	Other CMAPP Products	Business Presentations	Traditional Prose Paragraph
Front pieces			
Body: • Introductory information • Discussion • Conclusion	Body: • Introductory topic summary • Body text • Concluding summary	Body: • Introduction • Body • Conclusion	Body: • Topic sentence • Supporting sentences • Concluding sentence
End pieces			

FIGURE 8.4: Structure of Long/Formal Reports

Complex Content

The long report's message is more complex than that of the short report and thus requires considerable analysis and synthesis. Not only will a secondary message almost inevitably complement the primary message, but a long report frequently harbors a secondary purpose as well—for example, a report that attempts to convince the Tennessee Valley Authority that your environmental engineering company has fulfilled its contract specifications might have a secondary purpose of convincing the Authority to at least consider you for other contracts in the future.

Detailed Organization

Writing the engineering report just mentioned would require careful planning. You should use a multilevel outline (see Chapter 5) to develop its multiple focus and complex content.

Format

The long report never takes the form of a letter, a memo, or a prepared form; rather, it accompanies the transmittal memo (internal) or letter (external) that introduces it.

Content Type

The long report will always contain substantial text, often enhanced by visuals. It can never consist of visuals alone.

8–3b CONVENTIONS

Adherence to common conventions is perhaps the most recognizable attribute of formal reports. Most follow a three-part structure consisting of front pieces, body, and end pieces.

Front Pieces

Also referred to as *front pages* or *front matter*, front pieces assist the audience in dealing with the complexities of the report. They are numbered separately from the body of the report, usually with lowercase roman numerals.

Standard front pieces include:

- *Transmittal Letter* or *Transmittal Memo*: The transmittal letter or memo introduces the report to the audience. It may also direct the audience's attention to particular passages or recommendations. A transmittal letter or memo that runs longer than one page is numbered independently.

- *Title Page* or *Cover Page*: The title or cover page typically presents the title of the report, the name of the author, the identity of the primary audience, and the submission date. The title page is not numbered.
- *Abstract*: The abstract (discussed later in this chapter) is a brief description and assessment of the report. Most abstracts appear on a separate unnumbered page.
- *Executive Summary*: The executive summary (discussed later in this chapter) presents the main points of the report, often for the benefit of a nontechnical secondary audience. The executive summary always begins on a new page.
- *Table of Contents*: The table of contents lists the level heads of the report and the page number on which each appears. It provides a clear picture of the organization of the document, and allows the audience to easily locate particular sections. The table of contents always begins on a new page and, if more than a single page, is typically numbered independently.
- *List of Figures*: The List of Figures (also titled *List of Illustrations*, *Table of Figures*, or *Table of Illustrations*) indicates the page numbers on which the illustrations appear. It serves as a secondary table of contents in a report whose message relies significantly on its visuals (e.g., one that emphasizes comparisons and contrasts among tables of costs).

Body

The body constitutes the bulk of the formal report. Whereas lowercase roman numerals are typically used in the front matter, the body begins numbering afresh, usually with Arabic numerals. Mirroring the three-part structure of *introduction*, *body*, and *conclusion*, the body begins with an overview, proceeds to a description and analysis of the issues, and concludes with a short summation, and, in some cases, a set of recommendations.

End Pieces

Also called *end pages* or *end matter*, end pieces furnish relevant information not included in the body. The end matter typically continues the page numbering from the end of the body. Standard end pieces include the following:

- *Appendix*: A report may include one or more appendices. (An alternative plural is "appendixes," which, by the way, is also the standard plural for "appendix" when it refers to the small organ attached to the large intestine.) An appendix contains material not essential to an understanding of the report but of potential interest to the audience. Examples would include a table that is too long or cumbersome to be included in the main body of the text, a transcript of an interview, or the full text of a document referred to in the report.
- *Endnotes*: Like footnotes, endnotes provide additional information about particular items in the report. Unlike footnotes, endnotes appear at the end of the report rather than at the bottom of the relevant pages. If you use endnotes in your report, you should follow the format outlined in your chosen style guide (e.g., APA, Chicago, IEEE).
- *Glossary*: Sometimes used as a front piece, a glossary is an alphabetized list of words or phrases that the report's author believes may require definition or explanation. The choice of glossary items should depend on the results of a CMAPP analysis, particularly with respect to the audience's level of technicality.

- *Works Cited*: For reports containing quoted material, some style guides require a works cited page to present the bibliographic credits for those citations.
- *Bibliography*: More comprehensive than a works cited page, the bibliography lists all research sources consulted, including books, periodicals, interviews, websites, and so forth. The sources are typically arranged alphabetically, by the last names of authors. Again, consult your style guide.
- *Index*: An index is a list of alphabetically arranged entries and the page or pages on which they appear within the report. Its purpose is to allow the audience to locate key words or concepts. When developing an index, you should take care to ensure that it does not simply replicate the listing of level heads in the table of contents. You should also conduct a CMAPP analysis to determine which key words and terms your audience is most likely to want to look up—as well as whether you need an index in the first place.

8–3c SPECIFIC TECHNICAL REPORT ISSUES

Engineering reports, whether *short/informal* or *long/informal*, are, ultimately, just technical reports dealing with engineering topics, of which, of course, there is an almost unlimited variety. Nonetheless, you are likely to have to deal with certain specific details, including the following.

Equations and Formulas

Scientific equations and formulas are common in engineering reports. The way you actually produce them on the page will depend on your word processor. Simple items such as the Pythagorean theorem ($a^2 + b^2 = c^2$) can be entered using normal text, enhanced by the use of superscript for the exponents. For more complex formulas, Microsoft Word (2013), for example, offers the "ready-made" insertion of a few standard items, available in what it calls both *linear* and *professional formats*, as shown in Figure 8.5.

FIGURE 8.5: Quadratic Equation Formula

Word's *Linear* Format	Word's *Professional* Format
$x = (-b \pm \sqrt{}(b\wedge 2 - 4ac))/2a$	$x = \frac{-b \pm \sqrt{b^2 - 4ac}}{2a}$

Typically, equations and formulas in engineering reports are displayed using center justification and are captioned separately from other figures in the report. For example, if we wanted to include Sabine's formula for reverberation[1] at this point in our text, it would be captioned *Equation 1.*

Equation 1: Sabine's Formula

$$RT_{60} = \frac{24 \ln 10^1}{C_{20}} \frac{V}{Sa} \approx 0.1611 \text{sm}^{-1} \frac{V}{Sa}$$

[1] Expression of Sabine formula found at en.wikipedia.org/wiki/Reverberation.

References

Before you begin writing your report, you should check whether your audience requires you to adhere to a specific style guide. Remember that no specific style is "better" than others; however, all are slightly different. In the United States, an engineering report is most likely to conform to either the IEEE or the APA style. Make sure that you pay attention to the format issues involved.

Jargon

The Merriam-Webster Online Dictionary's primary definition of "jargon" is "the language used for a particular activity or by a particular group of people." Typically, there are two types of jargon: words that are used in common speech, but that have special meaning for "a particular activity" or "a particular group," and words that are used *only* for such an activity or by such a group. Figure 8.6 offers two examples of each type.

FIGURE 8.6: Examples of Jargon

Used in Common Speech and as Jargon	Occurring Only in Jargon
Boot Common speech = heavy footwear IT jargon = start up the computer	**Ferrimagnetism:** permanent and large magnetizations found in some ceramic materials; it results from anti-parallel spin coupling and incomplete magnetic moment cancellation[2]
Footing Common speech = placement of one's feet Civil engineering = a slab (usually concrete) designed to transfer loads into the surrounding ground	**Rheology:** the study of flow characteristics[3]

The important thing for you to remember is that jargon can be good or bad. For an audience that will understand it, engineering jargon is often the most efficient way to communicate specific ideas. For an audience that does not have the required level of technicality to be comfortable with it, however, jargon is counterproductive: It serves only to confuse and to irritate. Therefore, you need to ensure that your audience analysis has been as precise as possible, and that your message's "form" reflects that analysis.

Technical Report Examples

Case Study 8, at the end of this chapter, offers an example of a *short/informal* engineering report—presented in both an initial problematic version and in a more effective revised version. Since a *long/formal* engineering report might well be dozens of pages, if not more, we have not included an example. However, the two references under *Sample Technical/Engineering Reports* in the *Useful Websites* section at the end of the chapter do provide examples. A careful Google or Bing search could, of course, provide more.

---

[2] Definition taken from Materials Science and Engineering Dictionary, at engineering-dictionary.org/Materials-Science-and-Engineering-Dictionary/ferrimagnetism.

[3] Definition taken from Materials Science and Engineering Dictionary, at engineering-dictionary.org/Materials-Science-and-Engineering-Dictionary/rheology.

8–4 Summaries and Abstracts

Suppose you were a consulting engineer who had just completed setting up an automated manufacturing process for children's plastic pails. If you were asked to tell someone what you've done over the last two weeks, it's unlikely you'd even try to describe in detail every activity you could recall. (Such an account would also be excruciating for your audience.) Rather, you'd automatically select the points you'd consider most significant for your audience and briefly recount them. You would, in effect, be creating a summary.

In the context of engineering communications, a written summary is a concise document that conveys the important ideas and, in some cases, significant details found in the original document. What are significant details? The answer will derive from a CMAPP analysis. For example, imagine that you had to summarize a complex technical report on climate and weather conditions in the Raleigh, North Carolina, area over the past 20 years. If the audience for the summary—not necessarily the same audience as that for the report—were a group of graduate students in a university meteorology program, they would likely think significant such things as the annual mean temperature and the average yearly snowfall. However, if the summary's primary audience were the managers of a company that staged outdoor events in that city, they would have little interest in annual mean temperature; they would, though, want to know what the weather was like every July 4, a set of observations that would help them plan for next year's celebrations.

8–4a CATEGORIES

Unfortunately, conventional terminology for summaries tends to be used imprecisely. The same terms may be used for different things, and different things may be known by the same term. Nonetheless, summaries can be grouped into two broad categories: content summaries and evaluative summaries.

Content Summaries

Content summaries, also known as *informative summaries* or *informative abstracts*, sum up the important elements of the original document. As is the case in most engineering activities, objectivity is crucial in a content summary, which should not contain embellishment, opinion, or highly connotative vocabulary.

Two special types of content summaries are *minutes* and *executive summaries*. *Minutes* are the record of what occurred at a meeting; they document decisions made, actions to be taken, and, on occasion, comments made. An *executive summary* is typically found in the front matter of a formal report, usually targeting a secondary audience that is less technical than the primary audience. Most executive summaries do not exceed two pages, and many are limited to a single page.

Evaluative Summaries

Evaluative summaries, also known as *analytic summaries* or *assessment summaries*, provide not only the essential ideas of the original document, but also the author's opinions, often in the form of recommendations.

Two special types of evaluative summaries are *descriptive abstracts* and *abstracts*. Most common in an academic context, the *descriptive abstract* (generally no longer than a few lines) presents a synopsis of the original document, followed by a brief indication of its value or applicability to the project at hand. However, a descriptive abstract that follows an entry in a document's bibliography is often called an *annotation*—hence the term "annotated bibliography."

If you had written a term report on the use of Windows versus Linux in a small company, and your bibliography included the 2014 edition of *Linux for Dummies*, your annotation for that title might resemble the following:

> Part of the popular series of *Dummies* books, this is a 384-page primer on the operating system that still underlies a large part of the World Wide Web. Though the book is at times self-consciously populist and facetious, its 20 chapters provide a useful (and relatively nontechnical) introduction to UNIX.

A standard front piece in reports is the *abstract*, normally a short *evaluative summary* that presents a brief description of the report and offers the reader (typically the secondary audience) an evaluation of it (usually a very positive one). Abstracts tend to be shorter than executive summaries. Figure 8.7 compares examples of an executive summary and an abstract.

FIGURE 8.7: Executive Summary versus Abstract

Personnel Strategies for New RAI Dealership	
Executive Summary	**Abstract**
In January 2015, RAI requested that AEL identify personnel strategies for a potential RAI expansion into the St. Louis market. That investigation has yielded the following two potential strategies: Senior staff to be drawn from RAI's head office in Dallas. This strategy would involve the relocation to the St. Louis dealership of Vice president Alberto Chavez, General Manager George Finlay, and Sales Manager Mariela van Damm, with concomitant backfilling. Senior staff can be drawn from (a) RAI's Chicago dealership (involving the relocation of General Manager Tina Trann, who would become the St. Louis dealership's Vice president, and Sales Manager Ted Evans), and (b) RAI's Atlanta dealership (General Manager Phillip Osterhuis), again with concomitant backfilling.	AEL recently identified two strategies for effectively staffing RAI's planned St. Louis dealership. Both meet RAI's budget provisions and take excellent advantage of local conditions. While either is likely to prove successful, one relies primarily on RAI in-house expertise in Dallas, while the other would draw on RAI's Chicago and Atlanta locations for the necessary expertise.

8-4b CONTENT AND LENGTH

What you should include in, and exclude from, a summary depends on whether it is a content summary or an evaluative summary. In most cases, you can apply the criteria listed in Figure 8.8.

How long should a summary or abstract be? The answer is "as short as possible" and "as long as necessary." As you saw in Chapter 3, technical documents often require you to balance the need for precision against the need for brevity and concision. The more information you provide, the more you impose on your audience's time; the less information you provide, the more likely it is that your audience will not receive everything necessary.

Content Summary		Evaluative Summary	
Often Includes	**Typically Excludes**	**Often Includes**	**Typically Excludes**
Overall theme or goal Main points Significant details Reference to any conclusions in the original document Reference to any recommendations in the original document	Opinion New information Insignificant details Embellishments Technical jargon (if possible) Supportive examples or illustrations Visuals Quotations Citation credits	Overall theme or goal Main points Assessment/evaluation of original's content Assessment/evaluation of original's effectiveness	Since the purpose here is evaluation or assessment, it is impossible to identify any *a priori* exclusions.

FIGURE 8.8: Content of Summaries

Summaries typically run anywhere from 5 to 15 percent of the length of the original. However, keep in mind the *inverse proportion rule*, whereby the longer the original document, the smaller the percentage of its length that will be required for a summary. For example, a summary of a three-page report might be just over half a page (close to 20 percent), while a summary of a 300-page report might run two full pages (less than 1 percent). Remember, too, that most executive summaries still do not exceed two pages, regardless of the length of the original document.

8-4c CONTENT SUMMARY PROCESS

Creating an effective content summary can be a complex and painstaking undertaking. Following are suggestions for each of the stages of that process.

1. FAMILIARIZATION

Imagine, if you can, that you have never seen or even heard of what we call the Latin alphabet—the one used for English, Spanish, and German, for example. Now imagine you came across a cache of documents written in a script you'd never encountered. Among the thousands of characters you examine, you find the 70 shown in Figure 8.9.

How are you going to decide which symbols represent completely different meanings, or are simply variants of the same characters, or occur only at the end of a word (as is the case with some

FIGURE 8.9: Unknown Document Characters

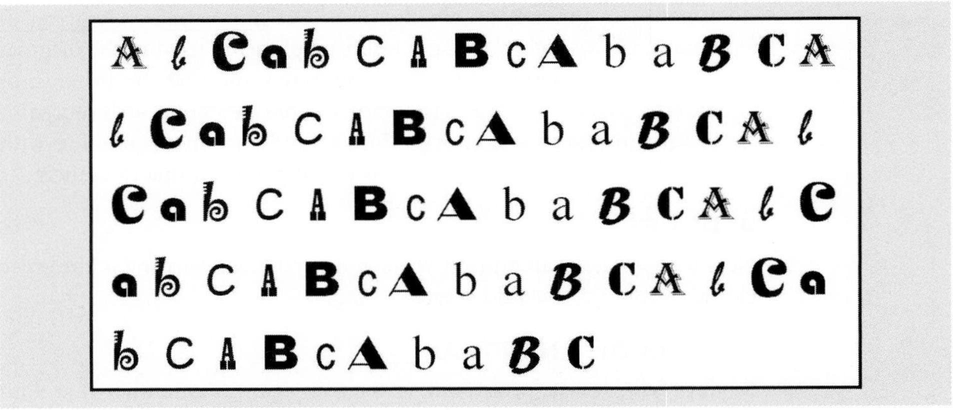

real alphabets such as Arabic and Hebrew)? Furthermore, these symbols are strewn across your documents in sequences that, as yet, mean nothing to you. (Remember, you're pretending you have absolutely no previous exposure to the Latin alphabet.)

To start with, you're going to have to get to know your material extremely well. The only way you're going to be able to look for patterns that will help you "decipher the code" is by first becoming very familiar with "everything you have." (Cryptographers, who work with real codes, take into account the differing average frequencies of use of different letters in different languages. You might want to look at Wikipedia's article on cryptography, available at en.wikipedia.org/wiki/Cryptography.)

Creating a summary requires an analogous first step: becoming thoroughly familiar with the original text. Doing so involves much more than the one cursory reading you might prefer. Even if you are the author of the original document, the rule still applies: Writing something is not the same as reading it. When familiarizing yourself with a document you have written, approach it as though you were not the author.

2. IDENTIFICATION AND MARKING

Identify the main points and significant details. While you will want to be as objective as possible, you are in a position not unlike that of a television news director who has to choose which items to broadcast, how much time to devote to each, and how to present each one. Selection is by definition a subjective process.

You can make a concerted effort to *try* to be objective by making use of verbal cues in the original document. You can look, for example, for words or phrases that suggest importance in the mind of the document's author. These words and phrases include the following:

Pointers: 1st, 2nd, 3rd, next, last

Causes: thus, therefore, as a result, consequently

Contrasts: however, despite, nonetheless

Essentials: moreover, in general, most importantly, furthermore

As you identify the points you need to include, mark them (you can star, underline, or highlight) *on your own copy* of the document. (You would not, of course, mark up other people's materials.) Try to identify single words or very short phrases; if you mark whole sentences, your task later in the process will be much more difficult.

3. COLLECTION AND (RE)ORGANIZATION

If you write out all the words and phrases you have identified, you will be faced with something that looks much like the results of the data collection that you examined in Chapter 5. You must now organize that data into usable information. In effect, you will be creating a kind of multilevel outline for what will essentially be a *new* document. This is an important point. A summary is an original document—not merely an abridgment. (See merriam-webster.com/dictionary/abridgment for an explanation.) As such, it will not replicate phrasing used in the original document.

As you develop your outline, you may sometimes find that in order to make your summary "flow" effectively, you have to reorganize some elements. Such reorganization is acceptable provided that your summary would not confuse an audience that is familiar with the original document, and that it does not take on a slant or bias not apparent in that document.

4. DRAFT

Now that you have an outline, write your first draft, treating it like any other engineering communication product you would create.

5. REVISION AND FINAL

Review, edit, revise, proofread, and check again. Remember that your summary, like every other CMAPP product, reflects on you and your credibility.

8–5 An Ethical and Cultural Aside

When summarizing, as when creating other communication products, you should try be aware of how your audience, because of their background, might respond to phrases and cultural referents you use.

Here is an example from author Dave Ingre's teaching experience.

> A student of mine once officially complained to the Dean about my use of a phrase that she felt was insulting: "rule of thumb." Her belief, one commonly held, was that the expression's etymology made its use demeaning to women, and that, consequently, I should never have used it in class. A "rule of thumb," she was convinced, derived from an ancient British law giving husbands the right to beat their wives, providing they used a stick of no greater thickness than that of a man's thumb.
>
> If the etymology she postulated had been accurate, she would have had a valid cause to object. However, many people have researched this question; for example, see Dave Wilton's *Word Origins* site at wordorigins.org. The consensus is that no such provenance can be confirmed, and that the phrase much more likely derives from rough measurements used during the Middle Ages, perhaps by tailors, brewers, and/or sailors. Nonetheless, some audiences, it seems, still believe (though mistakenly) that the phrase alludes to wife-beating.
>
> There's little to be gained by intentionally offending any audience. But, I doubt we can really ensure that what we say could not *ever* be misconstrued by *someone*. Consequently, I continue to use the expression when I think it appropriate. And, I often tell the story in class; it is, I have found, a useful object lesson regarding people's reaction to language.

It is important, therefore, that you be aware of the words you employ, and consider how your audience may respond to them. Once again, it is impossible to overstate the importance of effective audience analysis.

CASE STUDY

8 AEL Engineering Report

SITUATION

In January 2015, AAU's Department of Physical Plant, responsible for all university buildings, was notified by the university's principal real-estate insurance company, Vermont Life and Property, of a problem in Lincoln Square, a 55-year-old building now used primarily as a warehouse. The Vermont Life and Property inspector, who had observed fissures in the concrete panels around the truck loading bays, indicated that a policy renewal would not be forthcoming until the necessary repairs were completed.

AAU contracted with Accelerated Enterprises Ltd. to investigate the damage and submit an engineering report. The assignment went to Deborah Greathall, an AEL senior associate specializing in civil engineering. Greathall, who normally works out of the firm's New York office, spent the month of March 2015 working out of the Detroit office, but staying most of the time in Ann Arbor.

During her examination of Lincoln Square, she discovered that the truck loading bays, installed some 20 years earlier, had been carved from what had been an exterior wall. Unfortunately, the construction had not taken into account the long-term impact of large trucks backing up into the loading bays. The ongoing shocks, Greathall concluded, had eventually led to cracking in the concrete panels that made up the west wall of the building.

Greathall considered various options for repairing the wall. (The situation was complicated by the fact that segments of the panels were below exterior grade. As well, the university had no alternative loading bays that it could use during reconstruction.) She eventually decided on a course of action. She submitted, by email, her preliminary engineering report, introduced by a cover letter, to Dr. Joan Welstromm, chair of AAU's Civil Engineering Department, who had been appointed by Dr. Helena Paderewski, AAU's president, to be the university's liaison. Greathall sent a copy of the report to Frank Nabata, the AEL senior partner in Lansing, who had arranged for her to take the AAU contract.

Greathall's cover letter and preliminary report appear in Figures 8.9 and 8.10, respectively.

AEL ACCELERATED Enterprises Ltd.

Accelerated Enterprises Ltd.
25 West 36th Street
New York, NY 10017-8977
(212) 555-1111 Fax (212) 555-1222

Deborah Greathall, Senior Associate

March 30, 2015

Dr. Joan Welstromm, Chair, Civil Engineering
Ann Arbor University
5827 Dixie Road
Ann Arbor, MI 48103-0061

Dear Dr. Welstromm:

I have now completed my look at the troubles found on the west side of **Lincoln Square** on the University Campus on Dixie Road, which is where the building is.

On the following pages, please find my preliminary report regarding proposed repairs.

Once you have had the opportunity to examine my findings, please contact **AEL** to advise us of how and when you would like to proceed.

I'm soon heading off to go home to **New York**. However, here in the **Detroit** area, there's a really good **AEL** Associate you can talk to, of course. This is him.

Dr. Noam Avigdor
Senior Associate, AEL Detroit
(313) 555-8309

He will be pleased to assist you. Should you so wish, **Accelerated Enterprises** will be pleased to recommend competent construction firms to act as contractors.

Note that our terms are payment in full within **30 days**. Thanks.

Yours most sincerely,

Deborah Greathall

Deborah Greathall (Ms.)
Senior Associate, Civil Engineering
cc: Frank Nabata (AEL Lansing)

p.s. The sketches mentioned in the report are now being finalized and copied; I'll send them along as soon as they're ready.

FIGURE 8.9: AEL Engineering Report Cover Letter to AAU

Accelerated Enterprises Limited
213 Abbot Street Detroit, MI 48226-2521
Tel: (313)555-8309 Fax:(313) 555-8300

PRELIMINARY ENGINEERING REPORT

March 30, 2015

Short specification accompanied by procedures requested to rehabilitate lift-up panels at the industrial building called Lincoln Square, on the campus of Ann Arbor University, at 5827 Dixie Road, Ann Arbor.

The building was constructed on a pre-loaded site of approx. 4 acres in 2 phases:

1. 1982
2. 1992

The nine panels to be dealt with in this contract are of 5 ½" thickness of reinforced concrete. All panels are 32' high and approx. 16' wide. This refers to *nine* panels only.

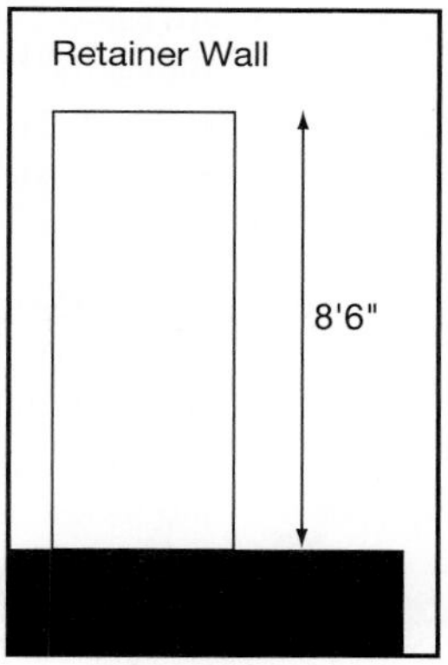

Many of the lift-up panels show signs of cracking and deterioration at the lower half. It is believed that the damage to these panels originated from

- trucks backing into the panels when maneuvering to load or unload
- stresses at construction time when being lifted by a crane into position onto their footing, possibly before having fully matured.

The 9 panels to be rehabilitated at the present time are on the western exterior wall in the 1992 construction phase. These refer to the **attached sketch** within building lines 2 to 5 and are shown thus in the original construction drawings.

Strengthening

It was decided to strengthen damaged panels by attaching them to newly-to-be-formed six-inch reinforced concrete retainer walls of approx. 8'6" height as measured from top of panel footing. As panel footings are competent and wide enough, they will also serve as footing for the retainer walls. However, the above holds good only for five panels marked **A**, and one panel marked A1. Also, one panel marked **D** is very similar but the height of the retainer wall is restricted by an exhaust duct opening.

Note

Please note that panel A1 is combined with panel B. A hollow steel rectangular column 4 inches wide by six and three eighths inches thick is attached to the footing by means of a ½" thick steel plate bolted to the footing and welded to the column (¼" weld) and is also attached to the south side of an existing reinforced concrete loading platform of panel B. It is intended to carry the weight of a horizontal retainer concrete beam attached to panel "B" above the door opening.

FIGURE 8.10 AEL Engineering Report to AAU

As panel "C" also contains an opening door with a cantilevered loading platform, its strengthening is in part achieved by another reinforced concrete beam 6 in. thick, attached to the panel and resting on concrete retainer walls at both ends.

No retainer wall is attached to panel B or panel C below the loading platforms. Instead, two 8-inch reinforced concrete walls at right angles to the panels are being provided for both and connected to the panels.

Retainer Walls

The retainer walls are to be reinforced with 20mm bars vertically at 2ft. o/c and horizontally with 10mm bars at 1ft. o/c. The retainer beams above the doors are reinforced in the entire length with two 20mm bars placed at the top and bottom. The reinforcement in the "retainer" walls is confined to its center by 10mm L-shaped anchors. These anchors are placed at 2ft each way into the panel. The anchor's legs are 6" long with 1 leg sticking in a hold drilled 3" deep into the panel with 12mm diameter, the other free leg parallel to the panel wall. The retainer beams are attached to the panels by the same type of anchors.

Procedures Requested

1. Reveal footing of panels of all 9 panels. (Approx. 144 ft in length)
2. Footing areas of panels "B" and "C" to be extended to the outline shown on the sketch.
3. All panels to be sandblasted to required height with "Black Pearl" coal slag to expose a visible profile and the aggregate of the concrete.
4. Holes of 12mm diameter 3 inches deep to be drilled at required spacing (2ft e/w into panels) to be required extent given by retainer walls or beams.
5. All sand blasted areas to be power washed.
6. Anchors to be placed into the drill holes after filling them with epoxy.
7. Soak panels for a minimum period of 24 hours with water and dry back to a "saturated surface dry" state immediately prior to placement of concrete.
8. Use dry to wet epoxy paint for bonding top 15 inches of space for retaining beams and walls/panel location just before concreting.
9. Use vibrator to place concrete against panels within form work.
10. Visible cracks in the panels up to top of retainer beam height and outside of retainer walls or beams should be cleaned out, washed and then filled with epoxy paint.
11. Paint all parts to the full height of the retainer beams with elastomeric paint to match existing light color.
12. Reinstate Grade and renew asphalt in the operation area.

Concrete Performance Requirement

1. Conform to CSA A231 type 10 normal Portland cement
2. Supplementary: class F Flyash
3. Exposure class F2
4. Minimum compressive strength: 28 M Pa

5. Maximum concrete aggregate: 14mm
6. Slump at point of discharge: 160 plus/minus 20
7. Air content 6 plus/minus 1
8. Minimum water ratio to cement: 0.55
9. Admixtures: normal dosage

Conclusion

This concludes my preliminary report.

Deborah Greathall

Deborah Greathall, Prof. Eng.
p.s. Sketches not quite ready; will be sent by courier within a day or two.

ISSUES TO THINK ABOUT

Let's assume that Greathall is a competent engineer, that she is conscientious about her work, and that she wants to make a good impression on both Welstromm and Nabata.

1. Is Greathall's letterhead effective in conveying an impression of the firm's competence and professionalism—why or why not?
2. Explain why you think her letter does or does not adhere to the CMAPP complementary attributes of *brevity*, *concision*, *clarity*, and *KISS*.
3. Discuss her use of *visuals* and of *document visuals* in the letter and in the report. What impact do you think they might have on the report's reception?
4. How do you think they might affect Welstromm's judgment of Greathall's expertise?
5. The sketches referred to in the report are not, in fact, attached. How might this oversight affect Greathall's audience?
6. Discuss her use of measurement systems in terms of context and message. (Hint: Look for examples of both metric and imperial measure.)

REVISION

Figures 8.11 and 8.12 represent improvements to the original correspondence. When you examine them, remember to consider the CMAPP complementary attributes. As well, note that the organization of some items has been changed, and that language use has likewise been modified for a consistent level of discourse. Incidentally, you should presume that, in this case, the sketches *are* attached to the report.

AEL ACCELERATED Enterprises Ltd.

213 Abbott St
Detroit, MI 48226-2521
Tel: (313) 555-8309
Fax: (313) 555-8300
AEL@AEL.com
AEL.com

March 30, 2015

Dr. Joan Welstromm
Chair, Civil Engineering
Ann Arbor University
Ann Arbor, MI 48103-0061

Dear Dr. Welstromm:

Having completed my examination of the west side of Lincoln Square on the University Campus on Dixie Road, I am pleased to present my preliminary report regarding proposed repairs.

Once you have had the opportunity to examine my findings, please advise AEL of your wishes. Since I must shortly return to New York, Dr. Noam Avigdor, AEL's Senior Associate in Detroit, will be happy to assist you. Should you so wish, he could also recommend competent construction contractors. You may reach him at the numbers shown above, or at his direct line, (313) 555-8109.

Yours sincerely,

Deborah Greathall

Deborah Greathall (Ms.)
Senior Associate, Civil Engineering
p.s. AEL will submit its invoice under separate cover.
cc: Frank Nabata (AEL Lansing)

FIGURE 8.11: Revised AEL Engineering Report Cover Letter to AAU

AEL ACCELERATED Enterprises Ltd.

213 Abbott St
Detroit, MI 48226-2521
Tel: (313) 555-8309
Fax: (313) 555-8300
AEL@AEL.com
AEL.com

Preliminary Engineering Report: 5827 Dixie Road, Ann Arbor, MI

Re: Specifications and procedures to rehabilitate lift-up panels at the industrial building called Lincoln Square, on the campus of Ann Arbor University, at 5827 Dixie Road, Ann Arbor, MI.

Background

The building was constructed on a pre-loaded site of approx. four acres in two phases, the first in 1982 and the second in 1992. At issue are the twelve 5½" thick reinforced concrete lift-up panels that make up the west wall. Each is 32' high and approx. 16' wide.

FIGURE 8.12: Revised AEL Engineering Report to AAU

The lower half of nine of the lift-up panels, all dating from the 1992 construction phase, show signs of cracking and deterioration. The damage likely derives from trucks backing into the panels when maneuvering to load or unload, and/or from stresses during construction while the panels were being lifted by a crane into position onto their footing, probably before the reinforced concrete had fully matured. These panels are marked on the attached sketches; note that in the original construction drawings that I consulted, they likewise appear within building lines 2–5.

Strengthening

It was decided to strengthen damaged panels by attaching them to newly formed 6" reinforced concrete retainer walls of approx. 8' 6" height as measured from top of the panel footing. As panel footings are competent and wide enough, they will also serve as a footing for the retainer walls. However, the above holds good only for the five panels marked A through E, and one panel marked F. The panel marked G is very similar but the height of the retainer wall is restricted by an exhaust duct opening.

Note

Please note that panel F is, in effect, combined with panel B: a hollow steel rectangular column, 4" wide by 6⅜" thick, is attached to the footing by means of a ½" thick steel plate bolted to the footing and welded to the column (¼" weld); this column is also attached to the south side of the existing reinforced concrete loading platform of panel B. It is intended to carry the weight of a horizontal retainer concrete beam attached to panel B above the door opening.

As panel C also contains an opening door with a cantilevered loading platform, its strengthening is in part achieved by another reinforced concrete beam, 6" thick, attached to the panel and resting on concrete retainer walls at both ends.

No retainer wall is attached to panel B or panel C below the loading platforms. Instead, two 8" reinforced concrete walls at right angles to the panels are being provided for both, and connected to the panels.

Retainer Walls

The retainer walls are to be reinforced with ¾" bars vertically at 2" o/c and horizontally with ⅜" bars at 1" o/c. The retainer beams above the doors are reinforced along their entire length with a ¾" bar placed at the top and at the bottom. The reinforcement in the retainer walls is confined to its center by ⅜" L-shaped anchors. These anchors are placed at 2" each way into the panel. The anchor's legs are 6" long, one ensconced in a ½" diameter hole drilled 3" deep into the panel, with the other free leg parallel to the panel wall. The retainer beams are attached to the panels by the same type of anchors.

Recommended Procedures

1) Footing of panels of all nine panels (approx. total length: 144") to be revealed.
2) Footing areas of panels B and C to be extended to the outline shown on the sketch.
3) All panels to be sandblasted to required height with "Black Pearl" coal slag to expose a visible profile and the aggregate of the concrete.
4) Holes of ½" diameter, 3" deep, to be drilled at required spacing (2" e/w into panels) as required by retainer walls or beams.
5) All sandblasted areas to be powerwashed.
6) Anchors to be placed into the drill holes after filling them with epoxy.
7) Panels to be soaked with water for a minimum of 24 hours and then let dry back to a "saturated surface dry" state, immediately prior to placement of concrete.

8) Dry-to-wet epoxy paint to be used for bonding top 15" of space for retaining beams and walls/panel location just before concreting.
9) Vibrator to be used to place concrete against panels within form work.
10) Visible cracks in the panels, up to top of retainer beam height and outside of retainer walls or beams, to be cleaned out, washed, and then filled with epoxy paint.
11) All parts to be painted to the full height of the retainer beams with elastomeric paint to match existing light color.
12) Grade to be reinstated and asphalt to be renewed in the operation area.

Concrete Performance Requirement

- Conform to CSA A231 type 10 normal Portland cement
- Supplementary: class F Flyash
- Exposure class F2
- Minimum compressive strength: 28 M Pa
- Maximum concrete aggregate: 14mm
- Slump at point of discharge: 160 ± 20
- Air content 6 ± 1
- Minimum water ratio to cement: 0.55
- Admixtures: normal dosage

This concludes my preliminary report.

Deborah Greathall

Deborah Greathall, Prof. Eng.
Attachment: Preliminary engineering sketches (5)

EXERCISES

8.1 Briefly describe four or more needs that are met by written reports.

8.2 Explain the difference between a lateral and a vertical report, and relate that difference to a report's usual level of technicality.

8.3 Identify and describe at least three of the six types of informal reports discussed in this chapter, and give an example of each.

8.4 Describe the differences between informal and formal reports with respect to focus, format, length, and content.

8.5 Distinguish between an informative report and an analytical report.

8.6 Identify at least two of the four common organization patterns found in informative reports.

8.7 Describe the organizational difference between the deductive pattern and the inductive pattern used in analytical reports. Explain the CMAPP implications for each pattern.

8.8 Assume that you have been asked to write an informal analytical report on possible changes to the parking situation at your institution. Identify the two organizational approaches you might use. Explain which approach you would choose if your primary audience was

a. the senior administration of your institution.
b. the student association executive.

8.9 Name the three main parts of a formal report. Identify and describe at least two elements in the first part and two elements in the third part.

8.10 Distinguish between a content summary and an evaluative summary.

8.11 Briefly describe each of the following types of summaries:

a. abstract (as front piece in a report)
b. annotation
c. executive summary

8.12 From AnnualReports.com, review and compare several annual reports; then, select one to analyze. How effectively does the annual report speak to its audience(s)? Is the structure of the report effective? How are visuals used to enhance the report's content? What changes do you think might make the report more effective?

8.13 Go to Penn State University's Writing Guidelines For Engineering and Science, at writing.engr.psu.edu. From the first row of options, choose "Reports"; then, choose one of the Sample Reports offered. Download the sample you have chosen, and summarize it.

USEFUL WEBSITES

Note: For ease of reading, we have removed the "http://www." from most of the URLs below. However, some browsers might still require that you insert it to precede some URLs so as not to receive an error message.

■ Austin Community College's David McMurrey provides an online *Internet Technical Writing Course Guide*, which includes an excellent chapter on technical report writing. Visit: prismnet.com/~hcexres/textbook

■ *Executive Summaries Complete the Report*, by Jay Christensen, is an easy-to-follow explanation of report writing in general, and Executive Summaries in particular. Visit: csun.edu/~vcecn006/summary.html

■ "Quoting, Paraphrasing, and Summarizing," a page from Purdue University's Online Writing Lab, offers straightforward advice on creating effective summaries. Visit: owl.english.purdue.edu/owl/resource/563/01

- Sample Technical/Engineering Reports:
  - State University's Writing Studio offers a sample engineering report for study.
    Visit: http://writing.colostate.edu/guides/page.cfm?pageid51594&guideid588
  - Midwestern Electrical Consultants, Inc., in Tinley Park, Illinois, provide a "generic" electrical engineering report.
    Visit: midwestelectrical.com/_resources/pdfs/mec_sample_engineering_report.pdf

- Montana State University's College of Engineering offers its *Engineering Communications Toolkit*, which includes both a Generic Report Template and a Generic Report Template (with comments).
  Visit: www.coe.montana.edu/StudentWritingAid/IECommunications/Reports/report%20template.html

- Wayne University's College of Engineering offers brief and worthwhile advice on "Writing Engineering Reports."
  Visit: http://www.eng.wayne.edu/legacy/MSE130/REPORT.html

CHAPTER 9

Mechanism Description, Process Description, and Instructions

Mechanism description is a common task for an engineer: to describe something's appearance in a very precise way. The object might be as simple as a paper clip, described for a patent application or a new manufacturer, or as complex as the avionics system of a military jet. A similar task, called *process description*, would be to detail how something works rather than what it looks like—for example, how carbon fiber is made, how a cyclotron works, or how the momentum forces of a bridge support are calculated. As an engineer, you're also likely to be charged with specifying how to do something—in effect, to generate instructions. You might be working with something as relatively simple as running a standard concrete durability test, or as complex as reverse engineering a programmable logic controller for a new factory assembly line.

9–1 Mechanism Description

A mechanism description explains and describes an object or a system. Although even a simple mechanism may have many parts, each one has a specific function. The parts work together toward a definite purpose. Any object or system whose parts function separately to achieve an overall effect can be called a mechanism. Mechanisms can vary from basic to elaborate, from minuscule to large, and from inanimate to living. Primary categories of mechanisms include the following:

- Tools and machinery—for example, a claw hammer consists of a handle, claw, cheek, neck, and face.
- Organisms—for example, a tree consists of roots, leaves, and bark.
- Substances—for example, automotive paint contains chemicals such as toners, reducers, and hardeners.
- Locations—for example, a construction site includes terrain and buildings.
- Systems—for example, avionics systems in airplane cockpits are likely to include communications and collision-avoidance systems.

The strategy for simple mechanism description usually involves the following steps and considerations.

1. *Introduce the item.* Provide a brief general description that includes the item's name and main use(s).
2. *Specify all relevant details.* Describe the characteristics of the item in detail. Include all relevant details such as length, width, height, depth, weight, density, color, texture, and shape. Choose vocabulary that is concrete rather than abstract, denotative rather than connotative, and precise rather than vague. For example:

Use	Instead of
The bus travels north on Main Street.	The bus goes up Main Street.
The sign is 750 meters straight ahead.	The sign is a fair distance ahead.
The marker lies at 48° 33'15" North.	The marker is just south of the border.
The temperature reaches 150°C.	It gets very hot.

Note, however, that two terms may be precise, while only one is accurate in its context. Thus, for example:

Use	Instead of
Spherical (for three dimensions)	Circular (for two dimensions)
Cubical (for three dimensions)	Square (for two dimensions)
Helical (for three dimensions)	Spiral (for two dimensions)

In consideration of your CMAPP analysis, you should choose an organization pattern that your audience will find logical and understandable. If you were describing the design of a building complex, you might choose a *spatial* organization pattern (each of the segments in order). In describing soil erosion, you might choose a *chronological* pattern of month by month, or a *geographical* pattern of sector by sector. In describing an industrial robot, you might choose a *topical* organization (component by component).

If appropriate, include visuals. (Be sure to apply what you learned in Chapter 7 about choosing or constructing effective visuals.) As well, pay particular attention to your use of document visuals: If your audience cannot readily follow and understand your description, your communication will not have achieved its purpose.

3. *Conclude.* A brief concluding statement may be used to sum up some key aspect or aspects of the item.

The description of the hearing aid in Figure 9.1 is an example of a simple mechanism description. Note that the conclusion implicitly indicates the item's purpose. From a CMAPP perspective, it gives little if any indication of a likely primary audience, primary purpose, or even context. Its most probable use would be, perhaps, as an entry in a brief encyclopedia intended for a lay audience.

FIGURE 9.1: Simple Mechanism Description of a Digital Hearing Aid

The Digital Hearing Aid

A digital hearing aid is an electronic device that amplifies sounds for the person wearing it, transforming sounds into digital signals and then converting these signals into analog, acoustic sounds for the listener. "In all hearing aids, sound enters through a microphone. It is then processed, amplified, and delivered to a receiver (loudspeaker). This sends the output either directly to the user's ear canal, via tubing to an ear mould and into the ear canal, or via a thin wire to a receiver placed directly in the ear." (http://www.oticon.ca/hearing/getting-help/hearing-aids/what-is-a-hearing-aid.aspx)

Monika Wisniewska/Shutterstock.com

A digital hearing aid's main components are the following:

- A battery compartment
- A power switch
- A volume control dial
- An ear hook
- A receiver
- An ear plug
- An earphone

After the amplified signal enters the user's ear, it is transmitted to the user's brain and processed into discernable, meaningful sounds.

9–1a LEONARDO'S MECHANISMS

As an engineering student, you have almost certainly heard of Leonardo da Vinci (1452–1519), whom many consider to have been the greatest European "renaissance man"—an artist, scientist, inventor, and ("unofficially") engineer.

Among his designs was what appears to be a precursor to today's parachutes. Along with factual information, Figure 9.2 offers a whimsical conjecture of a short "modern" mechanism description for that invention.

FIGURE 9.2: Leonardo's Parachute

Note: The da Vinci drawing shown is from *Il Codice Atlantico di Leonardo da Vinci nella biblioteca Ambrosiana di Milano*, Editore Milano Hoepli 1894–1904. The original drawing is housed in Milan's Biblioteca Ambrosiana. Da Vinci's own comment read: "Se un uomo ha un padiglione di pannolino intasato che sia di 12 braccia per faccia e alto 12, potrà gittarsi d'ogni grande altezza senza danno di se," which could be translated as "If a man has a seamless treated linen cloth of some 12 yards square, he will be able to descend without harm from any great height."

The Safe-Descent Machine

The safe-descent machine (see figure) is a semi-rigid, hollow quadrangular pyramid, constructed of a wooden frame, on which has been adhered linen cloth treated to reduce porosity, and from which descends a harness from which a man may safely dangle.

When falling, the safe-descent machine captures a large volume of air within its frame, thus significantly slowing its rate of descent to the ground. Consequently, having begun at any substantial height above the earth, the user, whether a gentleman seeking the exhilaration of flight, a military scout or other soldier seeking to escape from a high promontory or other lofty perch, or any man merely wishing to test through experimentation the validity of the science that underpins the machine may descend to the ground without injury.

Components of the Safe-Descent Machine

Wooden Frame

The frame is constructed of carefully cut lengths of strong wood, preferably ash or oak. Each of the diagonal struts is 11' 10" × 2' 5" × 2". Each strut used for the square base of the pyramid is 11' 6" × 2' 5" × 2". Six slats of 11' 8" × 1' 5" × 8" are attached to the top of the square frame, spaced evenly so as to provide support and strength while not interfering with the necessary entry of air into the enclosed space formed by the pyramid.

Sides

Each of the four sides of the pyramid is constructed of high-quality linen fabric, in the shape of a triangle of dimensions 11' 6" base × 12' 5" diagonal (height). Each piece of linen is treated with a mixture of linseed oil and resin, so as to prevent air from passing through it; each is attached to the base and diagonals of the frame with a combination of glue and sturdy, knotted gut; each is sewn to the other along the diagonals, with the finest of stitches possible.

Harness

The harness is made of fine horsehide leather straps, tanned and treated to withstand the strain of a large man being suspended from it while being buffeted by strong winds. The buckles that enable a man to fasten himself securely within the harness are made of polished brass, no weaker than that used in the manufacture of cavalry accessories. The harness is constructed so that the head of a man suspended in the harness, while dangling below the frame of the safe-descent machine, will be between 2' 6" and 3' 6" below the underside of the slats that comprise the base of the frame.

Conclusion

The value of the safe-descent machine to the prince or other ruler is inestimable, giving him, among other advantages, the means to have his men drop suddenly, stealthily, and safely to the place of his enemies. The safe-descent machine offers, to nobles, soldiers, and all kinds of gentleman adventurers, the ability to descend in comfort and safety from any high spot, whether it be the top of Milan's highest tower, or the edge of an Apennine cliff.

9-1b SIMPLE TECHNICAL DESCRIPTION

Sometimes, you'll need to describe something that is not technically (if you'll pardon the pun) a "mechanism." Rather, it is simply an object for which you require a technical description. Suppose, for example, that you were an engineer who had just completed the design of a decorative letter opener for your client (who is not enamored by email and Twitter). You might produce the brief technical description in Figure 9.3.

FIGURE 9.3: A Simple Technical Description of a Letter Opener

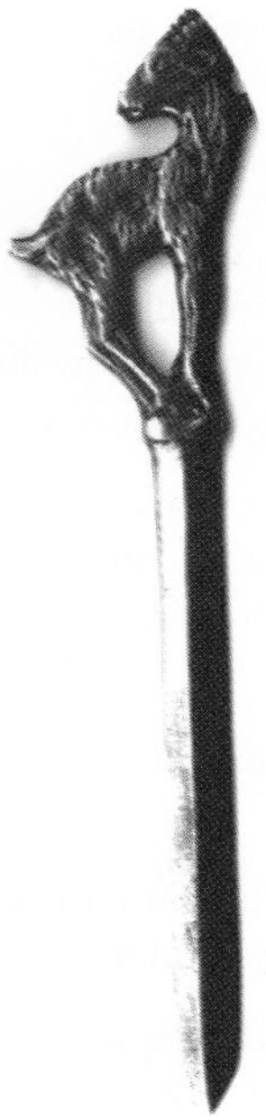

David Ingre

A Letter Opener

A letter opener is a common household or office tool designed to be inserted under the unglued portion of the diagonal edge of a sealed envelope and then used to slice open the top edge of the envelope without damaging any documents inside. Letter openers are commonly made of wood, metal, plastic, or a combination of materials. Their utility is often enhanced by decorative characteristics.

The decorative letter opener shown exhibits the following characteristics:

Shape:	Slightly convex, beveled, double-sided, double-edged sword blade surmounted by a curved-horned mountain goat whose four feet are planted on the base of the sword hilt. Mountain goat's features cast in relief on both front (shown in figure) and rear. Blade point and edges are sufficiently sharp to open an envelope efficiently, but not sharp enough to be dangerous for an adult user.
Composition:	Solid brass.
Texture:	Sword blade smooth, burnished; noticeable bevel along longitudinal axis of sword blade. Mountain goat's hide and horns emphasized by serrations.
Maximum height:	6.5" from point of blade to top edge of horns.
Height of sword blade:	4.1" from point to top of hilt.
Maximum width:	1.6" horizontally from tip of goat's tail to outer edge of its horn.
Width of sword blade:	0.4"

Depth of sword hilt:	0.2" (thickest point of letter opener)
Weight:	1.8 oz.
Appearance:	The style of the blade of the letter opener is reminiscent of medieval European sword blades, although lacking a suitably wide crossbar on the hilt. The relief sculpture of the mountain goat is slightly stylized, but readily recognizable and anatomically correct. Overall aspect is that of an antique style.

In conclusion, the blade of the letter opener is effective for its purpose. The mountain goat sculpture, however, presents an uncomfortable and unwieldy handle in the hand of an adult. Thus, this letter opener is designed more for decorative appeal than for utility.

9–2 Process Description

Process descriptions enable your audience to make informed choices and sound decisions, to work safely, and to operate professionally. Even though the audience may rarely carry out the described process, other actions may depend on a clear understanding of it.

A process description explains how things work or are done or made. Common examples might include how trans fats are reduced in the production of margarine, how side scan sonar works, or how aircraft "black boxes" function. Linear in nature, a process description consists of a series of steps that, typically, occur in a particular order. When creating one, you will likely want to consider the following.

9–2a PROCESS DESCRIPTION STRATEGY

1. Begin with an informative title and introduction. The introduction should define the process and provide an overview. It also might explain why or how the process is used, who or what performs it, and where or when it takes place.
2. Organize chronologically. Present steps in the order in which they occur.
3. Include each step in the body of the description. If you use paragraph form, include headings and subheadings to distinguish important details. A long description may require a heading for each main step. If you use bulleted and/or numbered lists, follow an outline of main points and subpoints, and use parallel structure—all to increase "accessibility" for your audience.
4. Although process descriptions often use the active voice, the passive is not uncommon, since it focuses the audience's attention on the process itself rather than on the person who is performing the process.
5. Identify the level of technicality of your primary audience. If necessary, define terms that may be unfamiliar to that audience.
6. Use a clear concluding segment. A long process description may require a full paragraph to summarize the process and, perhaps, remind the audience of its uses or advantages.
7. Drawings, diagrams, and flowcharts tend to be especially useful in process descriptions. As always, however, your CMAPP analysis will help you decide whether—and how—to use visuals.

9-2b COMPLEX DESCRIPTION

You will have noticed that the mechanism and technical descriptions given previously proceed in linear fashion: You expect that your audience will want to follow from "a" to "b" to "c," and so forth. More complex mechanisms or items, however, cannot be dealt with in the same linear pattern, because they involve several interdependent components. Examples might include the design of a new medical procedure, an automobile's automatic transmission, or the electric turbine generators in a dam. In developing your description, you will have to decide which ordering of the various components will best suit your audience, context, and purpose. Following the general guidelines for simple description, you can use this strategy:

1. Introduce the item.
2. Describe the overall relationship among the components.
3. Describe in turn each component and any subcomponents.
   a) Component A
   b) Component B
      (i) Subcomponent B1
      (ii) Subcomponent B2
   c) Other components, as required
4. Review the interrelationships.
5. Conclude.

Figure 9.4 depicts a hyperlinked version of a complex generic description.

9-2c COMPUTERIZED HYPERLINKS

You are more and more likely to be engaged in computer- or web-based mechanism description. For all but the very simplest of items, you will probably need to use a nonlinear approach, because your audience will expect to take advantage of hyperlinks. These familiar features of Web pages give your audience immediate access to key areas throughout the description. The addition of graphics, sound, animation, and/or video can further enhance the effectiveness of such description. Figure 9.4 shows the links that might be included in a generic description of an item containing three components, one of which includes four subcomponents. When creating such a description, you need to plan carefully: Think about which links between which components or subcomponents your primary audience would be most likely to want to follow.

FIGURE 9.4: Hyperlinked Generic Mechanism Description

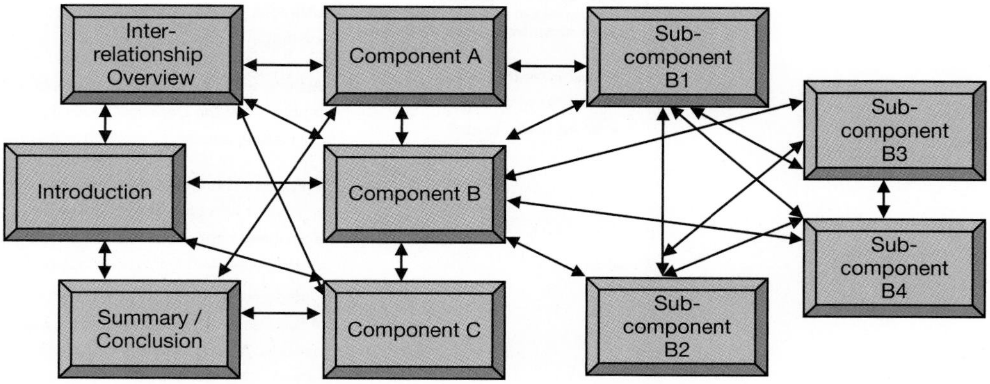

9-2d HYPERLINKED DESCRIPTIONS

Just as mechanism and process descriptions are more and more commonly published on the Internet, instructions are becoming more and more visible on the Web. If the instructions and descriptions that you create are meant to be available online, your audience will expect a series of hyperlinks to help them move around.

As you learned in Chapter 1, engineering communication is *audience centered*. It's your professional responsibility—as an engineer and as a communicator—to provide your audience with what they need and want. Again, your CMAPP analysis will help you decide what kinds of links from and to which elements would be most useful to your primary audience.

The authors of this textbook note that instructions and descriptions that contain too many hyperlinks become hard to read. Compare, for instance, the "clean" and easy-to-understand hyperlinks from HowStuffWorks regarding USB ports (Figure 9.5) to the visually distracting and "busy" Wikipedia entry for the iPad at en.wikipedia.org/wiki/ipad (Figure 9.6).

Instructions are covered in detail below.

FIGURE 9.5: Good Use of Hyperlinks in Technical Description

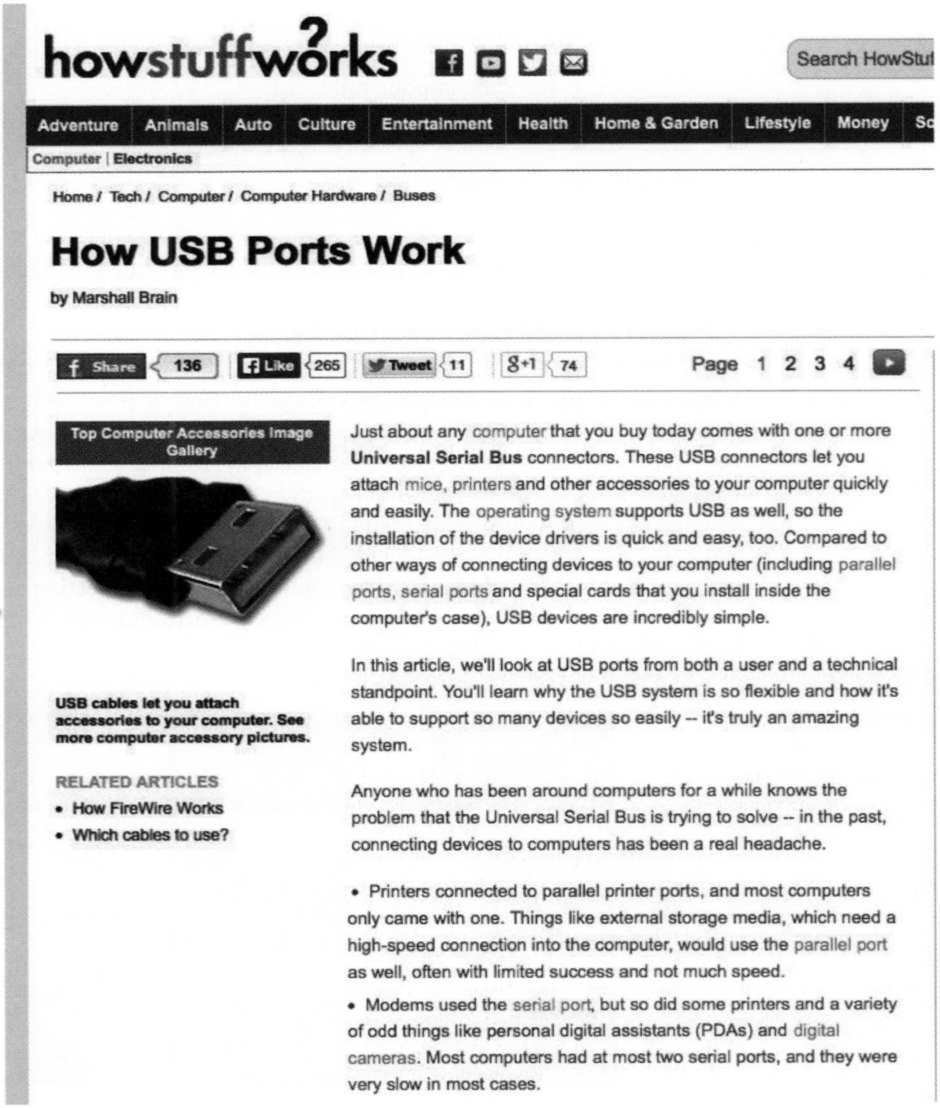

howstuffworks

Search HowStu

Adventure | Animals | Auto | Culture | Entertainment | Health | Home & Garden | Lifestyle | Money | Sc

Computer | Electronics

Home / Tech / Computer / Computer Hardware / Buses

How USB Ports Work

by Marshall Brain

Share 136 Like 265 Tweet 11 8+1 74 Page 1 2 3 4

Top Computer Accessories Image Gallery

USB cables let you attach accessories to your computer. See more computer accessory pictures.

RELATED ARTICLES
- How FireWire Works
- Which cables to use?

Just about any computer that you buy today comes with one or more **Universal Serial Bus** connectors. These USB connectors let you attach mice, printers and other accessories to your computer quickly and easily. The operating system supports USB as well, so the installation of the device drivers is quick and easy, too. Compared to other ways of connecting devices to your computer (including parallel ports, serial ports and special cards that you install inside the computer's case), USB devices are incredibly simple.

In this article, we'll look at USB ports from both a user and a technical standpoint. You'll learn why the USB system is so flexible and how it's able to support so many devices so easily -- it's truly an amazing system.

Anyone who has been around computers for a while knows the problem that the Universal Serial Bus is trying to solve -- in the past, connecting devices to computers has been a real headache.

- Printers connected to parallel printer ports, and most computers only came with one. Things like external storage media, which need a high-speed connection into the computer, would use the parallel port as well, often with limited success and not much speed.
- Modems used the serial port, but so did some printers and a variety of odd things like personal digital assistants (PDAs) and digital cameras. Most computers had at most two serial ports, and they were very slow in most cases.

Source: HowStuffWorks, http://computer.howstuffworks.com/usb.htm

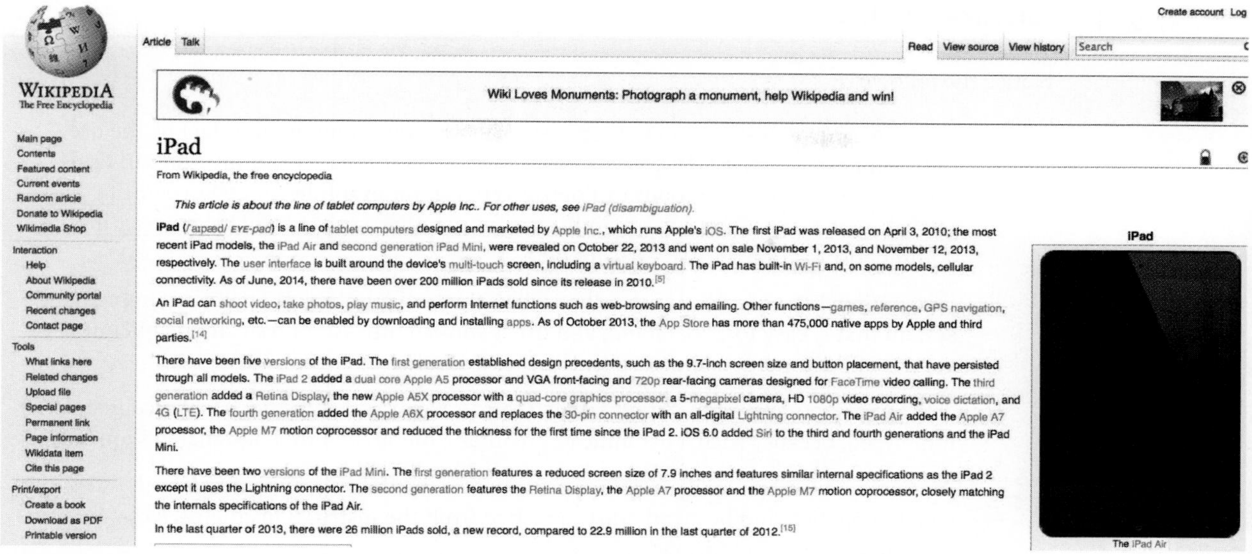

Source: Wikipedia, http://en.wikipedia.org/wiki/IPad

FIGURE 9.6: Distracting Overuse of Hyperlinks in a Technical Description

9–3 Instructions

In your professional and your private life, you will undoubtedly have to write simple instructions from time to time. Examples might include instructions a fellow employee should follow while you are on vacation, guidelines for a junior employee you hire, or directions to your home for an out-of-town guest. Depending on your position, you might also have to produce more formal sets of instructions—how to conduct a tensile strength test that you have developed, for example, or how to verify that construction work meets building code specifications. Since instructions by definition refer to processes, they might be considered a special class of process description. Their purpose, however, is different. In a standard process description, you want your audience to see or understand how something works; in instructions, you want your audience to be able *to do* something. This distinctive purpose, of course, affects the specific message.

Note the differences between these two examples:

> At this stage, it is crucial that the power light be green. Gear A now activates gear B. Immediately thereafter, cam C, turning on shaft D, lifts rod E 3.5 cm, causing it to make contact with plate F.

> CAUTION: Ensure that the power light shows green before undertaking the following steps. If it is not green, do not proceed; call the supervisor at once.
>
> **1.** Allow gear A to activate gear B.
>
> **2.** Turn shaft D clockwise, until cam C lifts rod E 3.5 cm, so that it makes contact with plate F.

The first example is appropriate for a process description; the second for instructions.

9-3a CHARACTERISTICS

Notice the following:

- We recognize an instruction (which is similar to a command or an order) by the use of the imperative mood. (If you're not sure what this means, look it up. You can find information on all four moods of English verbs at, for example, gmatclub.com/forum/english-verbs-have-four-moods-indicative-imperative-131341.html or dailywritingtips.com/english-grammar-101-verb-mood.)
- Instructional steps are easier to follow when they are numbered. Therefore, implement a system that is easy to understand, and use it consistently. A numeric or an alphanumeric system of the kind you looked at in Chapter 5 would be appropriate.
- A list of objects and/or materials needed to complete the instructions is typically not numbered, because such a list indicates neither a hierarchy nor a designated sequence.
- A warning must be readily visible (remember "accessibility"), appear before the step to which it applies, and be clearly distinguished from the instructions themselves—in terms of both language and format.

9-3b STRATEGY

The strategy for instructions is fundamentally the same as that for mechanism description.

1. Introduce the topic and provide any necessary preamble.
2. Specify all relevant details—the instruction steps themselves.
3. Conclude.

As you plan and create the instructions, ask yourself the following questions:

1. Who is my primary audience?
2. Do I understand the process? (It is unlikely your audience will if you don't.)
3. Is my language as precise and specific as possible?
4. Have I indicated how long it should take to complete the set of instructions?
5. Have I given a complete list of required equipment and/or materials? (Your audience will not appreciate discovering at step 8 that a specific tool is now needed.) And, remember that the list will likely be bulleted rather than numbered—unless the items must appear in a specific hierarchy or designated sequence.
6. Am I beginning at the appropriate spot or should I begin earlier or later in the process? For example, will my audience already know how to reach what I indicate as the first step? (Your CMAPP analysis will help you answer this question.)
7. Have I tested the instructions myself?
8. Even more important—have I tested them with someone who does not already know how to complete them, and would that "tester" actually reflect the intended audience for my instructions?
9. What level of technicality is appropriate? For example, suppose you are an electronics engineer, writing instructions for part of the manufacturing process of a new computer chip, and suppose that your primary audience will be experienced engineers, newly hired

into your company. Presumably, you should use a very high level of technicality with regard to electronics; however, your audience's level of technicality with regard to your company's procedures, and so forth, will be much lower.

10. How do I plan to reproduce these instructions? Think about such things as the legibility of photocopies and the use of color. For example, a diagram of electrical wiring may depend on color distinctions.
11. How are the instructions to be distributed? Are they to be posted on a wall, included with a purchased item, attached to a memo?
12. How should my audience deal with any confidential information, such as code numbers, passwords, or confidential technical innovations?
13. Have I distinguished between instructions and notes or other information? (Remember to use the imperative mood for instructions, and to make effective use of document visuals.)
14. Do I need visuals? Consider number, position, size, legibility, scale, and level of detail. Again, do a CMAPP analysis before beginning.
15. Is my numbering system clear and consistent?
16. Is the order of the steps the most logical for my audience?
17. Should I provide hints to help my audience recognize whether particular steps have been successfully completed?
18. Is this set of instructions amenable to linear progression, or are there too many interacting components, for example, to make this feasible?
19. Have I indicated how subcomponents relate to major elements, and have I expressed them in a logical and workable order?
20. Is it the safest sequence?
21. Are warnings or cautions properly placed, highly visible, and clearly worded?
22. Have I indicated when and how the audience could interrupt the process if necessary? (Consider "bathroom breaks" or quitting time, for example.)

CASE STUDY

9 Written Instructions for ShowRight

SITUATION

Paul Mahoney is a junior at Ann Arbor University, with an interest in electrical engineering. Hoping to gain some relevant experience, he has obtained a summer job at AEL's Lansing office, working with Sherry Grewal, an AEL consultant. One Monday morning, Grewal gives Paul a new task: producing a short document for one of AEL's long-time clients in Lansing, ShowRight, a small company that organizes local business conventions and trade fairs. Lyle Coss, ShowRight's owner, has sent the following information, which Grewal gives to Paul.

Opened just over 10 years ago, ShowRight has 11 employees: the owner (Lyle Coss), a financial officer (Ramona Mai), an office manager (Peter Bokharian), three public relations officers (Teresa Wong, Sam Schumann, Felipe Koi), two administrative clerks (Harinder Khan, Orley Ulsitt), and two contracts officers (Michelle Suttliffe, Laurent Beaudoin). The PR and the contracts officers often work past regular business hours. All on one floor, ShowRight has a main, street-level entrance double door (glass), and a side and a back door. As well, there are 10 windows, four of which look out onto the rear alley.

Phalanx Security recently installed a monitored alarm system for ShowRight's offices. Despite having taken a short training session from Phalanx, Bokharian has been unsuccessful in producing a brief, concise set of instructions to help other employees set the alarm should they be the last to leave the building.

The alarm's control panel is on the wall on the inside of Coss' office—all other employees have "cubicle offices." If all windows haven't been properly secured, the alarm cannot be armed. Similarly, the side and back doors have to be bolted, as must the right-hand side of the double door. The left-hand side must be fully open for arming. The deadbolt lock must also be set from the outside for the alarm to be fully armed.

After talking with Grewal, Paul contacts Coss and arranges to talk with him about how ShowRight's employees tend to use their offices. Afterwards, he conducts a brief CMAPP analysis. Some of his CMAPP questions appear in Figure 9.7. He then produces his draft instructions, shown in Figure 9.8.

FIGURE 9.7: Segments of Mahoney's CMAPP Analysis for ShowRight

Audience

1. Who is the primary audience for the instructions?
2. Is there a secondary and a tertiary audience? If so, who?
3. How and to what extent are my audiences homogenous (sharing relevant characteristics)?
4. How much do they know? How much do they *want* to know? How much do they *need* to know?
5. Are they receptive to technology?
6. Will they be willing to accept my expertise?
7. What will be the benefit(s) to the audience(s)?

Context

1. How should I consider office hierarchy at ShowRight?
2. How should I consider office hierarchy at AEL?
3. Will my audience respond favorably?
4. What might generate a negative response, and how might I best deal with it?
5. Will the addition of visuals help?
6. If so, how many should I use, of what kind, and where?
7. What level of discourse (formality of language) would be most effective here?
8. What level of technicality would be appropriate?
9. Should I show my draft to both Grewal and Coss? If so, in which order?

Message

1. What, precisely, are the instructions for?
2. How many steps are really involved?

3. Am I providing instructions about setting the alarm, about securing the premises, or about something else?
4. What should my starting point be? What should my end point be?
5. Is any part of this matter highly technical?

Purpose

1. Why am I really writing this: to help ensure building security, to impress Grewal, to satisfy Coss, to improve my resume?
2. What is my primary purpose (in CMAPP terms)?
3. What do I want my audience to remember most: How to arm the system? How to avoid the embarrassment of a false alarm? Something else?

Product

1. Would my audience be best served by
   i. correspondence to each of them?
   ii. a desk card for future reference?
   iii. a sign on the alarm panel or beside the door?
2. What should I use as a title?
3. Should I use subheadings—why, which, where?
4. Have I used the most effective and the safest order of steps?
5. What rationale do I consider when deciding whether the instructions should be reproduced in color, in grayscale, or in black and white?

FIGURE 9.8: Phalanx Alarm Instructions

Instructions for Setting the Phalanx Model R-245 Alarm

Notes:

a. These instructions are only for setting the Phalanx Model R-245 Alarm prior to leaving the building when it is to be unoccupied. The alarm should be set by means of the panel located on the inside wall of Lyle Coss' interior office, close to the door.
b. Before setting the alarm, familiarize yourself with the panel (see the following figure) and with the locations of all doors and windows in the office. Note that the main entrance is a double door; viewed from the inside, the right-hand part can be locked by securing the vertical throw-bolts (top and bottom) on the inside edge; the left-hand door can then be secured to it by the deadbolt lock. The back and side doors can be secured by their deadbolt locks, as can all 10 windows.
c. You must have a main entrance door key to follow these procedures.

Instructions

1. Ensure that the back and side doors and all windows have been properly bolted.
2. Verify that the vertical throw-bolts on the main entrance door have been thrown.
3. Ensure that the left-hand (viewed from within) door of the main entrance is open; use the foot-pedal door-stop to hold the door open.
4. Approach and face the Phalanx Security alarm panel.
5. If the panel cover is closed, open it by swinging it downwards on the two hinges. (See figure.)

Notes:

The panel comprises three sections.

a. The lettered rectangles in the column at the left indicate the armed status for each of the six zones set up for the office. The light in the rectangle will show solid red if the zone is armed, solid green if it is unarmed, and flashing red if there is a "problem" in the respective zone.

b. The light-buttons in the dark gray bar in the center correspond to the respective zones to their left. Detailed zone descriptions are in the Phalanx manual, kept under Manuals in the Administration file cabinet in the center of the office. Also, see the Caution, below.

c. The gray numbered keys on the right of the panel are for arming, disarming, and programming the alarm.

Caution:

The Zone A indicator represents the left-hand entrance door; it should be flashing red if that door is open. All other zone indicators should be green.

If there is any deviation from this pattern, check all windows and doors (see instructions 1 and 2, above). If the problem persists, call Phalanx Security's 24-hour number: (800) 555-5555.

6. Press and hold the * key until the panel beeps once.
7. Press and hold the # key until the panel beeps once.

Caution:

After completing instruction 7, you will have 90 seconds to complete instructions 8 and 9 and to exit the building and close the main entrance door. Otherwise, the alarm will report trouble to Phalanx.

8. Press the following keys sequentially: 1 9 9 0 1 9 7 9.

Note:

This is the arming sequence as of September 1, 2015; it may change. Check with the financial officer to verify. Note also that the *disarming code* is *never published*. The financial officer will tell you the disarming code if you are supposed to know it.

9. Disengage the door-stop at the main entrance door, exit the building, and close the door.
10. Using your key, lock the deadbolt that secures the two doors of the entrance together.
11. Wait for the remainder of the 90 second pre-alarm time. Then, verify that the small red light above the main entrance doors is flashing, indicating the alarm is properly armed.

Note:

Should that light not be flashing, unlock the door and reenter. Check the panel. (See the *Caution* following instruction #7, above.) If it appears to be in order, repeat the arming procedure. Otherwise—or if the arming procedure is unsuccessful a second time—call Phalanx immediately at (800) 555-5555, and follow their instructions.

ISSUES TO THINK ABOUT

1. Why do you think Grewal would have given Paul this particular task?
2. What do you think Grewal might or should have done differently?
3. How has Paul distinguished "instructions proper" from other items in his document?
4. How might he have handled the matter differently?
5. Are they any additional items that you think should appear in the instructions?
6. Are there any that you think should not appear?
7. How do you think that Coss is likely to react to what Paul has produced?
8. What do you think might be the effects—if any—of Paul's efforts on the relationship between AEL and ShowRight?
9. What changes—if any—do you see occurring in Paul's relationship with Grewal and/or with AEL?

EXERCISES

9.1 What are the three basic components of a mechanism description?

9.2 Give at least two clear examples to distinguish precise from vague terminology.

9.3 Give at least two clear examples to distinguish precise from incorrect terminology.

9.4 Reexamine the hypothetical mechanism description of Leonardo's parachute shown in Figure 9.2. Construct a brief CMAPP analysis that reflects what Leonardo *might* have considered when creating this fictitious description.

9.5 What is the principal difference between mechanism description and simple technical description?

9.6 What are the main complications introduced by creating descriptions designed for online use?

9.7 Give at least two examples of process descriptions that would apply to engineering fields.

9.8 Specify the components of the strategy for simple instructions.

9.9 How does an instruction (or a command) differ from a description?

9.10 Identify a process connected with your job (if you are working) or with your educational institution. (Examples might include how orders are filled, how standardized tests are conducted, how the staffing process works, or how you register for a particular program.) Using visuals if appropriate, create a relevant process description.

9.11 Locate a set of instructions used where you work or go to school. Assess its effectiveness against the 22 strategy questions listed in Section 9-3b.

9.12 After reviewing this chapter's Case Study, assume that you are Paul Mahoney, and complete his CMAPP analysis by providing answers to his questions.

9.13 Based on that CMAPP analysis, construct what would be an effective multilevel outline for Mahoney's instructions. Follow Chapter 5's guidelines for the creation of outlines.

USEFUL WEBSITES

Note: For ease of reading, we have removed the "http://www." from the URLs below. However, some browsers might still require that you insert it to precede some URLs so as not to receive an error message.

■ *The Free Online Textbook for Technical Writing*, by David McMurrey of Austin Community College, offers a worthwhile segment on producing instructions.
Visit: prismnet.com/~hcexres/textbook/models.html#instructions

■ HowStuffWorks was created by author and former computer science teacher Marshall Brain. This interesting and unusual site presents a wide variety of process descriptions (often enhanced by graphics and animation), with topics ranging from electric motors and steam engines to microprocessors and the effects of caffeine.
Visit: howstuffworks.com

■ Leonardo da Vinci's inventions are presented simply and effectively through the websites of Milan's National Museum of Science and Technology. Drawings and explanations have been made available in English as well as Italian.
Visit: museoscienza.org/english/leonardo

■ Oregon State University's Extension Service offers useful content on writing clear instructions.
Visit: http://extension.oregonstate.edu/eesc/how-to/how-write-clear-how-instructions

■ wikiHow's editors call their website "a collaborative effort to create the world's most helpful how-to guides. Like Wikipedia, wikiHow is a wiki, in that anyone can write or edit a page on the site." It covers thousands of technical topics as well as more whimsical ones such as "How to Make Grilled Strawberries and Bananas with Ice Cream."
Visit: wikiHow.com

CHAPTER 10

Persuasive Communication

This chapter examines common CMAPP strategies for persuasive communication: persuasion that targets the intellect (logical argumentation) and persuasion that targets the emotions (AIDA—for "attention," "interest," "desire," and "action").

Persuasive strategies are vital in a variety of contexts, from advertising and marketing to analytical reports and meetings. While some persuasive strategies depend on logic and argumentation and others more on emotional appeals, all such strategies are based on establishing a connection with an audience and providing something to which the audience will relate and respond. As an engineer, you will frequently focus on persuading an audience who possesses a high level of technical expertise; this will normally result in your targeting that audience's intellect. Nonetheless, your professional acumen may also depend, in part, on your ability to recognize and react to persuasion that targets the emotions.

10–1 Persuasive Strategies: Intellect versus Emotion

Many years ago, author Dave Ingre attended a talk given to an audience of professionals—mainly electrical and electronics engineers. The speaker was trying to convince them of the value of what was then a new and unproven technology—fiber-optic cable in the telecommunications industry. His basic approach was to provide a host of facts, examples, and statistics to bolster his message of the advantages of fiber-optic cable. The technical expertise of his audience was his rationale for persuading them by appealing to their intellect. This persuasive strategy attempts to convince through logic, supporting assertions through reason and evidence. Consequently, it relies heavily on the denotative value of its words, and on the precise structure of its arguments.

Now consider most television advertising. Instead of being presented with a wealth of information, you are invited to sense, to feel, and to believe. The message uses words of high connotative value and features dramatic images. Few facts intrude—after all, is one brand of sugar-coated cereal significantly different from another? Even in cases of real differences—between a small Chevrolet and a small Porsche, for example—the message relies primarily on intangibles: the "feel of the road," the "excitement of the drive," and so forth. Such persuasion is commonly expressed through slogans such as "whiter than white," "9 out of 10 doctors" (whoever they might be), "tried, tested, and true," and "built for the human race." All these slogans would collapse if subjected to logical analysis. Of course, this type of persuasion is not meant to be parsed or analyzed; it targets your emotions, not your intellect.

10–2 Strategies for Conveying News

However, before examining in greater detail the kinds of logical argumentation that will help you to effectively target your audience's intellect, let's look at two common writing tasks that, though you might not have been aware of it, are in fact *persuasive*. We use so-called *good news* (also called *direct*) and *bad news* (also called *indirect*) strategies to tell our primary audience things that are likely to elicit fairly predictable reactions; and we use them because we effectively want to *persuade* the audience to react in specific ways.

10–2a CONVEYING GOOD NEWS

The term "good news" applies to much more than, for example, "Congratulations! You've just won the Powerball Lottery!" Rather, it encompasses a broad variety of ideas, including the following:

- You have been appointed to the position.
- You have been accepted into the apprenticeship program.
- Your shipment will arrive on time.
- The repairs are covered by your warranty.
- Your materials have arrived in our warehouse.
- We are reinstating your membership.

In effect, good news is any information that your audience is likely to be pleased to receive. To convey it, and regardless of the particular CMAPP product involved, you will normally construct your message to fit the *good news* or *direct* strategy.

This strategy comprises three parts:

1. **State the good news:** Present the news directly, simply, and clearly. In doing so, you will communicate that part of your message that responds to what your audience wants to know.
2. **Explain the situation:** In this segment of the good news strategy, you explain the main points and significant details of your message. When preparing the "meat" of your message, remember to take into account the complementary attributes of the CMAPP model—particularly 5WH—discussed in Chapter 3.
3. **Conclude on a positive note:** Although the second phase essentially completes your message, the conventions of the direct strategy dictate that you finish with a brief, positive comment of some kind. Your conclusion might be something as simple as "Thank you for choosing AEL"; it might be a reiteration of the good news itself, as in "Again, I would like to congratulate you on your appointment"; or it might be an expression of continued interest, as in "We look forward to doing business with you again soon."

In the letter body shown in Figure 10.1, Angelos Methoulios, the coordinator of GTI's Internship Program, has used the direct strategy to notify Karima Bhanji of her acceptance into the program. Notice that the first paragraph—one sentence only—conveys the good news; the second paragraph provides the explanation; and the final paragraph offers the positive conclusion. As always, the specifics of your own message will be dictated by your context, your audience, and your purpose.

FIGURE 10.1: Direct Strategy

I am pleased to inform you of your acceptance into the MIS Internship Program, which commences in September 2016.

According to your application, dated April 12, 2016, you meet all the entrance requirements for the program. Within the next week, therefore, I shall be sending you the Internship Program Registration Kit. It will inform you of the registration procedures you should undertake between July 5 and July 9 of this year. It will also answer general questions you might have about the program. Should you require further information, please do not hesitate to contact my office at 802-555-4412 during regular business hours.

In the meantime, please accept my personal congratulations on your acceptance into a program that has proven itself highly successful over the years. I look forward to working with you next term.

How is this *persuasive communication?* Clearly, Methoulios wants his communication to *convince* Bhanji to proceed with entering the program. His *direct* strategy is designed to do so.

10-2b CONVEYING BAD NEWS

You are likely to use the bad news or indirect strategy whenever you believe that your audience will not be pleased to receive your message. An extreme case of bad news would be "You're fired." Less dramatic examples might include the following:

- We will not be contacting you for an interview.
- The new parts for your equipment have still not arrived.
- The photocopier I bought from you still does not work properly.

- We are unable to ship the goods at this time.
- Your account is overdrawn.
- The text you requested is no longer in print.

Sometimes, as in the photo copier example, your bad news would in fact be a claim letter—a statement of a problem with a product or service, coupled with a request for appropriate redress. The bad news conveyed in the delayed shipment example might be followed up with an offer, on the part of the communicator, of some sort of compromise or recompense.

More complex than the good news strategy (and thus often resulting in a longer product), the bad news strategy is typically composed of five parts:

1. **Describe the context:** Use your first paragraph to explain the overall situation.
2. **Provide details:** Expand on your introductory paragraph by offering the main points and significant details that will prepare your audience for what follows.
3. **Deliver the bad news—tactfully:** Indicate the real purpose of your communication, but try to do so diplomatically. This will involve using language that is not as clear or direct as it perhaps could be. Conveying bad news often makes use of passive-voice constructions (see Chapter 15), since they permit a more oblique expression. If your language is too circuitous, however, your audience, who would presumably prefer good news to bad news, may misinterpret your message.
4. **Provide supplementary details:** This fourth phase is optional. Depending on the circumstances, you may wish to include additional information that will have the effect of further "softening the blow," providing other avenues the reader might consider.
5. **Offer conciliation and/or encouragement:** In this type of communication, you will normally want to maintain a working relationship with your audience. Consequently, you will offer some kind of hope or "consolation prize" (e.g., a discount on repair charges or on goods shipped late or an offer of substitution). At times, the consolation will be less tangible—an expression of your commitment to service, for example, or a confirmation of your continued interest.

In the letter body shown in Figure 10.2, Methoulios is communicating rejection rather than acceptance. Many engineering professionals dislike the indirect strategy, in large part because its oblique presentation of information deviates from the CMAPP attributes of brevity, concision, and precision. Nonetheless, use of this strategy is so widespread that a departure from its conventions may cause your audience to see your message as brusque or even rude.

FIGURE 10.2: Indirect Strategy

Thank you for your application, dated April 12, 2016, for the September 2016 Internship Program.

You will recall that the Internship Program Criteria Bulletin specifies a minimum of 3.33 for the overall GPA for the year previous to the Internship intake, and a mark of at least 80% in the relevant core introductory course, in your case INFO 101. Our records show that although you received a grade of 84% in INFO 101 last term, your mark of 66% in INFO 104 brought your overall GPA down to 3.27.

The Bulletin also specifies that a student's application must be accompanied by at least three letters of reference, two of which must be from GTI faculty. Only one of your referees, Ms. Pat Hayakawa, is on our faculty. Your two other referees, Dr. Janice Fleming and Mr. Edward Skoplar, appear to offer personal references only.

In light of the above, therefore, I am not currently able to further your application for the September 2016 intake in the MIS component of our Internship Program.

I do note that most of your marks were entirely satisfactory, and that Ms. Hayakawa's recommendation was highly favorable. Unfortunately, the number of applications we receive far exceeds the number of Internship Program places available; thus, we have found that we must apply the selection criteria quite rigorously.

Your record at GTI suggests that you are enthusiastic about a career in MIS, and that you show some promise in the field. Consequently, I would urge you to make all efforts to raise your GPA this coming term, and to seek appropriate references for the January 2017 intake. Grandstone's MIS Internship Program would welcome your reapplication.

Once again, how is this *persuasive communication*? Here, Methoulios wants his communication to *convince* Bhanji that, despite the "bad news," GTI wants to maintain a good relationship with her. His *indirect* strategy is designed to do that.

10–3 Targeting the Intellect: Logical Argumentation

Logical argumentation, the essence of an appeal to the intellect, might be seen as the persuasive progression from the assertion of a claim to its proof. As an analogy, think of the way a trained engineer tends to approach all professional situations and many personal ones: obtain the facts, analyze them carefully and dispassionately, derive conclusions from that analysis, and apply those conclusions to synthesize the solution. (Incidentally, the reference to personal situations reflects Dave Ingre's own experience: His extended family has included two civil engineers, an electrical engineer, an electronics engineer, a marine engineer, a mechanical engineer, and—though not quite the same—an architect. They likely wondered what had happened to him.)

Persuasion that targets the intellect follows one of two strategies: *deductive* (also called *direct*) or *inductive* (also called *indirect*).

10–3a DEDUCTIVE STRATEGY

If your CMAPP analysis leads you to believe that your audience already has a good grasp of the issue under consideration and/or is likely to react favorably to your message, you should probably use the *deductive* (or *direct*) strategy. It comprises three parts:

1. *Make your assertion.* Provide clear statements. Remember accuracy, brevity, concision, and precision.

2. *Justify your assertion.* Your justification must consist of a series of cohesive points, normally organized chronologically, topically, or spatially. Each point should be the logical consequence of what has preceded it. (Think of your argument in terms of the construction of a building, with the safety of each successive story relying on what underlies it.)

3. *Conclude.* Briefly summarize your argument, and then refer your audience to the necessary conclusion—your initial assertion. (You have thus come full circle: from your claim, through your corroboration, to the restatement of that claim.)

Note that the length and complexity of the second component will derive principally from your context and your message, while your level of technicality will derive from your audience analysis.

10-3b INDUCTIVE STRATEGY

If your audience lacks the background necessary to follow your reasoning easily and/or is likely to be averse to receiving your message, you should likely use the *inductive* (or *indirect*) strategy, which comprises the following parts:

1. **Introduce the issue.** Briefly describe the issue under consideration, situating it in a way that will be relevant to your audience.
2. **Provide your arguments.** Make your points just as you would if using the deductive strategy. Each point must stem logically from what has come before.
3. **Draw your conclusion.** Briefly summarize your arguments, and then state your assertion as their logical conclusion.

Figures 10.3 and 10.4 are two versions of a persuasive memo that Melinda Shaw, Radisson Automobiles' Atlanta Vice President, might send to Griffin Radisson, the firm's Chief Executive Officer. Having received Radisson's memo regarding the reorganization (see Case Study 2b), Shaw would like

7011 Roswell Road
Atlanta, GA 30328
(404) 555-7397
Fax (404) 555-3378

Radisson Automobiles Inc.
Where all the cars are stars

Atlanta Dealership

Memorandum

December 9, 2015
To: Griffin Radisson
From: Melinda Shaw MS
Re: Reorganization
cc.

I have read your memo of November 12, 2015, to vice presidents and general managers, regarding RAI's reorganization for January 1, 2016. I would like to suggest you consider postponing implementation.

Over the years, vice presidents have developed good working relationships with the Board of Directors. While I personally feel that reporting to a single individual—you—rather than to a group will make my job easier, I am concerned that abruptly severing our long-standing ties with the Board may lead to the interruption of important projects across the country. Further, I suspect that many vice presidents will be resentful of the "surprise" nature of the announcement.

We vice presidents have come to rely on our general managers, who seem to trust and respect us in return. The success of these relationships is in large part due to their having ready access to us in their respective jurisdictions. I am certain they will be dismayed to unexpectedly find themselves accountable to Celine Roberts in Dallas, a great distance away for most of them. I think that many are liable to infer that the company suddenly lacks faith in their competence and loyalty.

Successfully implementing the reorganization will, I fear, require far more than the "administrative details" to which your memo refers. The new structure will impose fundamental changes in the way we manage our dealerships. Consequently, I would ask you to reconsider your January 1, 2016, deadline so that the vice presidents could meet with you for full discussion.

I would be grateful if you would communicate your decision within the next two weeks.

FIGURE 10.3: Sample of Deductive Strategy in Persuasion

7011 Roswell Road
Atlanta, GA 30328
(404) 555-7397
Fax (404) 555-3378

Radisson Automobiles Inc.
Where all the cars are stars

Atlanta Dealership

Memorandum

December 9, 2015
To: Griffin Radisson
From: Melinda Shaw MS
Re: Reorganization
cc.

I have read your memo of November 12, 2015, to vice presidents and general managers, regarding RAI's reorganization for January 1, 2016.

It seems to me that successfully implementing the reorganization may well require far more than the "administrative details" to which your memo refers. The new structure will, I believe, impose fundamental changes in the way we manage our dealerships.

Over the years, vice presidents have developed good working relationships with the Board of Directors. While I personally feel that reporting to a single individual—you—rather than to a group will make my job easier, I am concerned that abruptly severing our long-standing ties with the Board may lead to the interruption of important projects across the country. Further, I suspect that many vice presidents will be resentful of the "surprise" nature of the announcement.

We vice presidents have come to rely on our general managers, who seem to trust and respect us in return. The success of these relationships is in large part due to their having ready access to us in their respective jurisdictions. I am certain they will be dismayed to unexpectedly find themselves accountable to Celine Roberts in Dallas, a great distance away for most of them. As well, I think that many are liable to infer that the company suddenly lacks faith in their competence and loyalty.

Consequently, I would ask you to reconsider your January 1, 2016, deadline, and to arrange for all vice presidents to meet soon so as to allow full discussion.

I would be grateful if you would communicate your decision within the next two weeks.

FIGURE 10.4: Sample of Inductive Strategy in Persuasion

to persuade him to at least postpone the reorganization until discussions can be held with the vice presidents. The first version uses the deductive strategy. It reflects Shaw's recognition of Radisson's understanding of the issue and presumes that he will be amenable to feedback from his senior executives. The second uses the inductive strategy. The latter's premise is that Radisson may be unaware of all the implications of his decision and/or may view Shaw's remarks as inappropriate criticism from a subordinate. Note that the different strategies presume different contexts and thus generate quite different products, whose basic message, however, remains the same.

10–4 Syllogisms

Persuasion that targets the intellect must demonstrate logic that is apparent to the audience. One common approach is called deductive reasoning, which argues from the *general* to the *specific*. And one type of deductive reasoning is called a syllogism, which the Merriam-Webster Online Dictionary defines as "a formal argument in logic that is formed by two statements and a conclusion which must be true if the two statements are true." Although formal logic, including a full examination of deductive reasoning, is well beyond the scope of this book, persuasive communication that targets the intellect can often make profitable use of the syllogism.

10-4a EXPLANATION

The first part of the syllogism is called the *major premise*. You might think of it as a general statement that most reasonable people would accept as being usually true; thus, while you *might well* be able to find exceptions, most people would accept that the statement is accurate most of the time. The second part is called the *minor premise*. This is a statement that derives from the *major premise*; it is, in effect, a particular instance of the *major premise*. The third element is called the *conclusion*. It states the necessary result of applying the *major premise* to the *minor premise*. For example:

Major Premise: All published college textbooks are printed on paper.
Minor Premise: This book is a published college textbook.
Conclusion: This book is printed on paper.

10-4b SYLLOGISM PROBLEMS

An easily overlooked problem with the syllogism, however, is a flawed major premise, one that is in fact a statement about one specific case only. For instance:

Major Premise: The Philadelphia telephone directory is thicker than an issue of *Time* magazine.
Minor Premise: This textbook is thicker than my recent issue of *Time*.
Conclusion: This textbook is a Philadelphia telephone directory.

A related flaw occurs when the minor premise *seems* to stem from the major premise, but in fact derives from an *inversion* of the meaning of the major premise. Thus:

Major Premise: All dogs have four legs.
Minor Premise: My cat has four legs.
Conclusion: My cat is a dog.

Here, the problem resides in the minor premise's being based on interpreting the major premise as being "all animals that have four legs are dogs"—a preposterous statement. Another example could be the following:

Major Premise: Robotics technology can now undertake many dangerous manufacturing tasks formerly performed by workers.
Minor Premise: After his injury, my father could no longer work on the blast-furnace floor.
Conclusion: My father's job was cut because of the introduction of robotics technology.

Analogously, in this example, the major premise is about technology, while the minor premise derives from an assumed statement about danger.

10-5 Logical Fallacies

Logical argumentation can be undermined by other weaknesses as well, including what are referred to as logical fallacies. Here are some examples of common ones. (The Web references at the end of this chapter will direct you to others.)

10-5a FAULTY CONSEQUENCE

Faulty consequence, also known as *post hoc ergo propter hoc* (a Latin phrase meaning "after this, therefore because of this"), is based on the assumption that something seen to happen after something else is necessarily its consequence. For example, you could give reliable evidence that for several days last year, NPR's national news report began just after you watched the sunset from your window. If you were to try to make the case that the NPR news was triggered by the arrival of sunset in your area, you would be committing the error of faulty consequence. Your argument would be guilty of the same fallacy if you ran your toaster, microwave, electric kettle, and washing machine all at the same time, then turned on your food processor and noted that all the appliances (and your kitchen lights!) went off, and then claimed that your food processor (which was not defective) would turn off appliances.

10-5b HASTY GENERALIZATION

You would be guilty of hasty generalization if you postulate a general truth that is, in fact, based on insufficient evidence. An instance would be if you told your audience of two separate occasions on which an ATM dispensed less cash than the amount shown on the receipt, and then asserted that ATMs are therefore unreliable. Another example would be citing the tragic results of a head-on collision and then claiming that the city's roads are unsafe for public use.

10-5c UNDISPROVABLE THEORY

This is the logical fallacy of claiming that something is true simply because it cannot be proved false. For example, the statement "The designers of the Egyptian pyramids were inspired by telepathic communication from aliens" might sound absurd to you, but you cannot logically disprove it. Equally undisprovable is the claim that since the turnover to January 1, 2000, did not produce a computer catastrophe, American business and government should not have spent the fortune they did on Y2K readiness.

10-5d FALSE DICHOTOMY

Another logical fallacy is false dichotomy—the presentation of an issue as being either black or white, when in fact it could be any of a multitude of shades of gray. A true dichotomy (not a logical fallacy) could be exemplified by the statement "The light switch is on or off": It cannot be both on and off at the same time. Another example would be that it is either 8:48 a.m. or it is not: The time cannot be "partially" 8:48.

An example of false dichotomy, however, would be the claim that if someone is not tall, he or she is short. There are two problems with this claim: First, the terms "tall" and "short" are imprecise and highly subjective; and second, a significant proportion of the world's population is neither tall nor short. Another common example is the statement "If you're not for us, you're against us." Obviously, one can take any number of stands that are neither truly "for" nor truly "against" something. An analogous example might be the claim that "Engineering firms that refuse to computerize all their operations are obviously giving the profession a bad name."

10-6 A Note on Ethics

Remember that your reputation as a professional engineer will depend substantially on others' perception of your ethics. Arguments that incorporate logical fallacies can appear convincing, particularly to an audience of nonexperts. As an ethical communicator, however, you should guard against them. (And as a consumer, you should be alert to their use by others.) Remember that logical argumentation demands not only that your premise and your arguments *appear* sound, but that they in fact *be* sound.

10-7 Targeting the Emotions: AIDA

Persuasion that targets the emotions does not rely on logic; it does not try to appeal to the audience's sense of reason. Rather, it wants the audience to sense and to feel, to rely on emotional responses rather than logical ones. Consider, for example, the difference between the approach you would take to convince your boss to purchase thermocouples manufactured by Wilcon Industries rather than Geocorp, and the approach used by brewers to get you to drink Bud Light rather than Coors Light. In the former case, the audience will respond to arguments based on cost, reliability, just-in-time delivery, and so forth; in the latter, however, the brewers want you to "taste" the difference—they do not give you detailed specifications of their respective brewing processes. By the way, a successful example of persuasion that targets the emotions (though masquerading as targeting the intellect) is the series of Pepsi Next TV commercials that began in 2013. The images of happy, active, attractive young people and the Pepsi animations certainly targeted the audience's emotions; the "information" such as "30% less sugar" appears to target the intellect. You can likely still find this commercial at youtube.com/watch?v=PNU46swGtwU. Others in the series, such as the "performing baby" sequence at https://www.youtube.com/watch?v=E1KVn_XQlqA, rely on humor and irony to target the emotions, while still appearing to target the intellect by providing "information" about the drink.

Particularly apt for persuasion that targets the emotions is the already mentioned strategy called AIDA, an acronym that stands for attention, interest, desire, and action. Although you can examine the elements independently, they tend to function in a continuum, with each element leading into the next.

10-7a ATTENTION

Your audience is constantly being bombarded with persuasive material, whether on television, radio, billboards, or the Web. Thus, if you want your message to be noticed, it must catch your audience's attention. In documents—paper or electronic copy—you can make use of the techniques such as the following:

- **Color**—colored paper stock, colored type or graphics, sharply contrasting background or highlight colors
- **Shape**—nonstandard envelope and/or stationery, contrasting shapes on the page
- **Font**—decorative or otherwise unusual typeface or font size
- **Visual**—eye-catching graphics or unusual document visuals (but remember what you learned in Chapter 2 about the KISS principle, and in Chapter 7 about the use of visuals)
- **Wording**—words such as "free," "new," and "improved" are standards (look at the products on any supermarket shelf); words or phrases such as "Congratulations!" and "Are you a winner?" can also be effective (note that connotation plays a larger role than denotation).

As well, cultural preferences and cultural referents (discussed in Chapter 1) are likely to play a role in what catches your audience's attention.

10-7b INTEREST

After obtaining your audience's attention, you must find a way of sustaining that attention by generating interest. You might ask, for example, "How would you like to see the world from your very own yacht?" You generate interest in this sense by eliciting from your audience a feeling of curiosity about your message. Once again, connotation is important; you are not trying to inform your audience, but to elicit an emotional response from them. As is the case with the *attention* element, your audience's cultural background will in part determine the effectiveness of this step. Always ask yourself the appropriate CMAPP questions.

10-7c DESIRE

Now that your audience is willing to examine what you have to communicate, you need to convince them that they *must* have whatever it is you are offering. Advertisers traditionally create desire by suggesting that using their product or service will make us richer, stronger, healthier, happier, more beautiful or handsome, more sophisticated, and so forth. As mentioned above, we see this kind of message embodied in the television ads featuring young people whose happiness is portrayed as dependent on a particular brand of hair coloring, shaving cream, or cereal. The aim of such ads is not to provide concrete information but to generate emotions, often unconsciously.

Advertising taglines that have become famous have never tried to get their audiences to *think*; rather, the words and/or images are carefully chosen to generate powerful feelings. Examples include Apple's 2014 iPhone 5's "You're more powerful than you think," IBM's "Solutions for a small planet," Mazda's "Zoom Zoom," and Nike's "Just do it!" Note that politicians and government decision makers have generated slogans to promote their states, from Alabama's "Where America finds its voice," to California's "Find Yourself Here," to Hawaii's "The Islands of Aloha," to Wyoming's "Like No Place on Earth." None of these means anything specific in rational terms; in emotional terms, however, they are memorable and they work. In these and other slogans, connotation speaks more loudly than denotation, evocative abstractions replace specific detail, and precision yields to vague promise. Experience has shown these tactics to be highly effective; as consumers, we are not all susceptible to the *same* emotional stimuli, but we are definitely susceptible.

10-7d ACTION

Once you have persuaded your audience to desire your product or service, you must tell them how to satisfy that desire. You need a clear call to action, which could range from the somewhat pushy "Operators are standing by. Call now!" to the more restrained "We hope to hear from you soon." Unless you specify an action, your persuasive message may be lost.

10-7e A NOTE ON ETHICS

The AIDA strategy's effectiveness is reflected in the fact that American companies and governments spend billions of dollars each year creating highly successful persuasive messages that target the emotions.

But is this type of persuasion ethical in the sense discussed in Chapter 1? Is it a deliberate attempt to deceive and therefore an unethical practice? Advertisers might argue that it cannot truly be deceitful to suggest that a successful life depends on a floor that is "cleaner than clean" (whatever that means) because any thinking adult would be able to recognize the absurdity of the claim. In implementing AIDA, aren't advertisers merely asking us to whimsically *entertain an image or narrative* rather than *believe* what should be an obvious lie?

Many people, though, have pointed out that persuasion that targets the emotions is often directed at children. Particularly young children, it is thought, cannot distinguish the "obvious lie" from the "truth." Thus, it is claimed, ethics are badly breached in these cases. At what age, however, do children acquire the necessary insight? Should there be an overseer—perceived by many as a censor? And, if so, who should those people be, and who should appoint them?

These are questions without easy answers. What cannot be disputed is that learning how the AIDA strategy is applied makes us less susceptible to being unconsciously influenced by it, and that in itself is worthwhile.

10-7f CHOOSING A STRATEGY

Generally speaking, the higher the level of technical understanding possessed by the primary audience, the less effective will be persuasion that targets the emotions. For example, if you were a mechanical engineering consultant trying to persuade a potential client—a large testing facility—to award you a contract, your failure would be all but guaranteed if you talked earnestly and urgently about your family's social standing, or your ancestors' deeds, or your well-developed sense of contemporary clothing fashion, or the joy your smile would bring to the company's employees. You'd do better to focus on your credentials, your experience, your demonstrated expertise, and the like.

As already mentioned, engineering communication deals most often with what is factual, empirical, quantifiable, and verifiable. Consequently, its persuasion will almost always target the intellect. Again, your CMAPP analysis will help you determine how to construct an effective persuasive message.

CASE STUDIES

10a Acceptance Letter from GTI

SITUATION

Karima Bhanji is a 20-year-old student from the Seattle suburb of Redmond. She had moved from Washington State and registered at Grandstone Technical Institute because she wanted to take its highly regarded two-year Management Information Systems (MIS) Diploma Program with Internship option. Students who do well in their first year often gain a semester of real-world work experience with participating firms. After graduation, some of these students have been able to obtain full-time employment with the companies in which they served as interns.

Midway through her first year of the MIS program, Karima had applied for admission to GTI's Internship Program. On May 23, she received the letter shown in Figure 10.5. She was distressed to discover the following:

- The letter had been sent to her permanent address in Washington state rather than to her temporary address in Randolph, Vermont.
- Her parents had redirected the letter, but U.S. Postal Service delays seemed to have taken a toll.
- For whatever reason, GTI had, apparently, not notified her by email, as she would have expected.
- She had thus missed the deadline for confirming her acceptance of admission to the Internship Program.

Grandstone Technical Institute
Serving the Green Mountain State and beyond

Internship Office
400–11th Street
Randolph, VT 05060-4600
(802) 555-1212
Fax: 555-1313

May 2, 2015

Ms. Karima Bhanji
14832 NE 84th Street
Redmond, WA 98053

Dear Ms. Bhanji:

Congratulations on your successful application to Grandstone's Internship Program.

I have examined your transcripts and studied the recommendation report submitted on your behalf by your instructor, Pat Hayakawa. In the light of these documents, I have registered you in the MIS Group of the Internship Program.

Enclosed are profiles of the local firms among which we hope to find a two- to three-month work placement for you in the second semester of your second year. Please examine the company profiles and think about how one or more of them might benefit from the skills and personal qualities you feel you possess. Note that you will have to put your request for placement with particular companies in writing when you submit your Internship Placement Preference form in early June.

Note also that you must phone or fax me within the next two weeks to confirm your acceptance; otherwise, we may have to allocate your place to another student.

I look forward to working with you during your Internship Program and offer my best wishes for your success.

Yours sincerely,

Angelos Methoulios

Angelos Methoulios
Encl.
c. Mariana Lembo, Assistant Registrar

FIGURE 10.5: GTI Acceptance Letter

ISSUES TO THINK ABOUT

1. What communication strategy has Methoulios used in his letter to Karima? Give reasons for your answer.
2. What information that might be necessary to his primary audience is missing from Methoulious' letter?
3. What other aspects of his letter might be problematic? What changes would you make and why?
4. What would you suggest Karima do now? Explain the advantages, disadvantages, and implications of at least two options.
5. Justify the specific CMAPP product(s) you would, in fact, recommend Karima use in response to this letter.

10b Rejection Letter from RAI

SITUATION

Jerry Quelton grew up in Los Angeles but moved to Boston when he was 25. By his late 40s, Jerry was the divorced father of two grown children who both lived outside the country. With more free time on his hands, Jerry had begun to make occasional visits to Los Angeles, reestablishing old friendships. Eventually, he decided he would like to leave his well-paying job selling luxury cars at Slipstream Motors in downtown Boston and seek work in his old hometown.

It was around this time that Jerry's boss at Slipstream, Reg Planck, offered the name of a former acquaintance, Fran Jeffers, now vice president of Radisson Automobiles in Los Angeles. Jerry decided to write to Jeffers to ask for a position. About three weeks later, he received the letter shown in Figure 10.6.

Radisson Automobiles Inc.
Where all the cars are stars

Radisson Automobiles of Los Angeles
5544 Hollywood Blvd.
Los Angeles, CA 90028
Tel: (323) 555-4438
Fax (323) 555-1987

February 2, 2016

Mr. Jerry Quelton
277 Stuart Street
Boston, MA 02117

Dear Mr. Quelton:

I want to thank you for taking the trouble to write to me last month, and to apologize for my delay in responding.

I have taken the liberty of sharing your letter of January 3, 2016, with Michel Gagne, our general manager, and with Marty Duckworth, our sales manager. We were all impressed with your success at Slipstream, and with the considerable expertise you have acquired in 15 years of selling high-quality automobiles.

You were perhaps unaware of the fact that the great majority of our new car sales are in the mid-range: Our biggest sellers have traditionally been two Global products, the Minotaur and the Whirlwind. In recent years, we have reduced the already small number of more expensive vehicles that we offer our mainly middle-class customers.

Consequently, I must note that Radisson Automobiles of Los Angeles is not at present seeking a high-end specialist. Should conditions here change, I will of course think of you again.

In the meantime, thank you once more for your letter, and for the greetings from my old friend Reg Planck.

Yours sincerely,

Fran Jeffers

Fran Jeffers
Vice President
cc: G. Radisson

FIGURE 10.6: Rejection Letter from RAI

ISSUES TO THINK ABOUT

1. Which communications strategy has Jeffers used in this letter? How can you tell?
2. The letter contains no specific references to the fact that Jerry actually applied for a job—or that Jeffers has rejected him. What specific words or phrases in the letter would lead Jerry to conclude that his application was rejected?
3. What, specifically, makes you think that Jeffers thought about CMAPP when she was writing?
4. What do you think her primary purpose was? What makes you think so? What secondary purpose do you think she might have with her primary audience?
5. From what you already know about RAI's hierarchy, explain the "copy" indicator in terms of the CMAPP analysis that Jeffers might have conducted. (Hint: Think about what you already learned in Case Study 2b about RAI's executives.)
6. In that light, explain any purpose you think she had with regard to that secondary audience.
7. Explain how the tone of her letter is likely to affect her primary and secondary audiences and the context.

10c Persuasive Message from AAUSAEx

SITUATION

At a series of meetings of the Ann Arbor University Student Association Executive, the executive president, Dorothy Palliser, has argued in favor of a special Student Association "technology" levy of $25 per semester. This money would be used to purchase, exclusively for student use on AAUSA premises, a number of tablets and laptops with peripherals and software, including a variety of popular games.

Despite strong opposition to her proposal, Palliser has managed to convince the executive to put the issue to a student vote. AAUSAEx has also set aside a small amount for "campaign fees" to be used by both sides in the debate. Part of Palliser's strategy is to email a form letter to every student. Seeking advice, she contacts Amy Booth, AAU's senior development officer, and asks for assistance. Booth agrees to help. A week later, she provides two versions of text for a form letter, along with a note that says, in part, "Since you know your audience better than I do, Dorothy, you'll have to decide between these." The two versions that Booth created are shown in Figures 10.7 and 10.8.

FIGURE 10.7: Version A Text of Booth's Suggested Letter

Dear <Studentfirstname>,

As a mature university student, you will already have recognized the increasingly important role played by technology at our school. From the efficient production of professional-quality class materials such as essays and projects, to the research opportunities available through the Internet, computer literacy and ready network access have become a sine qua non for academic success.

If you have home access to today's generation of hardware and software, you are undoubtedly aware of your advantage and concerned about those among your fellow students who lack such access. Although university computers are available—on a first-come, first-served basis—in a few campus locations, access to the university network and to the Web is currently available only through the computers found on the second floor of the library. Many of you have been frustrated by the strong competition for use of those computers, or have had your grades lowered because of poor-quality printers.

FIGURE 10.7: (continued)

As a member of both the university community and the community at large, you are keenly aware of the increasing costs of higher education. Your fees have risen more than 23 percent over the last two years. Over the same period, the Student Association has managed to mitigate the impact of higher fees by imposing an 8 percent cap on increases to your Student Association allocation.

The fact that you have enrolled at Ann Arbor University demonstrates that you are committed to the long-term benefits of a university education, that you are willing to work hard to achieve those benefits, and that you appreciate their cost. Access to computers and the Web has become an indispensable feature of a quality education. A modest $25 increase in the Student Association levy is a small price to pay for increased access for AAU students.

When called upon to exercise your franchise, let reason and compassion prevail over parsimony: Vote in favor of the AAUSA technology allocation.

FIGURE 10.8: Version B Text of Booth's Suggested Letter

WHO CARES?

I think **you** do, <Studentfirstname>, because it's shameful!

Costs have soared, but you still need your education! Student fees keep skyrocketing, even though the Student Association tries to put some kind of lid on the pot by holding the line on AAUSA allocation. And what's the result? You still don't get to do all the work you need when you need to do it.

You can't succeed unless your work looks good. But what if you don't have the proper hardware and software at home? You can use the campus computer rooms, but how often can you actually get in? How often can you get the computers to work? How often do the printers actually have ink or toner?

And what if you have to connect to AAU's network or you have to do research on the Web? Well, the library lets you sit at one of its couple dozen stations—for less time than it takes to download just about any page or file you really need!

So what's the solution, <Studentfirstname>? AAUSAEx wants to have the equipment that students really need—laptops and tablets. And we want them here, in our own AAUSA rooms, hooked up to the AAU network, wired to the Web, and available 24/7/365 to any registered AAU student.

<Studentfirstname>, I know you care. Help us show AAU that we all do! Make sure that we all have the technology we need. When the ballot boxes come out, vote for the $25 ticket that'll buy everyone access to a decent computer!

ISSUES TO THINK ABOUT

1. Why do you think Booth created two different versions?
2. How would you explain her motivation in terms of context, audience, purpose, message (particularly the level of technicality), and product?
3. Examine each of the versions in terms of message. What differences can you identify?
4. How do the two versions differ with respect to level of technicality?

5. Does either version target the intellect? If so, would you say the argumentation is deductive or inductive? Give reasons for your answer.
6. Does either version target emotions? If so, what specific features suggest this? (Think about the AIDA strategy and its emphasis on connotation, abstractions, etc.)
7. Both versions include the indication of a mail-merge code, <Studentfirstname>. How do the two versions differ in their use of this item?
8. In what ways do you think those uses are effective or ineffective?

EXERCISES

10.1 Specify and briefly explain the components of the *deductive* or *direct* strategy of conveying news.

10.2 Specify and briefly explain the components of the *inductive* or *indirect* strategy of conveying news.

10.3 Briefly explain why many engineering professionals dislike the *bad news* strategy.

10.4 Explain the principal differences between targeting the intellect and targeting the emotions.

10.5 Explain when you would be likely to use the following for persuasion:

- deductive strategy
- inductive strategy

Give examples to illustrate your answers.

10.6 Identify and briefly describe the three components of a syllogism.

10.7 Provide an example of each of the following:

- valid syllogism
- invalid syllogism

10.8 Identify and briefly explain at least two types of logical fallacies, using an example to illustrate each.

10.9 Identify and briefly explain the four components of the AIDA strategy.

10.10 Look for a short piece of persuasive writing in your field of interest. You will be looking for something that targets the intellect. Then

- create what might have been the CMAPP analysis for that document.
- discuss the persuasive document in terms of strategies for logical argumentation.

10.11 Choose a magazine or online advertisement or a television commercial, and assess its use of each of the four AIDA components.

10.12 Discuss your chosen example of AIDA persuasion in terms of ethics.

USEFUL WEBSITES

Note: For ease of reading, we have removed the "http://www." from the URLs below. However, some browsers might still require that you insert it to precede some URLs so as not to receive an error message.

■ "Basic Helps for Practical Reasoning," by Dale Sullivan, of Northern Illinois University, offers concise explanations of various types of logical argumentation.
Visit: sullivanfiles.net/WID/assignments/discourse_field/logos.htm

■ Buffalo State College provides a pdf version of a course module on Persuasive Communication.
Visit: faculty.buffalostate.edu/smithrd/UAE%20Communication/Unit3.pdf

■ "Stephen's Guide to Logical Fallacies," developed by Stephen Downes, a Canadian academic and consultant who describes himself as an "information architect," has become a standard overview of logical fallacies.
Visit: onegoodmove.org/fallacy/welcome.htm

■ Tom Sant's Management Consulting News interview, "Persuasive Proposals," offers useful advice on constructing successful business proposals.
Visit: managementconsultingnews.com/interview-tom-sant/

■ "Understanding Persuasion," created by Joel P. Bowman of Western Michigan University, offers an excellent description and explanation of approaches to persuasion in the context of North American business.
Visit: homepages.wmich.edu/~bowman/c4dframe.html

CHAPTER 11

Proposals and Business Plans

Fundamentally, a proposal is a CMAPP product designed to offer a solution to a problem. Understandably, in your situation, you will be looking to offer a solution to an engineering problem, from CAD/CAM, to heat transfer, to geological investigation, to vapor-liquid equilibrium.

Solving a problem invariably costs time, money, energy, and so forth; thus, your proposal must persuade your audience (by targeting their intellect—see Chapter 10) that your solution warrants that expense. At the same time, your proposal must clearly and precisely explain the solution, its costs, and its benefits. Thus, a proposal usually combines persuasion with mechanism, technical, or process description (see Chapter 9).

Like informal reports (discussed in Chapter 8), a proposal may take the shape of other standard CMAPP products: a memo (for internal communication), a letter (for external communication), or simply a document titled, for example, "Proposal for . . ." (usually introduced by a transmittal memo or letter). Recall, as Chapter 8 also pointed out, that many people subsume proposals under the broad rubric of "reports."

11-1 Classification of Proposals

A common way to classify proposals reflects the CMAPP elements of message and audience. Consider the following four scenarios.

11-1a SCENARIO 1: SOLICITED INTERNAL PROPOSAL

Griffin Radisson, RAI's President, believes that the morale of senior managers in all seven dealerships has been declining, to the detriment of the company as a whole. He asks the company manager, Celine Roberts, to look into the matter and submit to him a proposal for raising morale.

Since Radisson has requested the proposal, we classify it as *solicited*. Since Radisson and Roberts are part of the same company, Roberts' proposal is termed *internal* and would likely be introduced by means of a transmittal memo.

11-1b SCENARIO 2: UNSOLICITED INTERNAL PROPOSAL

Jack Lee, the Functions Coordinator for AAU's Student Association, has thought of a way for AAUSA to make extra money when the new Student Association Building opens. He envisions renting some of the AAUSA space to campus clubs. He decides to write an implementation plan and submit it to the AAUSA President, Dorothy Palliser.

Because Palliser has not been expecting Lee's proposal, we would call it *unsolicited*. Because both individuals are part of the same organization, Lee's proposal is *internal*. Being relatively short, it would likely take the form of a memo.

11-1c SCENARIO 3: SOLICITED EXTERNAL PROPOSAL

Cal Sacho, GTI's Senior Facilities Manager, has been charged with studying the geological and geotechnical implications of building an Engineering Annex. Seeking a competent consulting firm for this major undertaking, he places in the local papers a call for proposals, often referred to as an RFP (request for proposals). Always on the lookout for interesting contracts, Mitchell Chung, an AEL senior associate, notices the call and submits AEL's proposal for the study.

In this case, Sacho requested bids; thus, we can label Chung's proposal *solicited*. Since two separate organizations are involved, the proposal is *external*. Chung would most likely use a letter of transmittal to introduce his proposal.

11-1d SCENARIO 4: UNSOLICITED EXTERNAL PROPOSAL

Nicolas Pleske, a junior associate with AEL, has experience in marketing and advertising. He has heard through personal contacts that RAI is considering opening a new dealership in his area. Eager to drum up new business, he obtains the name of RAI's President, Griffin Radisson, and creates and submits to him a proposal for bringing RAI's name to the attention of the car-buying public.

Since RAI has not requested anything, we would classify Pleske's proposal as *unsolicited*. Since AEL and RAI are different organizations, the proposal would be *external*. Pleske would frame his proposal either as a letter or as a document introduced by a letter of transmittal.

11–2 CMAPP Implications

One might assume that insider knowledge and easy access to information would make a solicited internal proposal the easiest to formulate, and that, conversely, an unsolicited external proposal would entail the greatest difficulties. For the proposal writer, however, the realities of workplace context and audience can upset that expectation.

11–2a SCENARIO 1: SOLICITED INTERNAL

A senior manager within the company, whom Roberts knows well has ready access to all relevant information. These advantages should make it easier for Roberts to construct her proposal.

Imagine, however, the reaction of the other senior managers when Roberts begins asking questions. Many are likely to be suspicious of her motives; some may even feel threatened. Their relationships within RAI are thus likely to be affected, regardless of the eventual proposal she develops. As well, the accuracy of the information they provide—and thus of her proposal—may be compromised by their own speculations regarding the use to which she will put what they tell her.

Having asked Roberts to propose solutions, Radisson is obviously aware of a "problem." Thus, she need not overcome audience skepticism that one in fact exists. What would Roberts do, however, if her research determines that morale has been in decline because of Radisson's own overbearing management style and his unwise operational decisions? In such a case, her own familiarity with her audience, and the fact that it is her boss who asked for the proposal, might subvert her desire to be factual and objective.

11–2b SCENARIO 2: UNSOLICITED INTERNAL

Because his proposal is unsolicited, Lee must first convince Palliser that there is a problem to be solved. Thus, his audience's initial skepticism may be high. However, his knowledge of his audience and his own position within AAUSAEx may well increase his credibility and thus his ability to persuade her.

A member of the AAUSAEx and an AAU student, Lee will benefit from specialized knowledge of audience and context to make his proposal more precise, relevant, and convincing. On the other hand, if the relationship between Lee and Palliser is not a good one, he will face an uphill battle: Being human, she may not be able to react objectively to his proposal and might feel that he is making use of his "insider knowledge" to undermine her position as president.

11–2c SCENARIO 3: SOLICITED EXTERNAL

Responding to a request for proposals (RFP), Chung need not convince his audience of the *existence* of a problem; he can concentrate on a workable solution.

However, he is likely unaware of, and thus will not respond to, specific GTI in-house concerns. Further, despite the public tender call, Sacho's decision might be colored by internal politics. For example, GTI's Board of Regents might have decided that small, local firms should be given priority, thus leaving AEL—a large national concern—at a hidden disadvantage.

11–2d SCENARIO 4: UNSOLICITED EXTERNAL

Pleske, an unknown "outsider," must first persuade his audience that there is in fact a problem. Radisson, not having requested anything, and believing his own situation to be well in hand, is likely to be highly skeptical and might even take offense. Not being part of RAI, Pleske cannot be certain that RAI is looking to open a St. Louis dealership soon. Further, he needs to convince Radisson not only to undertake marketing and advertising, but to hire AEL to develop and implement the program. In fact, he cannot even be sure he has chosen the appropriate audience: Radisson might be annoyed that a promotional proposal had not gone directly to Celine Roberts.

On the other hand, it is also possible that Radisson would be intrigued by an outsider's perspective, and might have perceived an unsolicited proposal from one of his own employees as a mark of disloyalty. It is also possible that Pleske, unaware of "how things have always been done" at RAI, might bring to bear new ideas whose originality will appeal to his audience.

11–3 Informal and Formal Proposals

Internal proposals are sometimes referred to as *informal*. Through such a proposal, for example, you might furnish your supervisor or a colleague with your solution to a specific, work-related problem—whether of a specifically engineering nature or not. Most informal proposals, therefore, will be relatively short, likely no more than a few pages, and will take the form of a memo. If the issue is complex, a longer informal proposal might be framed as a separate document, introduced by a transmittal memo.

External proposals may also be termed *formal*. Whether occasioned by a tender call or some other stimulus, they tend to be longer than informal proposals and more complex. If you, as a consulting engineer, are responding to a request for a specific solution to a client's engineering problem, for example, you would construct a *formal* proposal. Although some formal proposals may be included within a letter, most will be documents introduced by transmittal letters.

11–3a STRUCTURE

There is no prescribed structure for a proposal; its content and organization will be individually determined by the results of a CMAPP analysis. Typically, though, an informal proposal might include the following elements, although the wording of your headings may well differ. Notice the similarity with the direct or deductive approach to persuasion discussed in Chapter 10.

- **Introduction** Provide sufficient background for your audience to appreciate the rationale for your proposal.
- **Recommendations** Provide a brief, concise list of the steps or actions that you are proposing.
- **Justification** Discuss, in detail, your arguments to support your recommendations. Don't forget to consider potential costs and potential benefits, including any alternatives, and some kind of implementation time line. Remember that this section is the "meat" of your persuasion.
- **Summary** Briefly and concisely recapitulate your proposal. If you think your case requires it, seek authorization to proceed. Include a "call to action": As in any persuasive communication, you must specify exactly what you want your audience to do next.

If your informal proposal is more complex, or if you are preparing a *formal* proposal, you might use the following structure. Again, note the deductive approach and recall that this list does not specify required headings.

Introduction

Indicate the background. If the proposal is solicited, specify the request to which it is responding. If it is unsolicited, explain your rationale for "intruding."

Proposed Solution

Describe precisely the steps or procedures you are suggesting. Indicate the benefits that should derive from implementing your solution.

Budget

If appropriate, you might use headings such as Costs, Staffing, Personnel, and/or Requirements, either instead of Budget or in addition to it. Present these expenses clearly and precisely.

Benefits

Specify the benefits to your audience of implementing your proposal. Use your CMAPP analysis to help define your content.

Schedule

If your proposal incorporates several sequenced steps, detail when each should be accomplished. If timeliness is important—it usually is—specify the deadlines you think are necessary.

Authorization/Action Request

If appropriate for your context and your audience, request approval to begin the project. In all cases, however, include a clear call to action.

11-3b SAMPLE PROPOSAL

If Jack Lee follows through on his proposal to Dorothy Palliser, his informal, unsolicited internal proposal might look like Figure 11.1.

Ann Arbor University Student Association
Memo

Arbor Hall 243 **Ann Arbor University Student Association Executive** **Local 4334**

From: Jack Lee, Functions Coordinator
To: Dorothy Palliser, AAUSAEx President
cc.
Date: Monday, October 19, 2015
Subject: Profitable utilization of new AAUSA premises

Introduction

As you know, the new Student Association Building (SAB) is scheduled to open in January 2016. Despite its name, the building will not be owned by AAUSA, but by the University. AAUSA has, however, committed to leasing most of the main floor.

AAUSAEx members remain aware of the possibility of financial difficulty in light of our contracted lease payments, particularly since we have been averse to raising Association fees for AAU students.

I would like to propose what I believe would be an effective solution, one that I feel would bring social as well as financial benefits.

FIGURE 11.1: Lee's Proposal to Palliser

Proposed Solution

We obtain permission from AAU administration to sublet part of the premises that we have contracted to lease.

We subdivide this section into five "club rooms" and contract with AAU clubs to lease them from AAUSA.

Costs

What has so far been set aside as the "General Purpose Room" comprises some 37' × 34', or some 1258 square feet. Since we will be leasing space from AAU administration at $0.14 per square foot per month, the prorated cost to us of the General Purpose Room will be approximately $176 per month.

Robust Construction has given me an informal estimate of approximately $4000 to convert the General Purpose Room into five club rooms, each of approximately 76 square feet: divider walls, doors, electrical renovations, and so forth. Prorated over one year, this would entail a cost of $333 per month.

From my informal discussions with the Engineering Club, the Foreign Students Association, and the Athletics Club, I predict that each of the five club rooms could be rented at $85 per month. Since AAU operates on a standard trimester system, the five rooms should be occupied for 11 months per year, providing a total annual income of $4675, and thus a yearly prorated income of $389 per month.

The financial implications of the first year would thus be:

Item	Amount
Monthly Rental to AAU	(176.12)
Monthly Renovation	(333.33)
Monthly Sublet Income	389.58
Monthly Total	(119.87)
First Year Total	(1,438.44)

The financial implications for subsequent years would be:

Item	Amount
Monthly Rental to AAU	(176.12)
Monthly Sublet Income	389.58
Monthly Total	213.46
Yearly Total	2,561.52

From the above tables, you can see the following:

- The first year would show a loss of $1,438.44 that could readily be covered by our "rainy day" fund, which currently stands at $2,000;
- The second year would show a profit of $2,561.52 − $1,438.44 = $1,123.08;
- Subsequent years would show an annual profit of $2,561.52.

Benefits

AAUSA would actually begin to see financial benefits before the end of the 3rd quarter of the second year of operation, by which time the renovation costs would have been paid off.

Subletting the space to campus clubs would foster greater intermingling of both clubs and individual students, and would likely result in greater cooperation between AAUSA and independent campus clubs.

Student members of the clubs that sublet from AAUSA would be able to accomplish "one-stop shopping" when they came to the SAB.

FIGURE 11.1: (continued)

Schedule

We should attempt to receive sublet approval from AAU administration before December 1, 2015.

Immediately upon securing this approval, we should contact the presidents of campus clubs and should try to have contracts with them signed by December 15, 2015. Such contracts should be for a minimum of one year.

Authorization

I request that within the next week you give me your approval to begin implementing this proposal so that I may initiate contact with both AAU administration and AAU club presidents.

Please respond by return memo or by email (leej@aau.edu).

11–4 General Considerations

Whether your proposal is internal, external, solicited, or unsolicited, you should bear in mind several things.

1. A proposal offers a solution to a problem. If you cannot clearly and precisely describe both the problem and your solution to it, your likelihood of persuading your audience to pay for implementing your solution will be negligible at best.
2. When developing your proposal, consider each of the complementary attributes of the CMAPP model that were discussed in Chapter 3. To succeed, your proposal must be an effective "total package," as discussed in that chapter.
3. When preparing your proposal, think of ways to counter the inevitable audience skepticism. If your proposal is unsolicited, your audience will probably question the very existence of the problem you are claiming to be able to solve. If your proposal is solicited, you will have to combat concerns about such things as the cost of your solution. In this case, you will normally have to convince your audience that your solution is superior to those proposed by others.
4. Conduct the necessary research to ensure that your CMAPP analysis answers questions such as the following:
   a. What are your audience's goals and objectives?
   b. Is your audience a small company with limited resources or a larger company with extensive and varied options?
   c. Has the company funded other projects like the one you are proposing?
   d. Have you sought out copies of any previously successful proposals of this type? They might be available on public record, on file in libraries, or on the Internet.
5. Work through your solution in sufficient detail to permit your audience to make an informed decision. Remember that you are targeting your audience's intellect and must thus provide thorough, cogent arguments. Conversely, you must be brief and concise: Extraneous detail tends to prompt a negative response.

6. Ensure that your solution is reasonable. If Celine Roberts had determined that the cause of RAI's morale problem was Griffin Radisson himself, would it have been reasonable for her to propose to her boss that he remove himself from the company?
7. Ensure that your solution, particularly your costing, is feasible. If Nicolas Pleske had suggested a 20-million-dollar marketing and advertising program for RAI, it is unlikely that Radisson would have considered that budget practical. However effective Pleske's proposal might otherwise have been, it would probably have been rejected out of hand.
8. Pay careful attention to the accuracy of your entire message. Numerical errors, faulty terminology, verifiably false assertions, and misspelled names will cause loss of credibility for both you and your proposal.
9. Similarly, as a professional engineer, your reputation for accuracy, ethics, honesty, integrity, and objectivity will play an important role in the success of your proposal.

11–5 Writing a Business Plan

As an aspiring engineer, your career could end up taking an entrepreneurial turn: You might want to establish your own business or make an investment in the idea of a friend, family member, or colleague. If this happens, you will need to prepare a business plan to guide yourself and your colleagues through the start of your business; you will also need a business plan to persuade a bank or investors to invest in your project.

It is more likely, however, that as a technical professional, you will contribute your research and expertise to a business plan that is being coordinated among several industry professionals. The plan could be written to set goals for an entirely new enterprise, or it could be put together to set out the practical matters and challenges of a major new initiative in the company for which you are currently working. Either way, the business plan itself would be organized in a very similar format.

A business plan is a complex report in that it contains numerous discrete sections that often can stand alone as simple, short reports. Because it is a report that attempts to persuade, and one that suggests a solution to an identified problem, it can readily be seen as a proposal.

Here are the sections found in virtually all business plans.

11–5a EXECUTIVE SUMMARY

This section provides the reader with an overall picture of your business plan. Typically no more than two pages, the executive summary is meant to convey all of the essential elements of the plan, allowing key decision makers to assess it without having to read every detail. Executive summaries are described at length in Chapter 8.

11–5b BUSINESS OVERVIEW

This section describes the products and services that will be the basis of your initiative, as well as the industry environment you will be entering. Here also, you will need to describe your customers: what kind of people they are, how many there are, and how you expect to convert them into clients. Finally, here you will concisely and precisely describe what your company's objectives are and how you will measure them. Note: Your overview is not meant to be a "best case scenario." Rather, it should be a realistic, hard-headed—even skeptical—depiction of your research. People will be making financial decisions based on your work. Don't leave out "inconvenient" facts that seem to make your initiative less daunting than it really is; potential investors need to know these facts *before* they take out their wallets. If they come to feel that you've been "hiding" things that might have influenced their decision, they'll quickly put their wallets away again—without giving you anything.

11-5c OWNERSHIP

Potential investors need to know the legal structure of a new business entity. Is it a public or private company? Does it have limited legal liability? Who are the current shareholders, and what percentage of the company do they currently own?

11-5d SALES AND MARKETING PLAN

Malcolm H. B. McDonald in "Ten Barriers to Marketing Planning" wrote, "Marketing planning is simply a logical sequence and a series of activities leading to the setting of marketing objectives and the formulation of plans for achieving them." This might make this part of the business plan seem more straightforward than it really is. The sales and marketing plan consists of several subsections:

- a description of the marketing environment into which you will be entering;
- a SWOT (Strengths/Weaknesses/Opportunities/ Threats) analysis;
- a clear and enumerated description of your marketing strategies and objectives;
- an action plan that engages each part of your company; and
- a rubric by which you can measure the success (or failure) of your marketing plan.

This section also often has an analysis of the organization's probable competition.

11-5e OPERATIONS PLAN

Here you explain how you will make your products and provide your services, in terms of research and development as well as in the production process. You will describe the resources, technology, and personnel you need. You will also provide a contingency plan for your organization in case you run short of these resources—or if their costs go up much higher than you had predicted.

11-5f MANAGEMENT AND HUMAN RESOURCES PLAN

A business idea can be "beautiful" and solid in most respects, but still amount to nothing without the right people putting the idea into action. This part of the business plan demonstrates to potential investors that the people running the company have sufficient experience, education, and leadership abilities to be successful. If there are "holes" in your management structure—let's say you haven't been able to sign on a chief financial officer—indicate that explicitly. Your reader needs to know that part of your plan is to fill these holes. The human resources part of this section is also very important: Do you know where you will be able to find—and then hire and retain—employees with the expertise and experience to do the work your initiative requires? Do you have sufficient financing to pay these employees at a market-competitive rate?

11-5g TIME LINE

It is important that investors and employees know what is supposed to happen and when in terms of meeting important company goals. A time line shows stakeholders that the business plan is, in fact, a real plan, with deadlines and benchmarks.

11-5h RISKS AND CONTINGENCY PLAN

We typically ask our own students what was the number one risk in investing in Microsoft in its early days. Most say that superior competition from Apple or IBM could have presented a big risk. In fact, the biggest risk would have been the untimely demise of its chief shareholder, Bill Gates. Without that wily genius, it is doubtful that the company he cofounded would have grown, or even survived. Early investors needed to be apprised that Microsoft's business plan relied largely on Gates' remaining at the company's helm, inspiring and leading his employees and intimidating his competition. No doubt, Gates and his management team also provided a "contingency plan"—that is, a strategy to handle unexpected or problematic issues that might adversely affect a company's performance—for when he departed the scene. (He retired as Chief Software Architect in 2008 and stepped down as CEO in 2014.)

11-5i FINANCIALS

This section comprises numerous types of monetary data: past, current, and expected future financial performance; start-up costs; break-even projections; profit-and-loss statements; loans and debt; cash flow analysis; and use of proceeds. All analysts will tell you that in the preparation of financial statements, you should err on the side of caution and skepticism. You can open yourself up to legal liability if your statements prove to be overly and willfully optimistic.

11-5j PUTTING IT TOGETHER

A key reason many businesses fail is because individuals' business plans never live up to their names: actual *plans*. They remain hopes or putatively "good ideas." Each of the preceding sections needs to be addressed diligently, fully, and skeptically. Only then will you have a plan on which you can base intelligent decisions.

It is not a quick process. We have found that even plans for simple start-up businesses can take a lot of time to write: up to 500 person-hours. There are no shortcuts. As an engineer contributing technical data and analysis to such a document, you will be expected to put in a lot of those person-hours, and to know how your contribution fits into the whole of the document.

Business plans present a particular challenge when it comes to editing. As noted previously, they are often composed over a long period of time, longer than most other documents described in this book. This means that facts that are current and correct early on in the drafting process might be out of date by the time you are finishing up your executive summary (the last thing you write in a business plan). You need to double-check literally every number, every fact one last time before you submit your plan.

Another revision challenge is that *business plans are typically multiauthored documents*. Senior management might draft the company overview; a marketing specialist might compose the section on competition and marketing plans; you, as the engineer, might be called in to draft all or part of the operations section; and an accountant might be responsible for putting together the financials. You need to make sure that each section is congruent with the others; for example, the marketing initiatives proposed in the marketing section should "square" with the amount of funds allotted to these initiatives in the financials; and the management and senior staff described in the management and human resources section should be capable of overseeing the activities, research, and technology described in your operations discussion.

How long are business plans typically? We have found that most business plans that address all of the sections described here run from 25 to 40 pages—that is, as short as possible, considering all of the research that has gone into it, but as long as necessary for the context, message, audience, purpose, and product.

11–5k CRITICAL POINTS

a. A business plan is an immensely complex CMAPP document, because it needs to address numerous and disparate audiences and other stakeholders: potential investors (and perhaps bank managers), current and future employees and management, as well as outside contractors like engineers, accountants, and suppliers.

b. All of these people rely on the accuracy and currency of the business plan in order to make sound decisions and act appropriately.

c. Your plan should be written in plain and clear language, avoiding jargon as much as possible. Remember that a wide variety of people need to have a precise understanding of your plan. You can't afford to confuse anyone.

d. The fundamental purpose of a business plan is to foster and maintain relationships—relationships that together enact a new, often expensive initiative. The business plan lets the key figures know what to expect from the organization and what is expected of them.

e. Avoid unwarranted optimism. While certain stakeholders might be very excited about the "idea behind the plan," your goal as business plan author/contributor is to be skeptical, reasonable, and diligent. It is more important for you to protect your clients' wallets or purses than it is to make them happy. If you are writing a plan for a business that will face very stiff competition, don't hide that fact; it is critical your readers know what might be in store for them.

f. Make sure your research is up-to-date. Author Bob Basil once submitted to a client a business plan—the result of several months of research, planning, and writing—that was sound and current at the time but out-of-date the next day. That next day was September 11, 2001. All assumptions regarding finances, the market, and even management had to be reassessed following the attack on the World Trade Center.

CASE STUDY

11 Proposal to AEL from Superlative Design

SITUATION

In 2014, Leila Burke, AEL's senior associate in Lansing, Michigan, hired a small, local design firm called Superlative Design to remodel the company boardroom. Within months of the job's completion, Burke began to receive complaints from office personnel about the quality of the workmanship. She conceded that her choice of designer had been a mistake. At the same time, she recognized that Superlative Design had, technically, fulfilled the terms of what she now realized had been a too vaguely worded contract.

This year, AEL plans to renovate the Lansing consultants' offices. Consequently, Burke requested proposals from three design firms with whom she had had contact. However, Jessica Greyland, a partner with Superlative Design, happened to learn about this through an acquaintance; she decided to submit her own proposal, which is shown in Figure 11.2.

Superlative
Design, Ltd.

231 Devonshire Way
San Francisco CA 94131
415/555/7475 Fax: 555/7476

Ms. Leila Burke, Consultant
ACCELERATED ENTERPRISES
801-b market Street San Francisco CA
April 20, 2014
Dear L. Burke,

We have heard of your call for proposals of the 3rd inst., this kind of short proposal is our response. Thank you for giving Darrell and I the opportunity to submit this proposals. AEL is a firm we are always happy to have good business relations with.

The quality and excellence of every office facility htat we design reflects our commitment to our clients. Our innovations are often cited as excellent. All ready this year we have won an award. We pride ourselves in our ability to quickly and efficiently design well engineered systems. We have yet to miss a deadline in our many years of operation, our ten-year anniversary was celebrated last year.

We are familiar with your staff's needs due to the fact that we worked with you last year on the bordroom project. If you may recall, that interior, which Darrell actually designed, required less changes and cost less than expected. And Darrell is confident he can achieve similar results on current project.

Our Human Factors study report (we did one in absentia so to speak) shows that the absolute best use of your office space is to divide that humungous room into nineteen cubicles, 4 cute private offices, and 2 shared offices. We'll get great affects from the high-res chromatography analysis, the shades and tints and hues will be awesome and Statistics indicate that most workers prefer private spaces that are in close proximity to the people they work with. It is clear that the same principle applies to managers and chief executives. According to our research, the number of Worker Compensation claims among employees who suffer from a from of repetitive motion syndrome is greatly reduced by using high-quality, adjustable chairs and desks.

We have forwarded a copy of this proposal to Frank Nabata in Lansing who I remember your secretary said really makes all the big decisions. We will be very interested in what he recommends.. If you have any questions, please feel free to contact me whenever. You can either reach me at my office or can call me at my home number, (417)5550099.

We look forward to starting soon. Thank you for taking the time to review this proposal.

Very very sincerely yours,

Jessica Greyland ☺

J. Greyland, Human Factors Consultant
SUPERLATIVE DESIGN
JG/jg

FIGURE 11.2: Superlative Design's Proposal to AEL

ISSUES TO THINK ABOUT

1. If you were Leila Burke, what would be your first reaction to this proposal in light of your previous dealings with Superlative Design?
2. Evaluate Greyland's proposal in terms of the following:
   a. letterhead
   b. inside address
   c. salutation
   d. grammar
   e. proofreading
3. How would you assess the format of Greyland's proposal? In your answer, consider her use of document visuals and the extent to which she follows (or fails to follow) standard business letter conventions. (Note: These conventions will be illustrated in Chapter 12; so, yes, you'll have to read ahead.)
4. How might Burke respond to Greyland's reference to Frank Nabata and Burke's secretary?
5. Evaluate the proposal in terms of
   a. level of discourse
   b. level of technicality
6. Which type of proposal organization do you think Greyland was trying to use? How well did she succeed?
7. Greyland has made several assumptions in this proposal. One appears in her first paragraph, another in the action request. What are the assumptions and how do you think Burke would react to them?
8. How would you evaluate the proposal in terms of brevity, precision, and persuasiveness?
9. Is Greyland's proposal solicited or unsolicited? Justify your answer.

REVISION

Had Greyland been more of the professional she pretends to be, her unsolicited proposal would have looked more like Figure 11.3. You'll note that

1. she recognizes that her letter might not be warmly received and thus chooses an inductive strategy and a high level of discourse;
2. her mention of changes within her own company is implied acknowledgment of AEL's likely dissatisfaction with Superlative's earlier work;
3. since she understands that she does not have—in fact, is not supposed to have—the information required to produce a formal proposal, her letter is, in effect, a request to submit one; and
4. Greyland tries to create a "total package" that adheres much more closely to the complementary CMAPP attributes described in Chapter 3.

Superlative Design Limited

231 Devonshire Way, San Francisco, CA 94131
Phone: 415-555-7475 Fax: 415-555-7476

April 20, 2014

Ms. Leila Burke, Senior Associate
Accelerated Enterprises Ltd.
801B Market Street
San Francisco, CA 94102

Dear Ms. Burke,

Superlative Design has undergone a number of changes since our contract with you last year: a new management structure, new quality control and quality assessment procedures, the addition of two highly experienced, accredited interior designers, and a renewed and more clearly articulated commitment to client satisfaction. I am delighted to be able to tell you that, over the last several months, our efforts have brought us compliments from both private- and public-sector clients such as Bentley Industries, WireCo, and the California Department of Transportation.

Our policy is to work with our clients, listening carefully and consulting with them consistently and openly. Thus, we are now able to pride ourselves on accurate estimates, competitive pricing, quality design, and strict adherence to both budget and deadline.

Through a chance meeting with a colleague, I have recently been led to understand that you are planning further renovations to AEL's San Francisco offices. Although I recognize that you have not gone to public tender, I would ask you to consider giving Superlative Design the opportunity to submit a detailed bid. I am certain that we would be able to meet your needs quickly and effectively.

I hope you will allow us to discuss your requirements, and I would ask you to call me to arrange a meeting at your convenience.

Thank you for taking the time to consider the new Superlative Design.

Yours truly,

Jessica Greyland

Jessica Greyland
Principal

FIGURE 11.3: Revised Proposal from Superlative Design to AEL

EXERCISES

11.1 Identify and briefly describe the four types of proposals discussed in this chapter.

11.2 Using examples, describe the CMAPP implications for each type of proposal identified in question 11.1.

11.3 Distinguish between informal and formal proposals.

11.4 Explain why proposals employ primarily a persuasive communication strategy.

11.5 Explain why proposals target the intellect rather than the emotions.

11.6 Indicate whether proposals employ the *direct* strategy or the *indirect* strategy (analogous to the *direct* and *indirect patterns* discussed in Chapter 8 with respect to evaluative reports) and explain why.

11.7 Assume that you and several of your classmates believe it would be more beneficial for your education in a particular course you are taking to work on a term project rather than take the currently scheduled final exam. Construct a CMAPP analysis for the proposal you would submit to your professor.

11.8 Using the guidelines provided in Chapter 5, construct a formal multilevel outline for your term project proposal.

11.9 Write that proposal.

11.10 For the following scenario, you should invent all required specifics; make sure, however, that they are reasonable and cohesive.

The electrical engineering company you work for will soon be moving its offices to a nearby town, resulting in your having a one-hour commute to and from work every day. Because your job requires much individual work at your computer and less than average daily interaction with other employees, you would like to telecommute.

After researching telecommuting, prepare a proposal in memo format for your supervisor. Be sure to

- propose a specific and feasible solution supported by your research;
- highlight the potential benefits (e.g., finances, morale) to your company of telecommuting;
- take into account the probable audience skepticism and find ways to counter it; and
- pay attention to the CMAPP complementary attributes discussed in Chapter 3, and to the logic and effectiveness of your arguments and counterarguments (see Chapter 10).

11.11 As a young engineer you have been asked to contribute to a business plan being prepared by a privately owned natural resources exploration company headquartered in Chicago, Illinois. The president (and owner) sees an opportunity to enhance the company's operations in Argentina, where it has several long-term clients, by opening up a permanent facility in Buenos Aires. You have been asked to contribute to the operations plan. You need to determine what resources, personnel, and technology this new office would need to function on its own, without logistical support from the Chicago headquarters.

- What primary research would you conduct? That is, to whom would you speak or write to conduct sound research regarding these operational issues? What questions would you ask? What details would be best discussed in "real time," face-to-face, or over the phone? What questions should be listed in writing?
- What secondary research would you conduct? That is, what published sources would you consult? How would you determine the credibility and currency of these sources?
- What type of document would you submit to your supervisor, summarizing your findings? Would you supplement this document with a presentation? Why or why not?

11.12 Following the scenario above, you have been asked to contribute to the company's SWOT Analysis, a helpful marketing exercise that helps a company identify internal issues (Strengths and Weaknesses) and external ones (Opportunities and Threats) that might have a bearing on your plan's success. You are very familiar, of course, with your company's internal operations, and you have access to several secondary research sources to examine external issues. Making up reasonable details regarding your company's internal situation, and performing some secondary research regarding the economic and competitive environment in Argentina, brainstorm and then compose a SWOT Analysis that you might give your company's marketing director. Note that it is typically presented as a box divided into four quadrants, as follows:

Strengths	**Weaknesses**
• Strength 1	• Weakness 1
• Strength 2	• Weakness 2
• Etc.	• Etc.
Opportunities	**Threats**
• Opportunity 1	• Threat 1
• Opportunity 2	• Threat 2
• Etc.	• Etc.

USEFUL WEBSITES

Note: For ease of reading, we have removed the "http://www." from the URLs below. However, some browsers might still require that you insert it to precede some URLs so as not to receive an error message.

■ Bplans.com has hundreds of sample business plans for you to peruse.
Visit: bplans.com/sample_business_plans.php

■ Bplans.com also has an excellent section on SWOT analyses.
Visit: articles.bplans.com/how-to-perform-swot-analysis

■ CapturePlanning.com provides a useful article called "How to Write a Business Proposal," which offers links to additional information.
Visit: captureplanning.com/!hc_how_to_write_a_proposal.cfm

■ Colorado State University offers an Engineering Proposals overview through its Writing@CSU | The Writing Studio guides; the page provides specific and highly relevant advice.
Visit: writing.colostate.edu/guides/documents/proposal

■ Entrepreneur.com provides a handy template to help you research and write a simple business plan.
Visit: entrepreneur.com/how-to-write-a-business-plan

■ Proposalworks.com features a wide variety of resources for proposal writers, including articles on all aspects of the proposal development process. It also has a library of 165 sample proposals you can review.
Visit: proposalworks.com

■ Delaware Community College English instructor Alice Reid provides "A Practical Guide for Proposal Writing" that includes a planning sheet for a proposal and its presentation.
Visit: facstaff.gpc.edu/~ebrown/pracguid.htm

■ University of Toronto's Engineering Communication Centre offers a worthwhile overview of the requirements of a good engineering proposal. Although the focus is Canadian, the principles are equally applicable to the United States.
Visit: ecp.engineering.utoronto.ca/online-handbook/types-of-documents/proposals

CHAPTER 12

Correspondence: Letters, Memos, Email, and Faxes

In 1773, Benjamin Franklin wrote to "Messrs. Abel James and Benjamin Morgan" to discuss the "Introduction of Silk-weaving into America." Franklin closed that letter with the then common phrase "I . . . am, with great esteem, Gentlemen, your most obt. Humb. sert." Two years later Franklin wrote to a Mr. Timothy to submit a "Lecture on Office Seeking." That letter's complimentary close was "I am ever, dear Sir, your faithful & most obedient Servant." (Through its online Making of America publications, available at http://ebooks.library.cornell.edu/m/moa/, Cornell University has made these and many other historical materials available to the public.)

Although mail wagons did make fairly regular runs among the colonies' major centers around that time, having a letter conveyed to your intended recipient was a somewhat haphazard affair.

On July 26, 1775, the members of the Second Continental Congress agreed

> [t]hat a postmaster general be appointed for the United Colonies, who shall hold his office at Philadelphia, and shall be allowed a salary of 1000 dollars per an: for himself, and 340 dollars per an: for a secretary and Comptroller, with power to appoint such, and so many deputies as to him may seem proper and necessary. That a line of posts be appointed under the direction of the Postmaster general, from Falmouth in New England to Savannah in Georgia, with as many cross posts as he shall think fit.

(This quote and others are available at http://memory.loc.gov/.) Later that year, Franklin was elected Postmaster General of the United Colonies. And, according to the United States Postal Service, in 1788, "President Washington appointed Samuel Osgood as the first Postmaster General under the Constitution. A population of almost four million was served by 75 Post Offices and about 2,400 miles of post roads."

12–1 Letters

Of course, both society and technology have changed dramatically since then. And you might be tempted to ask whether today's engineers even need to write effective letters. The short answer, of course, is "absolutely." For elaborations, you might want to look at "Writing Letters to the Editor" (ieeeusa.org/policy/guide/lettertotheeditor.html), from the Engineer's Guide to Influencing Public Policy, at Engineering Career Services at https://ecs.engr.wisc.edu/public/student/letters.php, or at Cover Letters and Letters of Transmittal at www.cewriting.ling.pdx.edu/images/PDFs/Genre%20Unit%202.pdf, to cite only three good websites.

Here is an anology. Despite continuing change in most areas of automobile design over the last 100 plus years, there has been a contrasting movement toward standardization. For instance, in the early part of the last century, not every car manufacturer placed the clutch pedal on the far left and the gas pedal to the right of the brake pedal—or, for that matter, even used the same distribution of controls. At http://www.reddit.com/r/AskHistorians/comments/2otbx2/how_did_the_control_system_of_automobiles_develop/ and at https://en.wikipedia.org/wiki/Car_controls, for instance, you'll find interesting examples of the changes in the way cars were controlled. Over time, however, uniformity of design became the rule. The same tendency has been reflected in the development of standard formats and common conventions for business letters and other forms of correspondence, like memos and, more recently, emails.

The effectiveness of your correspondence will in part be in inverse proportion to the extent that your audience is consciously aware of these formats and conventions. When they are in some way violated, your audience is likely to notice the violation and to react negatively to the entire message. For example, if you were to receive a letter in which the sender's letterhead and your name and address appeared at the end of the letter rather than on the first page, you would be distracted, perhaps to the point of suspecting something was wrong with the letter's content as well. Similarly, if you received a letter that concluded with the phrase "I remain your humble and obedient servant," you would be bemused and possibly suspect an attitude of sarcasm on the part of the sender.

Notice the consequent dichotomy: You should adhere to letter conventions . . . precisely so that your audience does *not* notice them. If your readers do in fact notice that the organization, design, and verbal composition of your correspondence seem somehow "off," they will *also* suspect that you are not fully a professional person. And then you might start frustrating colleagues and losing clients.

12–1a BLOCK AND MODIFIED BLOCK

Two standard formats for business letters are the *block* and the *modified block*, shown in Figures 12.1 and 12.2, respectively. (Note that in both letters the body text is devoted to an explanation of the format.)

Grandstone Technical Institute
Serving the Green Mountain State and beyond

400–11th Street
Randolph, VT 05060-4600
(802) 555-1212
Fax: 555-1313

April 3, 2015

Ms. Olivia Barclay
Executive Assistant
Radisson Automobiles Inc.
6134 Bank Street
Dallas, TX 75204-1010

Dear Ms. Barclay:

This letter illustrates the **block format** with **mixed punctuation**.

You will notice that all lines begin at the left margin. Paragraphs are not indented in this format. You should also note the following common conventions.

Key the inside address four lines below the date. Use only one line space to separate the inside address from the salutation as well as the salutation from the first paragraph of the text.

If you have not been able to adjust your paragraph spacing to approximately 1.5 times line spacing (see Chapter 7), you should single-space the body of the letter and double-space between paragraphs. Place the letter appropriately on the page using top, bottom, left, and right margins of between 1" and 1¾".

The complimentary close is keyed two lines below the body of the letter. Only the first letter of the first word is capitalized. A company name may appear after the complimentary close if company letterhead is not used. (Note, however, that *not* using a company letterhead may well be seen as amateurish. If used, though, the company name is usually keyed in all capital letters, two lines below the complimentary close.) Key the writer's name four lines below the complimentary close. *Enclosure* or *Copy* notations should be separated by two lines.

Yours truly,

P.R. Hayakawa

(Ms.) Pat Hayakawa, Instructor, MIS
Enclosure

c. Boris Milkovsky

FIGURE 12.1: Block Format

12–1b LETTER CONVENTIONS

The conventions for the formats shown are discussed in this section.

Letterhead

The individual's or organization's letterhead (or return address) appears on the first page of the letter. (Note that this page does not normally include a page number, while succeeding pages are numbered but include no letterhead.) When designing a letterhead, you should keep in mind the document visual guidelines discussed in Chapter 7.

Ann Arbor University
Seeking Excellence
In the Great Lakes State

Ann Arbor University
5827 Dixie Road
Ann Arbor, MI 48103-0061
(734) 555-9989

April 3, 2015

Mr. Flavio Santini
Senior Consultant
Accelerated Enterprises Ltd.
342 Center Street West
Lansing, MI 48980-1720

Dear Mr. Santini:

This letter uses the modified block style with open punctuation. Please note the following conventions.

In formal letters the salutation will be followed by a colon; in less formal ones, by a comma.

In this format, all lines begin at the left margin except for the date, the complimentary close, and the signature block, which begin at the center of the page. This format is efficient because only one tab setting at the center is required. Place the letter appropriately on the page using top, bottom, left, and right margins of between 1" and 1¾".

Paragraphs may begin at the left margin, or—as in this example—the first line of each paragraph may be indented five spaces. If you have not been able to adjust your paragraph spacing to approximately 1.5 times line spacing (see Chapter 7), you should single-space the body of the letter and double-space between paragraphs. Key the inside address four line spaces below the date. Use only one line space to separate the inside address from the salutation and the salutation from the first paragraph of the text.

The complimentary close is keyed two line spaces below the body of the letter. Only the first letter of the first word is capitalized. Key the writer's name four lines below the complimentary close. Enclosure or copy notations should be separated by two line spaces.

Yours sincerely,

Roger Concorde

Roger Concorde
Registrar

Encl.

c. Nancy McDonald

FIGURE 12.2: Modified Block Format

The letterhead should always include the company's name, a mailing address, and a phone number. Most commonly, it will now include, as well, a fax number, an email address, and a Web address.

Date

We often write dates in numerical format, such as 10/24/15 for October 24, 2015; we are accustomed to using the sequence of MM/DD/YY, and, because there are only 12 months, the meaning is usually clear. Not all nations—or organizations—use that numerical format, however; they use a YY/MM/DD or YYYY/MM/DD sequence instead. So, although in the United States, 03/02/15, for example, would likely be read as March 2, 2015, it could be misunderstood elsewhere as February 3, 2015. In a letter, the date should be unambiguous. Therefore, it is best practice to write it *in full*.

Inside Address

The inside address specifies the letter's primary audience. Here is what remains the most common standard formula for an inside address:

> Honorific + First Name or Initial + Last Name
>
> Professional Title
>
> Company
>
> Street Address
>
> City + State + Zip Code

Notes

An honorific is a "title of politeness" such as Mr., Ms., or Dr. If you do not know the appropriate honorific, you may omit it—although some audiences will notice and react adversely to its exclusion.

Analogously, if the information in the first line as shown above is unavailable, it has become more common and more acceptable to leave it out and use simply the professional title.

Thus, it will no longer raise many eyebrows to write, for example, "Manager of Student Affairs" rather than the specific name of the woman or man who holds that particular position.

Salutation

The salutation acts as a greeting to the receiver. Its formality depends on the relationship between the sender and receiver of the letter. As a general guide, use the name that you would use if you were addressing the person face to face. Thus, for most business letters, for example, you might use the following:

> Dear Ms. Ramirez:
>
> If you were already on a first-name basis with her, you might instead use the following:
>
> Dear Violeta,
>
> If the letter is addressed to a job title only, use that title in the salutation—for example:
>
> Dear Manager of Student Affairs:

A colon or a comma follows the salutation if **mixed punctuation** is used; no punctuation follows the salutation if **open punctuation** is used.

Complimentary Close

The complimentary close is the formal closing. Only the first letter of the first word is capitalized. Use a comma after the complimentary close when using mixed punctuation; omit it when using open punctuation. Frequently used complimentary closes include

> Sincerely,
>
> Yours truly,
>
> Cordially,

Note that each ends with a comma.

Signature Block

The signature block contains two elements: the signature itself and, immediately below it, the signature line, in which the sender's full name is keyed. The name is often followed, either on the same line or the subsequent line, by the sender's professional title. See Figure 12.2, for example. Note, however, that the signature block should not include the company name: That should be in the letterhead.

12–1c SIMPLIFIED FORMAT

Many engineering professionals who want to adhere to the preceding conventions run into difficulty when confronted by the fact that we do not (as yet) have a standard honorific that does not show gender. Author David Ingre's article in Figure 12.3, which first appeared in a Vancouver,

FIGURE 12.3: Honoring Thy Signee

A long time ago, English solved a social problem that many languages still have. It did away with the distinction between formal and informal second-person pronouns.

If you were a friend of Will Shakespeare, you would have been used to using "thou," "thee," "thy," and "thine." As well as indicating that you were addressing only one person, the words denoted definite social familiarity. You said "thou" to a child, to your spouse (though perhaps not to your mother-in-law), to a close friend, or to a "social inferior." But you always said "you" to a superior. (Just to make things more complicated, the "informal" pronoun was also used to address the deity.) Occasionally, it was difficult to decide what to call whom . . . and many an insult resulted.

Things seem a lot simpler now. But we still have the uncertainty in business correspondence of how to address women. When you address a letter to "Mr. John Smith," the "honorific" indicates gender and number: "Mr." refers to a man and is singular. But every time you write "Dear Mrs. Brown" or "Dear Miss Jones," you are also stipulating marital status. And in a business letter, a woman's marital status is—unless she feels otherwise—no one's business but her own.

The last decade or so has witnessed the introduction of a new creation—"Ms." Although some people don't like it, I find it a highly useful and appropriate addition to the language. Although the sound of "Ms." may not please a lot of traditional ears, the word itself succinctly transmits a necessary message.

Just as everyone should have the right to choose a form of address, it's also good business etiquette to let people know what you prefer. Let's suppose I received a letter signed "J.R. Smith." If I don't know who "J. R. Smith" is, what should I put in the salutation in my reply?

If I use "Mr.," "Mrs.," or "Miss" (or simply "Dear Sir" or "Dear Madam"), I'm making an assumption I might later regret. "Dear Sir or Madam" always seems a trifle clumsy, while "To Whom It May Concern" makes me think of a new service charge or a warning against toxic gas. And although "Dear Mr./Mrs./Miss/Ms. Smith" does cover all the bases, it might not impress a potential client.

Old J. R. could have saved me a lot of frustration by letting me know in the signature line how he/she likes to be addressed.

It's a courtesy I try to extend to others. Although my actual signature—which is quite illegible—has only my initials, I use my given name in the typed signature line. Up to now, almost everyone has assumed "David" means I would prefer "Dear Sir" or "Dear Mr. Ingre."

But what if your first name is Pat or Les—or Charlie, for that matter? If you have found some variety in the honorifics people use when they write you, you might want to express your preference a little more clearly.

How—or whether—you do so is entirely up to you. If you're male, chances are you won't need to worry. Most men—and women—still assume an unidentified business correspondent is male and write accordingly. (That's not right, but I'm afraid it's still largely true.)

If you're female, you could ignore the issue and smile at "Dear Sir." You could include your given name in full if it is considered gender specific in our society (Mary or John, for example). Or you could put the honorific you prefer between parentheses before or after your name.

As our population becomes increasingly diverse, this issue will grow more acute. While we can (usually) guess the gender of names we're familiar with, the number of names we encounter that many of us don't easily recognize as masculine or feminine is growing. Both courtesy and efficiency dictate that we make allowances—either in what we write or in the way that we write it.

Adapted from *Hats Off to English* by David Ingre, published by David Ingre Written Image Services, 1990. © David Ingre.

Canada, business publication, addresses the problem. Interestingly, although the item was written over 15 years ago, the issues do not appear to have been resolved.

Some years ago, the burgeoning direct-mail industry needed to deal with the gender-specific honorific problem referred to in Honoring Thy Signee. The companies that compiled and maintained the mailing lists often did not know their audiences' preferred honorifics, but it was to their commercial advantage to offend as few people as possible. Consequently, the industry helped popularize a very practical solution: a different style of business letter called the simplified format. The letter shown in Figure 12.4 is an illustration of the simplified format. (Again, the letter's body text explains its format.)

AEL ACCELERATED Enterprises Ltd.

Accelerated Enterprises Ltd.
San Francisco Office
801-B Market Street
San Francisco, CA 914103
Tel: (415) 555-8708, Fax: (415) 555-0907

April 3, 2015

Tina Trann
General Manager, Radisson Automobiles Inc.
4965 Madison Avenue
Chicago, IL 60602-0057

SIMPLIFIED FORMAT

The simplified format omits the salutation—but uses a subject line—and the complimentary close. As in the block format, all lines begin at the left margin. The spacing between letter parts is the same as that in the block or modified block format.

In general, place the date approximately two inches from the top of the page so that the inside address is positioned for use with a window envelope. Key the subject line in ALL CAPS or in uppercase and lowercase letters two line spaces below the inside address. Position the writer's name and title four line spaces below the body in either ALL CAPS or in uppercase and lowercase, as you prefer.

When writing a letter in this format, Ms. Trann, many people "personalize" it by incorporating the receiver's name within the body of the letter.

Leila Burke

LEILA BURKE
SENIOR ASSOCIATE

FIGURE 12.4: Simplified Format

The main disadvantage of the simplified format is that some audiences respond negatively to it. Perhaps because of the format's origins in the advertising industry, many audiences equate its use with someone wanting to sell them something. Again, consider your CMAPP analysis: *What you know about your audience should condition your product.*

12–2 Memos

A few years after establishing Accelerated Enterprises in Lansing, Sarah Cohen and Frank Nabata rented a small office on Center Street. When Frank had a message he wanted to leave for Sarah's comments, he would find a blank sheet of paper, put the date at the top, write *Sarah*, scribble his message, add his name, and leave the sheet in Sarah's in-box. After reading the message, Sarah would append her comments, initial or sign the message, and leave it in Frank's in-box.

But, the transmission of memos could become considerably more complicated, particularly in large organizations. For example, one government department in which author David Ingre worked many years ago employed two different memo forms—handwritten and typed. Exchanged between audiences at the same hierarchical level, the handwritten memo form was divided into originator's and respondent's segments, and had two carbonless copies attached. You wrote your message, kept copy 2, and sent off the other sheets. The recipient wrote a reply, kept copy 1, and returned the original. The typed memo was used for a senior management audience. The form for this memo was not divided, and had only one copy attached. You carefully typed your memo (on a typewriter—this was before desktop computers and shared printers), kept the copy, and sent off the original. If the department's "big guns" deigned to respond, you would receive their original.

We believe that this unwieldy format is no longer common. Nonetheless, today's memos have acquired a set of conventions to which you need to adhere.

Note that a few students have questioned whether memos are even used anymore. It is certainly true that they are less common within organizations than they used to be even a few years ago: Instant messaging, texting, and email have, in large part, supplanted them. However, many companies *do* still circulate memos, more likely in electronic rather than paper format and customarily for "top-down" communication.

Here, then, are a number of conventions you should follow when using memorandums. (Note: You will occasionally see the alternative plural, "memoranda.")

- Memos are used *within* an organization. If you are writing to an audience that is not part of your organization, use a letter instead.
- The word "Memo" or "Memorandum" should appear at or near the top of the form, often accompanied by the organization's name.
- The memo will show the date. (Some memo forms preface the date with the word "Date," an element many brevity-minded engineers consider redundant.)
- There will be a line headed "To" on which you stipulate the audience or audiences. Many memo forms also have a line headed "cc." or just "c.", on which you can indicate any copies. (In effect, you may be identifying secondary audiences.)
- There will be a line headed "From" on which to identify the author of the correspondence.
- There will be a line headed "Subject" or "Re" (Latin for "with regard to") that should indicate in concise terms the topic of the memo. Note that most memos will address one topic only. Note also that the information in the Subject line is not the same as that given

in the introductory sentence of the memo. The former specifies the issue that the memo addresses; the latter introduces the situation, often by summarizing it.

- Memos are more likely than letters to include bulleted and/or numbered lists.
- The memo will be initialed or signed, often beside the typed name of the originator. Some house styles, though, prefer that memos be signed or initialed after the memo's final paragraph.
- Memos do not contain an inside address, a salutation, a complimentary close, or a signature block.

Memos are designed to convey information quickly. In creating a memo, therefore, you should adhere to the CMAPP principles of brevity and concision. You should also regard a memo as a CMAPP product to which CMAPP strategies and analysis should be applied. The somewhat tongue-in-cheek memo shown in Figure 12.5 provides further information about this CMAPP product.

Formidable Forms, Inc.

Internal Memorandum

April 1, 2015

To:	A. Lert Manager, Memo Division, Regional Office
c.	G. Whiz, Comptroller
From:	Peter Carborundum PC Supervisor, Effectiveness Branch, Headquarters
Re:	Company Memoranda

For some time now, we have been considering the development and production of "in-house" memo forms. You will recall that:

- the meetings were interminable;
- the points were irrelevant; and,
- the results were inconclusive.

Consequently, I have decided to take this bull by the horns and barge full steam ahead. (Please excuse my mixed metaphors, which conveniently double as worn-out clichés.)

The word "Memo" or "Memorandum" will appear prominently. A standard memo form contains provisions for indicating the date, "To," "From," and the "Subject." These items, by the way, are often referred to as the memo's *masthead information*.

The message usually deals with one issue only, and the style is often point form.

Unlike a business letter, a memo does not display an inside address, a salutation, a complimentary close, or a signature block.

Nonetheless, you *must* initial a memo (unless your house style requires a full signature) to indicate that you have examined it before sending it out. You will normally do so beside your name on the "From" line; some companies prefer you do so after the memo's final paragraph.

Unless your company house style requires otherwise, the memo's body text should use left justification.

FIGURE 12.5: A Sample Memo

12–3 Email

Email has become one of the most essential day-to-day communication products in the workplace. It has largely supplanted letters (facetiously referred to as "snail-mail"), memos, and faxes as the preferred vehicle for rapid written communication. Not only can you now expect all but instantaneous delivery of your message, regardless of where on the planet your audience may be, but you can readily attach other electronic files (documents, graphics, sound, video clips, etc.) to your message.

At first this increasing dominance at work was matched by the rise in email's use in a social context. But among students and young people today, email's use in the latter arena has dwindled almost to nothing. Nearly instantaneous messages are now conveyed via SMS ("short message service," or texting) and social-media platforms like Facebook and Twitter.

Although email's use in the workplace has remained paramount for almost two decades, its primacy is now being challenged by various social media like the two platforms mentioned in the last paragraph, and by various intranet data-management and communications systems. Nonetheless, email is still in its relative infancy, when compared to letters or to other document genres discussed in this book, like proposals and informal reports and business plans. Because of this, the etiquette and protocol of email are still relatively fluid: Different organizations have widely divergent expectations, in particular toward tone (informal vs. formal) and audience analysis. Figure 12.6 presents an anecdote told in *The Essential Guide to Email for Office and Home*, by David Shipley and Will Schwalbe (Knopf, 2007). Although the story is several years old, the divergent expectations it illustrates remain timely.

FIGURE 12.6: An Email That Backfired on Management

Shipley and Schwalbe hold up an email exchange between an executive and a secretary at a large American company in China. The executive nastily wrote:

"You locked me out of my office this evening because you assume I have my office key on my person. With immediate effect, you do not leave the office until you have checked with all the managers you support."

The secretary wrote back:

"I locked the door because the office has been burgled in the past. Even though I'm your subordinate, please pay attention to politeness when you speak. This is the most basic human courtesy. You have your own keys. You forgot to bring them, but you still want to say it's someone else's fault."

She then performed the two-click operation that sent copies of her and her boss's emails to the entire staff of the company. Before long the exchange appeared in the Chinese press and led to the executive's resignation.

Source: *The Essential Guide to Email for Office and Home*, by David Shipley and Will Schwalbe. Knopf, 2007.

12–3a THE PRIVACY ISSUE

The anecdote above demonstrates a key fact about workplace emails: They are not private. When you press "Enter" or "Send," your message can be forwarded to 10 or to a thousand other people with a single keystroke. So remember to look before you take that leap.

Even more concerning to some, in the United States the National Security Administration (NSA) can access emails on American Internet servers using warrants that need not be made public. So, in short, email *nothing* that you would mind making public.

12–3b MAKING A "PAPER TRAIL"

The characteristics that can make email a volatile medium—it can be saved indefinitely and reproduced infinitely and without monetary cost—are what make it so valuable to professionals. Engineers, for example, can save and access every single email, sent or received, with its attached diagrams and specifications and maps, related to a particular project. You can show your colleagues and supervisors that you have done your work. You have the electronic documents—the "paper"—to prove it.

Additionally, you can access this paper trail anywhere there is Internet access, as email platforms like Gmail serve as a "cloud" making your workplace documents exist . . . anywhere and everywhere.

12–3c NETIQUETTE AND PROTOCOL

Although email's conventions to some extent remain in flux, several conventions have emerged. For example, some corporations now require a signature file in every outgoing message. A signature file is the email equivalent of a letter's signature block. A company may stipulate, for example, that the signature file contain the firm's name and the employee's email address, full name, and office phone number.

More entrenched are conventions—informal organizational and industry protocols—that are often referred to as *netiquette*. The rules of netiquette are still comparatively loose; rather than being issued by a central authority, they constitute the preferences of a multitude of unconnected (and often uncooperative) users.

In Figure 12.7 your authors list rules that do indeed seem endorsed by the great majority of good communicators in American industry.

FIGURE 12.7: Eleven Keys to Good Email Hygiene

1. For every one message you need to convey, write one email. If you need one person to respond to five requests, write five emails. (Here is why: Email in-boxes are to-do lists. If somebody responds to two of your five requests that you have sent in a single email, your email might be archived without the other three requests being addressed.)
2. Subject headers should describe your email's topic in a short phrase.
3. Your first paragraph should convey the email's context, its main message, and the response you want from your reader.
4. Your emails should never be more than a single computer screen monitor can contain. If your message is longer, attach that message as a file.
5. Write as if you are writing. Text messaging is wonderful. Email is not text messaging. Capitalize the first person singular pronoun as "I," not "i,"and write "you" not "u."
6. Remember that for your audience, what you see is what you get. Subtleties like irony generally don't travel well.
7. Do not overuse "emoticons" such as :-) or abbreviations like ROTFL (look it up if you don't recognize this one!) unless you are on very informal terms with your audience.
8. Write using complete words that are spelled correctly and are in standard sentence format that is properly punctuated. (Why? Your reader is judging you all the time. You might be thousands of miles away from your reader, but your email is inches away from your audience's eyes.)

FIGURE 12.7: (continued)

9. If you are sending a file as an attachment, consider whether your audience has the software to permit proper opening, viewing, or editing of that file.
10. Don't email when you are very emotional; don't email when you are intoxicated or hung over; and don't email when you should really be sleeping.
11. Finally, never send out an account's username and password in the same email. If you must convey these items to a recipient digitally, use one platform/medium for the username (say, SMS) and another one for the password (say, Twitter's Direct Message).

12–4 Faxes

Surprisingly, fax machines were invented in the mid-1800s. But they became commonplace in the mid-1980s. Since the advent of email, though, they are now used infrequently for workplace purposes. More commonly, fax machines these days are jammed with pages of promotional items and PR announcements.

Perhaps their main important current use is in the signing and verification of contracts and other legal documentation when the parties are separated by geography and when digital signatures are not acceptable.

Their other use today is in the transmission of confidential proposals or other sensitive information that either the sender or the recipient—often both parties—don't want stored digitally on a hard drive or in a "digital cloud." When sending a confidential fax, make sure that no one other than the intended party has access to the fax machine, or that no one else is authorized to collect the fax at a particular machine. And, of course, be *very* careful that you have keyed in the right fax number before you press "Send."

12–4a FAX CONVENTIONS

Fax conventions include the frequent use of bulleted or numbered lists and the clear indication of:

- product (e.g., fax transmission)
- date of transmission
- company name
- primary audience, including phone and fax numbers
- sender, including phone and fax numbers
- number of pages being transmitted
- topic of the fax.

Use the most professional-looking, uncluttered fax cover page that your organization offers.

CASE STUDIES

12a AAU Interdepartmental Memorandum Situation

SITUATION

In late July, 2015, Kaz Lowchuk, a promising civil engineering student, received a letter from AAU's Admissions Office informing him that he would not be allowed to complete his registration for his third year because he had failed two of his compulsory second-year courses, Architectural Design 223 and Drafting Principles 200. Stunned, Kaz went to the Admissions Office and pointed out to the admissions officer that his marks before the final exams in those courses had been 78 percent and 82 percent, respectively. He stressed that he was certain he had done well on the exams, and that his other course marks were all very good.

The admissions officer checked the computer record and informed Kaz that the two engineering professors had submitted his marks as required. According to the system, Kaz's marks for AD223 and DP200 were 38 percent and 48 percent respectively. The admissions officer told Kaz that both professors were on vacation, but that he could launch a formal appeal of grades, although the process normally took six weeks.

Not wanting to lose admission to his third year because of what he was sure was an error, Kaz set up an appointment with Joan Welstromm, chair of the Civil Engineering Department. To prepare for the meeting, Welstromm examined Kaz's records and managed to contact the professors at their homes. At the meeting, she was able to reassure Kaz that he would be permitted to register.

After Kaz left, Welstromm composed the memo that is shown in Figure 12.8.

Interdepartmental Memorandum
Office of the Chair—Civil Engineering

Ann Arbor University
Seeking Excellence
In the Great Lakes State

Date: August 15, 2015
To: Roger Concorde, Registrar
From: Joan Welstromm, Chair
cc. John Vanetti, Hans Scart
Subject: Kaz Lowchuk

Welstromm

It is unfortunate that another lapse on the part of your staff has again created a situation that causes a student worry and work, that two faculty members must examine while on vacation, and that senior administrators must resolve despite their heavy workload.

Your *Admissions Office* has denied **Kaz Lowchuk** (Student Number **100888938**) admission to his 3rd year of **Engineering** because of their own error in recording his grades. Their best advice to him was to "appeal"—a process that would have put him well beyond the registration deadline for this **September**.

I telephoned the two professors involved, **Vanetti** and **Scart**, interrupting their vacations, and received assurances from them that **Kaz** had, in fact, continued to do very well indeed in the courses in question. Their records—which, by astounding luck, they both had at home—confirm that **Vanetti** had entered a mark of **83 percent** and **Scart** a mark of **84 percent**. I am sure I need not further detail the sequence of careless events that led to **Kaz's** being refused admission by your office.

I have assured the student that his registration is assured without further difficulty, and would appreciate your confirming this to me by memo.

FIGURE 12.8: Welstromm's Memo

ISSUES TO THINK ABOUT

1. Why do you think that Welstromm used a memo for this communication?
2. What other product might she have used instead? Why?
3. To what extent does the memo reflect the conventions outlined in this chapter?
4. What would you say about Welstromm's use of language, and what do you think it says about her?
5. What is your assessment of her use of document visuals, and what changes would you make and why?
6. After conducting a CMAPP analysis of the memo, answer the following questions:
   a. Who is her primary audience?
   b. What other audience(s) might be suggested by her memo?
   c. What kind of relationships with her audience(s) does Welstromm appear to presume?
   d. How well do you think she has gauged the context here?
   e. What effect do you think her content will have on the context?
   f. What appears to be her primary purpose?
   g. Indicate how you think that purpose reconciles with what you think she should be trying to accomplish.
   h. What secondary purpose or purposes does her memo suggest?

REVISION

A revised version of Welstromm's memo is presented in Figure 12.9. Consider it the version she would have sent had she waited until anger no longer clouded her normally strong appreciation of good CMAPP usage.

12b Email: Personal Communication at RAI

SITUATION

Some time ago, the Board of Directors of Radisson Automobiles Inc. had approved the creation of a Radisson intranet, connecting all Radisson dealerships and including an email gateway to the Internet. Thus, Radisson employees could now easily send messages to each other, and could take advantage of standard email as well. Salespeople began to give clients and potential customers their Radisson email addresses, and found that this additional communication option increased sales.

For the last 11 years, Walter Stephenson has been selling cars for Radisson Automobiles in Chicago. Several times, he has won an award as salesperson of the month. He is popular with his colleagues and with his customers. His "repeat" sales are consistently among the highest in all Radisson's dealerships.

When Victor Liepert, the Chicago dealership Vice President, received Griffin Radisson's curt memo announcing the January 2016 reorganization (see Case Study 2b in Chapter 2), he decided to share it immediately with his staff. Many felt offended, particularly the people who had been with the firm for a long time. Not surprisingly, Walter Stephenson was one of those people offended.

Interdepartmental Memorandum
Office of the Chair—Civil Engineering

Ann Arbor University
Seeking Excellence
In the Great Lakes State

Date: August 15, 2015
To: Roger Concorde, Registrar
From: Joan Welstromm, Chair
cc. John Vanetti, Hans Scart
Subject: Kaz Lowchuk

Welstromm

The above-named student was recently advised by the Admissions Office that his final grades in AD223 and DP200 were 38 percent and 48 percent, respectively, and that if he disagreed with them, his only recourse would be to appeal. Since the appeal process is time-consuming, he would likely be precluded from imminent registration for his third year of Engineering.

I have managed to contact Professors Vanetti (for AD223) and Scart (for DP200). Having checked their records, they maintain that they submitted grades for Kaz Lowchuk of 83 percent and 84 percent, respectively. It would appear the numbers were inadvertently transposed during the data-entry process. Professors Vanetti and Scart will shortly be faxing their relevant grade sheets to your office.

Please consider this memo my official request, on behalf of Professors Vanetti and Scart, to correct Kaz Lowchuk's grades, to do so without delay so that he may complete his registration before the deadline of August 27, 2015, and to send me a copy of the grade correction form by return campus mail.

If you have any questions, please contact me at local 4887 or by internal email at welstrommj@aau.edu.

Thank you for your prompt assistance.

FIGURE 12.9: Welstromm's Revised Memo

The more Walter thought about it, the angrier he got. As the reorganization deadline approached, he found himself really incensed: This was, he felt, a typically high-handed power grab by people in the head office who hadn't the faintest idea of what things were like in local dealerships and had probably never sold a car in their lives. (Surprisingly, Walter was unaware of Griffin Radisson's experience in every sector of the firm.)

Checking his email at work one late December evening, and still angry about the changes coming within a couple of weeks, Walter sent the message that appears in Figure 12.10.

On New Year's Day 2016, Stephenson was at home and logged on to the Radisson intranet to see if any customers had sent him messages. To his astonishment, he found a message from Celine Roberts, the Company Manager at Head Office in Dallas. This message is in Figure 12.11.

ISSUES TO THINK ABOUT

1. In terms of ethics, how do you react to Stephenson's message?
2. How well did Stephenson consider the "Eleven Keys to Good Email Hygiene" presented in this chapter?
3. In the same terms, how do you feel about Roberts' reply?
4. How would you describe Stephenson's use of email as a product? In particular, how clear is his message, and how many people might make up his intended audience?
5. What do you think his primary purpose was?

FIGURE 12.10: Walter Stephenson's Email

Date:	December 21, 2015, 20:25:22 CST
To:	dwang@infra.com, fharmnn@nosuch.com; harmon2@nosuch.com
From:	"Walter Stephenson" <wstephenson.radiss.reg@cars.com>
cc.	listserv@auto.int.new.com, listserv@beanies.com
NEWSGROUPS:	alt.business.harangues.personal; alt.complaints.management.people
Subject:	Dilbert's cohorts strike again!

MESSAGE

Thought you might like to know what's still happening in the world of big business . . . However, KEEP THIS CONFIDENTIAL!!! cause I don't want to get kicked in the you-know-what yet again by the bosses. ;-)

Heard of Radisson Autos, right? I've been there almost a dozen years now and thought I'd seen everything. Guess not. Seems that that cartoon character Dilbert, the one that shows up all the idiocy in business, has been walking around Radisson at Head Office in Dallas. . . . aren't those Texas hot shots all alike? . . . just told us that they're reorganizing the whole lousy company. Our boss here at least had the guts and the decency to tell us all. HEAD OFFICE DIDN'T EVEN BOTHER! ☹

Is this going to help sales? NOPE! Is this going to help morale (which was OK anyway)? NOT A CHANCE! Is this going to get the unions in a snit? WHAT DO YOU THINK? Are we going to take it lying down? Probably. GOOD JOBS ARE HARD TO FIND AND THEY KNOW IT, TOO!

Yeah, I'm mad. At least writing this gets it off my chest. Now I guess I have to hope none of the management types subscribe to the lists or newsgroups I do. :-) Unlikely. They're probably too busy trying to find new ways to screw us working slobs.

Have a nice day from "Supercar"!

W. Stephenson
Radisson Automobiles Inc.
Chicago, IL
wstephenson.radiss.reg@cars.com

FIGURE 12.11: Celine Roberts' Email to Walter Stephenson

Return-path:	<radiss@cars.com>
Date:	December 29, 2015, 20:25:22 CST
Sender:	owner-<radiss.cars.com>
From:	"Celine Roberts" <roberts.radiss.ho@cars.com>
To:	"Walter Stephenson" <wstephenson.radiss.reg@cars.com>
cc:	"Victor Liepert" <vliepert.radiss.reg@cars.com> "Tina Trann" <ttrann.radiss.reg@cars.com> "Griffin Radisson" <gradisson.radiss.ho@cars.com>
SUBJECT:	Your misuse of company intranet email and consequences of your action

---

On December 21, 2015, wstephenson.radiss.reg@cars.com wrote:

Thought you might like to know what's still happening in the world of big business . . . However, KEEP THIS CONFIDENTIAL!!! cause I don't want to get kicked in the you-know-what yet again by the bosses. ;-)

Heard of Radisson Autos, right? I've been there almost a dozen years now and thought I'd seen everything. Guess not. Seems that that cartoon character Dilbert, the one shows up all the idiocy in business, has been walking around Radisson at Head Office in Dallas. . . . aren't those Texas hot shots all alike? . . . just told us that they're reorganizing the whole lousy company. Our boss here at least had the guts and the decency to tell us all. HEAD OFFICE DIDN'T EVEN BOTHER! ☹

Is this going to help sales? NOPE! Is this going to help morale (which was OK anyway)? NOT A CHANCE! Is this going to get the unions in a snit? WHAT DO YOU THINK? Are we going to take it lying down? Probably. GOOD JOBS ARE HARD TO FIND AND THEY KNOW IT, TOO!

Yeah, I'm mad. At least writing this gets it off my chest. Now I guess I have to hope none of the management types subscribe to the lists or newsgroups I do. :-) Unlikely. They're probably too busy trying to find new ways to screw us working slobs.

Have a nice day from "Supercar"!

MESSAGE

You are no doubt aware that

- the Radisson intranet is company property;
- all email originating through that intranet must be work-related;
- all such email *is and remains the property* of Radisson Automobiles Inc.

You may *not*, however, recall that Head Office makes spot checks on all intranet internal, incoming, and outgoing email, so as to ensure its efficient and economical use, and to verify that the hardware and software are functioning as expected.

As it happens, your message of Dec. 21 was one of those chosen for scrutiny. In the opinion of the Board of Directors, you have

- misused Radisson property for your own benefit;
- brought shame on the Radisson good name;
- without foundation, accused the Company of improper, unethical, and perhaps illegal activities;
- disparaged blameless individuals within the Radisson family.

I am now in communication with your Vice President, Victor Liepert, and with Tina Trann, your General Manager. Please report to Ms. Trann at 9:00 a.m. on Monday Jan 4/16. She will inform you of the action we have decided to take.

Note that I am sending copies of this message to Mr. Liepert, Ms. Trann, and Mr. Griffin Radisson, President of the Board of Directors of Radisson Automobiles Inc.

Celine Roberts
Company Manager
Radisson Automobiles Inc.
radiss.ho@cars.com

6. Do you think he had more than the one purpose in sending the message?
7. What kind of response might he have been expecting from his audience?
8. How does his message conform to general rules of netiquette?
9. Explain why you think his use of company email to criticize the company was or was not ethical.
10. Explain what you think Roberts' primary purpose is.
11. Discuss any secondary purpose you think she might have.

12. What changes in the context do you think these messages will have generated? (Hint: Think of various relationships within the company.)
13. How do you think the messages will affect the context of the communication to take place in person in Chicago a few days later?
14. What do you think of the ethics of RAI's monitoring of email?

12c Student Association Cooperation: Fax from AAU to GTI

SITUATION

In November 2015, the Ann Arbor University Student Association held a Thanksgiving party, at which two popular bands, Hodgepodge and Really Me, appeared. On behalf of the association, Jack Lee, the Functions Coordinator, had signed the contracts with the bands.

At the end of the evening, Hodgepodge's leader, Tommy Eldridge, had complained that the sound system had broken down twice during their performance, and that as a consequence, they had not finished their arranged sets until over an hour later than scheduled. He demanded that the association pay them an extra $500 to cover their time.

Jack, while recognizing that Hodgepodge had, in fact, stayed later than intended, had countered that their playing had not met the contract for two reasons:

- Their well-known lead singer, Buzzy, had been absent, replaced by a relative unknown, whose caliber, according to Jack, was below par.
- Despite repeated requests from the audience, Hodgepodge had not performed their "signature" number, "Night Is Nowhere."

After what had become a heated argument, Hodgepodge had finally packed up and left. A week later, however, Dorothy Palliser, the Student Association President, had received an angry letter from Eldridge, claiming that he was going to sue, and that the association would be blacklisted and would never be able to hire another band of any quality.

A month after that—and with no further word from Hodgepodge—Dorothy was talking on the phone with an old friend, Susan Trebanian, a student in the Electronics program at Grandstone Technical Institute in Randolph, Vermont. Susan happened to mention that the GTI Student Union was planning a big "pre-Christmas bash," and that her boyfriend was delighted that one of the bands performing would be Hodgepodge.

Taken aback, Dorothy told Susan about her own experience with Hodgepodge, and they decided that Dorothy should warn the GTI Student Union that they might have trouble. The following day, after obtaining the name and number of GTI's Student Union president from Susan, Dorothy sent the fax that appears in Figure 12.12. (Note that she used photocopied fax cover pages on which she hand-wrote her message.) Searching through the AAUSA records, she found a copy of the contract with Hodgepodge and Jack's handwritten notes describing his argument with Eldridge. To support her warning to GTISU, Dorothy faxed these sheets as well.

ISSUES TO THINK ABOUT

1. What is your reaction to the "mixed" product: a hand-written message on a printed fax cover sheet? Why might AAUSA be using it? Do you think it has any particular advantages or disadvantages?
2. Today, faxes are used principally by businesses and other organizations to (a) verify legal contracts or (b) convey confidential information (when individuals don't want, for example, an electronic "paper trail"). Is Dorothy's use of a fax appropriate?

Fax Transmission

Dominion 243	**AAU Student Association Executive**	**Local 4334**

Phone: (734) 555-4334 Fax: (734) 555-1421
Email: aausaex@wnet.com

Date: Dec 14/15
To: Vlad Ridescu
President, GTI Student Union
Fax: (802) 555-0505
From: Dorothy Palliser
President, AU Student Association
Subject: Warning re band Hodgepodge
Number of pages in total: 7

You don't know me, but I'm the President of the AAU Student Association. A friend of mine, Susan Trebanian, just told me you're planning to hire Hodgepodge to play at GTI, and I wanted to warn you about doing that.

We hired them a while ago, and it really went sideways. They didn't really give us what they were supposed to, and they wanted more money, and now they're threatening to sue us.

I'm sending some things I think will help you know what you're probably up against, including a copy of our contract with them, and some notes I got from our Functions Coordinator, Jack Lee, who dealt with them personally.

I think it's really important for student associations to talk to each other and help each other. So, when you've looked through this stuff, you can get in touch with me if you want more details.

Good luck!

Attached: 6 pages

FIGURE 12.12: Fax from AAUSA to GTISU

3. Could Dorothy have accomplished her purpose by sending an email and scans of the documentation instead of a fax? Which communications product do you think the recipient, Vlad, would have preferred?
4. Do you notice anything missing in the fax's "header" section (the printed segment that precedes the message itself)? How might that omission affect the message? What impact might it have on context?
5. How would you describe the way Dorothy appears to see the context here? To what extent do you think she undertook a CMAPP analysis? What might she have done to get more information before proceeding?
6. If you were Ridescu, how do you think you would react? What impact would your actions have on the ongoing context of this scenario?
7. To what extent does Dorothy's fax conform to the conventions discussed in this chapter?
8. What strategy has Dorothy used in her message? Do you think she used it effectively? Why?
9. In terms of communications ethics, do you think she should have sent a copy to anyone else? If so, to whom and why?
10. What is your reaction to Dorothy's sending Vlad copies of the AAUSA-Hodgepodge contract and Jack's notes? (Hint: Think about ethics and, perhaps, about legalities.)

REVISION

If Dorothy had taken more time for a CMAPP analysis, her fax—if indeed she had still chosen that product rather than an email and scans—might have resembled the one in Figure 12.13. Note that

- it has been word-processed, making it easier to read and more professional in appearance;
- information is more precise and accessible;
- the approach is more even-handed;
- the language is less connotative, more denotative; and
- she does not attach copies of the contract or of Jack's notes.

Fax Transmission

Dominion 243 **AAU Student Association Executive** **Local 4334**

Phone: (734) 555-4334 Fax: (734) 555-1421
Email: aausaex@wnet.com

Date: December 14, 2015
To: Vlad Ridescu President, GTI Student Union
Fax: (802) 555-0505
From: Dorothy Palliser President, AAU Student Association
Subject: Hiring of band Hodgepodge
Number of pages in total: 1

Susan Trebanian, a GTI Electronics Program student and a personal friend, recently mentioned to me that your Student Union is looking to hire Hodgepodge to play at your "Pre-Christmas Bash." I felt that I should apprise you of the AAU Student Association's dealings with them.

Last October, our Functions Coordinator, Jack Lee, contracted with both Really Me and Hodgepodge to play at our Thanksgiving dance in November. Despite the fact that our sound system broke down twice and the event thus concluded about 40 minutes late, Really Me's performance was highly professional, and we would recommend them without reservation.

Our experience with Hodgepodge, however, was different. When its leader, Tommy Eldridge, demanded an additional $500 because of finishing late, Jack Lee pointed out that

- Buzzy, their lead singer, and one of the main reasons we had hired Hodgepodge, had been replaced by someone called Harley Monk, a new member of the band;
- Monk's performance had led to several audience complaints during and after the show; and
- despite repeated requests, Hodgepodge had refused to play "Night Is Nowhere," the signature song we had been expecting.

After a few minutes' unresolved argument, Hodgepodge packed up and left. A week later, I received a letter from Eldridge informing me that he intended to sue the Student Association for breach of contract, and that he would attempt to blacklist us to prevent our being able to hire well-known bands in the future. I did notify our lawyer; but, to date, we have heard nothing about a lawsuit, and have had no difficulty dealing with other musicians.

Our experience, therefore, leads me to suggest you might wish to exercise caution in your dealings with Hodgepodge. If you wish to discuss this further, please don't hesitate to contact me.

FIGURE 12.13: Revised Fax from AAUSA to GTISU

EXERCISES

12.1 Give brief definitions of the following components of business letters:

- inside address
- salutation
- complimentary close
- signature block

12.2 Name two common business letter formats and give the defining characteristics of each.

12.3 Describe the principal differences between letters and memos; refer both to typical usage and to format conventions.

12.4 Review Case Study 10a, including Welstromm's original memo, and imagine you are Roger Concorde.

- Construct a brief CMAPP analysis for a reply to Welstromm.
- Create the product to reflect that analysis.

12.5 Briefly describe the principal conventions ascribed to:

- a fax
- an email message.

12.6 Locate the person responsible for setting house style within the company you work for, or, if you are not working, within your educational institution, and obtain the guidelines for

- letter format and conventions
- memo format and conventions
- fax cover pages.

Assess the material in terms of

- consistency of image
- KISS
- accessibility
- potential audience reaction.

12.7 If you work for yourself and have already set out your own house style, assess it as above. If you are not self-employed or have not yet developed a house style, set out the criteria you would choose, again in terms of the above issues.

12.8 Find out your organization's/institution's policy with regard to employee email privacy. Defend your opinion of its ethics.

12.9 Find an educational institution that provides email accounts for both faculty and students, and discuss whether you think the ethical issue of email privacy differs from one group to the other? For example, could one consider faculty "employees" but students "clients"?

USEFUL WEBSITES

Note: For ease of reading, we have removed the "http://www." from most of the URLs below. However, some browsers might still require that you insert it to precede some URLs so as not to receive an error message.

- "Netiquette Home Page" provides links to just about everything you might want to know about this topic.
Visit: albion.com/netiquette/index.html

- Dave Barry's *New York Times* review of Shipley and Schwalbe's book on email, *You've Got Trouble*, presents an amusing take on new communications technologies. "Imagine, for example, how useful [email] would have been for Paul Revere. Instead of having to climb onto a horse in the middle of the night and ride through Massachusetts spreading the alarm, he could have simply whipped out his BlackBerry, fired off a quick message to the patriots in Lexington and Concord, then gone to sleep (unless he also had TiVo)."
Visit: http://www.nytimes.com/2007/05/06/books/review/Barry.t.html

- George Mason University's Virginia Montecino has created an organized set of Web links on a page called "Computer Ethics, Laws, Privacy Issues." It functions as a highly effective portal to information concerning how our society looks at our own dealings with technology.
Visit: mason.gmu.edu/~montecin/eprivacy.htm

- Public Broadcasting Services' "Do You Speak American?" pages offer a wealth of interesting and, often, unusual information on how electronic communication has been changing American English.
Visit: pbs.org/speak/words/sezwho/wiredwords

- Rensselaer Polytechnic Institute's Writing Center features an excellent discussion of memos.
Visit: www.ccp.rpi.edu/resources/memos/

- The U.S. Postal Service offers an entertaining history of its mail delivery.
Visit: about.usps.com/publications/pub100/welcome.htm

- The resource website "Writing and Speaking Guidelines for Engineering and Science" offers a wealth of information and advice on creating a wide variety of documents, including letters and memos.
Visit: writing.engr.psu.edu

CHAPTER 13

Effective Presentations

Inevitably, you will have to speak in public. In your personal life, you might be asked to propose a toast at a wedding, act as MC at a social function, or deliver a eulogy at a funeral. As an engineering student at college or university, and, later, as a practicing engineer, you will at times be required to speak to classmates, colleagues, clients, public authorities in charge of licenses and building codes, and strangers. In college or university and in the workplace, you might be required to present reports or projects, or demonstrate your company's product or service—or your own.

The first group of activities listed above are examples of *speeches*. Speeches are designed to amuse, console, persuade, or entertain. Their function is primarily social rather than pragmatic, and though they might be constrained by the rules of etiquette, their approach is fundamentally casual. The second group of activities are examples of *presentations*. To develop and deliver them effectively, you will be concerned with the characteristics of technical communication introduced in Chapter 1: necessity for a specific audience, integration of visual elements, ease of selective access, timeliness, and structure. Further, your effectiveness as a presenter may well be enhanced by the judicious application of the CMAPP approach. Studies suggest that, at least in part, your success as a professional engineer will be linked to your effectiveness as a presenter.

13–1 The CMAPP Approach to Presentations

You can look at your presentation in terms of its CMAPP components, which might include the following.

13–1a CONTEXT

1. The reasons you have been asked to be a presenter
2. The physical setting of your presentation, including
   a. the size of the room and the distance between you and those members of your audience seated farthest from you
   b. the size of your audience
   c. the necessity for a microphone or sound system
   d. the length of time you have been allotted
   e. the technology available for integrating visuals
3. The time of day, including whether your audience will have just finished a meal, for example
4. The expectations your audience has of you, and vice versa
5. The relationships that exist among you, your audience, and anyone else involved (instructor, classmates, boss, colleagues, and so forth)
6. The impact you think your presentation will have

13–1b MESSAGE

1. Your words (recall the effect of both denotation and connotation)
2. Your visuals (recall the impact of both information and visual impression)
3. Your delivery (think about the paralinguistic/nonverbal considerations discussed later in this chapter)

13–1c AUDIENCE

1. The specific identification of your primary audience (note that in CMAPP terms, the *primary audience* for a presentation refers specifically to those people actually in attendance.)
2. The type of audience you are addressing, including
   a. their level of technicality
   b. their likely bias toward you (whether positive or negative) before you even begin

3. The expectations of your audience
4. Your understanding of
   a. what they likely already know
   b. what they likely need to know
   c. what they likely want to know
5. The specific identification of any secondary audience(s) (Note that in CMAPP terms, a *secondary audience* will not have been in attendance. However, in developing and delivering your presentation, you will have taken any secondary audience into serious consideration. An example might be your own boss, even if he or she was not at your presentation.)

13-1d PURPOSE

1. What you are trying to achieve by delivering your presentation
2. How you want your audience to react to your presentation

13-1e PRODUCT

Which of the four types of presentations (discussed later in this chapter) you will use:

- manuscript
- memorized
- impromptu
- extemporaneous

13-2 Audience Analysis

As part of your CMAPP analysis, you should consider the following seven audience analysis factors.

13-2a AGE RANGE

A group of "twenty-somethings" will respond differently than a group of seniors; their interests will be different, as will the cultural referents to which they will likely relate.

13-2b CULTURAL BACKGROUND

Different cultures respond differently to particular stimuli, references, topics, types of humor, presentation approaches, and so forth. When developing your presentation, you should take into account the cultural background of your audience. Recall Chapter 1's discussion of cultural impact.

13-2c LEVEL OF EXPERTISE

Whatever your presentation's topic, the reactions and expectations of a highly knowledgeable audience will likely be different from those of a less expert group. Your levels of discourse and technicality should vary accordingly. Think, for example, of a presentation about developments in neutrino detectors. What you include and how you explain it would vary considerably if your audience were professional engineers at a scientific colloquium or a citizens' group concerned about the proliferation of nuclear technology.

13-2d OCCUPATION

An audience composed of professional electronics engineers will be different from an audience composed of lawyers. You should be sensitive to the interests and expectations of whichever audience you address.

13-2e POLITICAL AND RELIGIOUS AFFILIATION

There are many hot-button issues associated with politics and religion. To be effective, your CMAPP analysis should take into account the potential views and affiliations of your audience. This does not mean that you should automatically avoid controversial topics; however, there is almost never an advantage to insulting or offending your audience. Remember, a primary purpose of all professional communications is to foster and maintain relationships: Being on friendly terms with potential adversaries can lead to mutually accommodating and profitable initiatives down the road for both you and for those who don't share many of your views.

13-2f GENDER

Chapter 1 refers to gender in terms of how vocabulary has changed to reflect new cultural expectations and norms. With reference to audience analysis, the issue is a contentious one that many people feel strongly about. You might have heard the quip, "It's always dangerous to make a general statement." Perhaps so; nonetheless, while developing your presentation, you shouldn't ignore some observed tendencies with regard to large groups—not, we stress, regarding *individuals*. For example, a predominantly female audience may tend to have somewhat different interests, preferences, and expectations than a predominantly male one. And the larger the audience involved, the more likely it is that such variations will exist.

13-2g SOCIOECONOMIC STATUS

People from different income groups tend to have different priorities, interests, and expectations. For example, as groups they are likely to respond quite differently to such issues as taxation, medical insurance, and subsidized housing.

13-3 Purpose

You must have a solid idea of what you wish your presentation to accomplish, of how you want your audience to react, and of what you want from it. For instance, imagine that you are to deliver a presentation regarding proposed changes to the code of ethics of your professional engineering association. Depending on the other CMAPP aspects, you could choose as your primary purpose one of the four described next.

13-3a DESCRIPTIVE PURPOSE

If your primary purpose is to describe, you will want members of your audience to see (in their mind's eye) or to feel something they have not seen or felt before, or to see something more clearly or feel it more strongly than before. You might, for example, talk about personal reactions to a former client's accusations of unethical behavior.

13-3b INFORMATIVE PURPOSE

If your main purpose is to inform, you will want your audience to know or understand something it did not before. You might discuss the differences between the existing code and the proposed one, and the implications to the association of making these changes.

13-3c INSTRUCTIVE PURPOSE

If you want primarily to instruct, you will want your audience to know how to do something that it could not do before. You might, therefore, want your audience to know how to interpret the changes, or how to obtain an official interpretation of them.

13-3d PERSUASIVE PURPOSE

To persuade, you must effect a change in your audience, normally in terms of a belief, attitude, or behavior. Perhaps you will want to persuade an undecided audience to vote in favor of the proposed changes to the code of ethics; or maybe you will need to convince some members of your association to take a more active role in developing the specifics of the proposed changes.

13-3e OVERLAP

It is impossible to describe without informing, to instruct without describing, and so forth. Inevitably, your presentation will incorporate elements of more than one purpose. Nonetheless, in order for your message to be effective, you must choose a single purpose as the foundation of your presentation. Otherwise, you and your audience will have difficulty relating to each other, and it will have trouble accepting you as a credible speaker.

13-4 Conquering Stage Fright

The technical term for fear of speaking in public is "glossophobia." According to glossophobia.com, at least 75 percent of the population suffer from it; and, "comedian Jerry Seinfeld used to joke that at a funeral, most people would rather be lying in the casket than delivering the eulogy." Incidentally, there are no documented cases of anyone actually dying of stage fright.

Dealing with that fear is important. Your presentation will not be effective if you don't appear calm and confident. Note our use of the word "appear." In the context of a presentation, what counts is not how you actually feel, but rather how you *appear* to feel. Your heart may be thudding, but your audience can't hear it; nor can it see the perspiration on your sweaty palms or feel the dryness in your throat. As far as your audience is concerned, if you *appear* to be in control of yourself and your presentation, you *are* in control.

There's some truth to the old saw, "Fake it till you make it!" Each time you "fool" an audience by appearing confident, you fool yourself a little too. Do this often enough and you will find that you are actually becoming more confident. As another cliché stipulates, nothing succeeds like success.

Remember, of course, that we are not really countenancing deceit, which is unethical by definition. Recall as well that while the "fake it till you make it" dictum might apply to coping with stage fright, you should not really apply it to other aspects of your career as a professional engineer.

Here are some tips for dealing with stage fright.

13-4a AVOID STIMULANTS

Someone might have told you that the best way to conquer stage fright is to have a stiff drink. Don't. It might make you feel more confident for a moment or two, but it won't make you or your audience feel better when your performance suffers as a result. The same advice applies to other chemical crutches, from pot to forms of "speed," and even to high doses of caffeine (whether in coffee or soft drinks) or taurine (found in many of the so-called energy drinks).

13-4b PRACTICE BREATHING AND VISUALIZATION TECHNIQUES

Get into the habit of taking several slow, deep breaths just before you walk up to deliver your presentation. Long before you reach that point, you can use a technique known variously as visualization, self-visualization, positive visualization, or positive imaging. Fundamentally, it involves imagining yourself—as vividly as you can—delivering a successful presentation and reaping the rewards of having done so. Taken seriously and practiced assiduously, the technique seems to work for most people.

13-4c IMAGINE THE WORST-CASE SCENARIO

From time to time, even the best, most experienced presenter flops badly. So accept the fact that, despite your best efforts, your presentation will occasionally be a washout. Then, consider what is the worst that could possibly happen. The answer could be anything from your feeling quite embarrassed to losing a coveted contract. And then what? Those of us who have had the experience have somehow managed to reassemble the shattered pieces of our personalities and continue to live productive and fairly contented lives. You will, too. So imagine the worst possible result, think of how you will deal with it, and—almost always—then watch it *not* happen.

13-4d BE PREPARED

The Scouts seem to be right: For most things in life, proper preparation pays off. Thus, you're investing a great deal now to prepare yourself to be a professional engineer. But, if you're lazy about developing your presentation, your lack of effort will be evident and your audience will be alienated. Take or make the time to prepare thoroughly: Conduct a CMAPP analysis and apply its results long before you stand before your audience.

13-4e MAINTAIN A POSITIVE ATTITUDE

If you believe that you're going to do poorly, that your audience will dislike you and your presentation, that the experience will be frightening and uncomfortable, and that you have no chance of success, then all those things may indeed come to pass. On the other hand, if you can convince yourself that you're going to do well, that your audience wants you to succeed (and that is usually the case), that the experience will be rewarding for both you and your audience, and that success is within your grasp, these convictions will likely come true. Remember that your attitude toward a task has a lot to do with how well you accomplish that task. Ask any engineer.

13–4f YOU *SHOULD* FEEL A LITTLE DIFFERENT WHEN YOU ARE SPEAKING IN PUBLIC

It is, for most of us, a relatively rare experience for five, 10, or 150 fellow professionals to show up in a single room just because *we* are there. That these professionals are in that room because they want to hear what we are going to say *honors us*. Good speakers acknowledge that they are being honored by the presence of their audience. They know that this is a special occasion and there should be a feeling of excitement in their stomachs. We tell our students that they should regard this excitement not as "nervousness" but as "an appropriate sense of anticipation."

13–5 The Development Process

In 1998, Martha Sloan, then Chair of the American Association of Engineering Societies, expressed the opinion that "The essence of engineering is design and making things happen for the benefit of humanity." Over many centuries, of course, engineers developed myriad ways to accomplish this goal. Perhaps analogously, and also over centuries, people have constructed various processes for developing presentations.

Deciding which process to use is less important than actually using one. However, the process outlined next is both simple and effective. Note that you will often work on the first four steps simultaneously, since they tend to be interdependent. The other four, a direct reflection of the multilevel outlining process you learned in Chapter 5, do tend to be sequential.

13–5a STEPS

1. Conduct your CMAPP analysis.
2. Decide on the specific topic you wish to cover. If you are not able to express your topic in a short and clear phrase, you have not got it firmly fixed in your own mind.
3. Formulate a *thesis statement*. The thesis statement is a single sentence that identifies your primary audience, your primary purpose, and your topic. It will serve as the foundation for step 6.
4. Conduct your research.
5. Determine your goals or objectives—the main points of your presentation. These will turn into the level-one heads you create in step 6.
6. Construct a multilevel outline. It should contain sufficient detail (e.g., specific references to any quotations and visuals you will be using) to permit you to develop a new set of speaking notes (see step 7) months later, should you need to do so. If you are preparing for a manuscript or memorized presentation (discussed later in this chapter), your outline will serve as the skeleton for your text. For an extemporaneous presentation (also discussed later in the chapter), you will "condense" your outline into the speaking notes described next, in step 7.
7. Develop your speaking notes (also called speaker's notes) for your extemporaneous delivery. Speaking notes—not your outline—are what you actually use while delivering your presentation. They are designed to remind you of the issues you will discuss,

issues with which you have already made yourself very familiar. Unlike your outline, speaking notes are brief and concise, often to the point of being cryptic. If you were to look at them several months after your presentation, you might well fail to understand them.

8. Because you want your audience's attention focused on you rather than on your notes, you will want the notes themselves, as well as your use of them, to be as inconspicuous as possible. Here are some suggestions:

  a. Write legibly and in large, clear letters: You want to be able to glance quickly at your notes and find your next point without difficulty.

  b. Use single words or very short phrases; clauses or sentences take longer to "find" when you are trying to glance at your notes without drawing your audience's attention to the fact. You might make use of symbols—for example, a $ to remind you to discuss financial issues, or a symbol character such as 🔒 to prompt you to mention security concerns.

  c. Use white space generously: You will be able to find your points more readily.

  d. On letter-size paper, write on one side only. This will allow you, during your presentation, to move from page to page without making your audience aware that you are doing so.

  e. Many presenters put speaking notes on small cards (3" × 5" index cards cut in half, for example) and hold them in the palm of their hand while delivering. If you decide to use cards, you should allow for plenty of white space (put only three or four items on each card), print on one side only, and number your cards so that they can be quickly reordered if dropped.

9. Rehearse your presentation. We know of no better way to find out whether your presentation is likely to work. Note that rehearsing does not mean going over your presentation in your mind. It means finding some kind of audience and delivering your presentation as though "for real." Among other things, rehearsal allows you to test your arguments and your visuals and to verify your time and timing. While rehearsing can feel a bit awkward, most presenters would tell you that the private discomfort of rehearsal is preferable to the public embarrassment of a flawed presentation. (Author Dave Ingre once rehearsed a presentation in front of his wife, a family friend, and his dog—so he could identify its weaknesses before delivering the talk to a paying audience.)

13-6 Types of Delivery

Broadly speaking, there are four types of presentation delivery. Each has advantages and disadvantages. Which you choose will depend on the results of your CMAPP analysis.

13-6a MANUSCRIPT PRESENTATION

A manuscript presentation is the delivery of a carefully prepared text that you refer to while presenting. The main advantage of this delivery type is that your presentation can be meticulously crafted: You can take pains to ensure that you always have the right word and/or visual in the right place at the right time. As well, if you have to deliver the presentation again, you can do so accurately and with a minimum of effort. One more advantage is that in the event of an interruption (recall *external interference* discussed in Chapter 2), you can easily find your place and continue.

The main drawback of manuscript presentations has to do with the differences between written language and spontaneous, spoken language. It's enormously difficult to write a presentation that, upon delivery, does not sound recited—and thus potentially boring to your audience. (Through a combination of training and experience, a good TV news anchor, for example, has learned to surmount this problem. Almost no engineers are also news anchors, alas.)

Another disadvantage of manuscript presentations is that it is more difficult to engage and interact with an audience while referring to a written text. A presenter who rarely strays from the written text, or who seldom looks up from the lectern, runs the risk of becoming monotonous.

13-6b MEMORIZED PRESENTATION

A memorized presentation is the delivery of composed material that you have memorized. Like manuscript presentation, this delivery type allows you to craft your presentation carefully. As well, it allows for greater interaction with an audience because it removes the barrier created by the manuscript itself. On the negative side, it's no easy task for most people to memorize more than a small amount of text. For those who can, it is a challenge to make memorized text sound spontaneous and natural rather than rehearsed and recited. And "picking up where you left off" after an interruption can be difficult and awkward. Finally, repeat presentations can exacerbate the problem: Like average actors in a long-running play, presenters may find their delivery becoming increasingly stale.

13-6c EXTEMPORANEOUS PRESENTATION

The extemporaneous presentation is the most widely used of the four delivery types. It involves diligent preparation that makes possible what then appears to your audience to be a spontaneous presentation.

The extemporaneous presentation allows you to prepare carefully, tailoring your presentation (including any visuals) to your context, audience, message, and purpose, and to adjust your presentation accordingly "on the go." Moreover, since you will be making use only of speaking notes, you will be able to talk spontaneously and interact extensively with your audience. On the negative side, inadequate preparation will be painfully obvious to both you and your audience. As well, should you lose your focus during the presentation, you will have only the cryptic content of your speaking notes for support.

When preparing your extemporaneous presentation, you should follow the development process outlined earlier in the chapter. Despite what you might have learned elsewhere, we must stress that you should never write out, word for word, any part of your intended presentation. Doing so will almost inevitably make your presentation as a whole sound uneven and make parts of it sound recited.

13-6d IMPROMPTU PRESENTATION

The least formal of the delivery types is the impromptu presentation. You will use it when you are (more or less) unexpectedly called upon to speak, for example. In class or in the workplace, you may be asked for your opinion or your analysis. Although you might have guessed that you would be called upon, and although your audience will expect you to have some familiarity with the topic, you will not have had the opportunity to thoroughly prepare what you are going to say or how you are going to say it.

An obvious advantage of the impromptu presentation is that, because you are given little, if any, time to prepare, your audience is disposed to cut you some slack; their expectations are comparatively low. Further, since you are likely to have no notes, your presentation will appear to be what it is—spontaneous. In addition, you are free to interact extensively with your audience.

The main disadvantage is the flip side of the first advantage: Because you have negligible time to prepare, your success is much more dependent on your ability to think on your feet. The more complex your arguments, the greater your risk—and your audience's—of becoming confused. Remember the KISS principle, and always try to apply the three-part rule for perfect presentations given in the next section.

Interestingly, although the impromptu presentation seems, and often is, the most spontaneous of presentation types, it requires the most content expertise on the part of the speaker. A neophyte in the field of landmark construction specifications, for example, can deliver a manuscript or memorized presentation on that topic with great grace, at least until the question-and-answer period. With enough research and presentation, our neophyte could probably get away with a nifty extemporaneous presentation as well. An impromptu talk on this topic would be impossible, though, for anybody without considerable education and experience in building codes, materials, and design, for example.

13–7 Elements of the Presentation

For your presentation, you will use the same tripartite structure found in all technical communications: an introductory segment, a body, and a concluding segment. One representation of this configuration is the following three-part rule for perfect presentations:

1. Tell 'em what you're gonna tell 'em.
2. Tell 'em.
3. Tell 'em what you told 'em.

Although this whimsically expressed rule may not appear scientific or empirical—attributes preferred by engineers, of course—it has proved effective for many generations of presenters. (And, if it seems overly simplistic, remember the enormous complexity that you derive from the apparent simplicity of two little digits—0 and 1.) To apply it, of course, you need to amplify each of its elements. Here are some suggestions for doing so.

13–7a INTRODUCTORY SEGMENT

The introductory segment should normally include the following elements, though not necessarily in the order given here.

Attention-Getter

You'll want to start things off by gaining your audience's full attention. To get it, don't explode firecrackers or do or say anything else that is totally unrelated to your presentation—loudly asking, "How y'all doin'?" for example. For the ethics topic mentioned earlier, you might gain your audience's attention by saying something like, "How many of you agree that an organization can change its formal code of ethics without changing its ethical stance in the marketplace?" You might also make a reference to a recent news story surrounding a controversy involving a company or professional organization in your industrial sector.

Self-Introduction

If someone introduces you to your audience, you need not reintroduce yourself—although doing so might be useful in increasing your "recognition factor." When you do introduce yourself, however, you must clearly state your first and last names and, if appropriate, your position or title as well. Remember: Your audience probably doesn't know your name as well as you do, and if it doesn't catch your name immediately, it will at once begin to lose interest in who you are. And then your presentation will likely fail.

Initial Summary

The initial summary serves as a roadmap to your presentation. You prepare your audience for what is to come by stating your purpose and your main points (typically the "goals" or "objectives" that became the level-one heads in your multilevel outline). This element is the most noticeable expression of "tell 'em what you're gonna tell 'em."

Speaker Credibility

Typically, we can identify two kinds of credibility—*a priori* (meaning "coming from before") and *earned*. If you are a recognized expert in the field you're going to discuss, your audience might grant you substantial *a priori* credibility before you even begin. Otherwise—and, often, in addition—you'll have to gain *earned* credibility from them. Normally, you'll try to do this by explaining to your audience why you are the right person to talk about the topic. But don't confuse celebrity with deserved credibility: Obviously, being a world-famous singer or actor should not automatically confer credibility with regard to innovations in mechanical engineering, for example. Most of the time, your credibility as a speaker has to be renewed each time you give a presentation. You might mention your experience in the field, your credentials and/or qualifications, your research on the topic, and so forth.

Audience Relevance

Relying on your CMAPP analysis, indicate why the topic should be of interest to the audience. Audience relevance is not necessarily obvious or straightforward. For example, suppose that you are speaking to a group of urban condominium owners and your topic is the possibility of a new professional football franchise team locating in the downtown core. Talking about the excitement of the game will likely score few points, so to speak, with this group. A better strategy would be to focus on things like construction jobs, tax revenues, parking, and street noise. This part of the introduction can be the most daunting to compose, especially if you have a large and diverse crowd in attendance. A thorough audience analysis requires time and effort—but it is also worth it. Attendees who know that your content will be meaningful and helpful to them personally will overlook some ragged spots that occur subsequently in the body of your talk.

13-7b BODY

The body of your presentation represents the bulk of your CMAPP message. Its content, of course, depends on your context, audience, purpose, and product (your delivery type). Among the techniques you might wish to include are the following.

Rhetorical Questions

Rhetorical questions (ones that do not really expect an answer) can be effective in the introductory segment as well. An example is the sample attention-getter mentioned earlier. Within your presentation body, they can likewise be useful devices to focus your audience's attention on a particular point. An example: "Does anybody here believe you can get by in your field without seeking professional development opportunities at least once a year regarding digital-media applications?"

Signposts

Signposts indicate to your audience where you intend to go, and thus what they can expect. Signposts include words and phrases such as "first," "second," "immediately following," "that is," and "furthermore."

Transitions

Transitions are related to signposts, but indicate a change from one point or idea to another. They include phrases such as "now that we have . . . ," "let us now turn to . . ." and "having looked at this, I will now . . . ."

Emphasis Markers

Showing your audience that you consider your next words important, emphasis markers heighten attention. They include phrases such as "most importantly," "without a doubt. . . ," "I'd like to draw your attention to . . . ," and "please note that . . . ."

Repetition Markers

You can at times refocus your audience's attention on your point by repeating it through paraphrase. When doing so, you can make use of repetition markers such as ". . . and, as I've already mentioned . . . ," and "let me repeat that . . . ."

Segment Summaries

Particularly if your message is complex and/or lengthy, you will help your audience (and thus yourself) by providing brief summaries of individual segments. A segment summary includes a transition into the next segment. For example, at some point in an informative presentation on the proposed changes to your code of ethics, you might say, "So we have looked at how our association came to develop a code of ethics, how changing market forces caused us to consider making changes to parts of that code, and how such changes could benefit both our members and the public we serve. Let's now turn our attention to what our bylaws would require us to do in order to effect those changes."

13-7c CONCLUDING SEGMENT

Your concluding segment should include four elements, incorporated in the order shown below. Note that the terms we use here are jargon; other people using other communication models may well use other words.

Closure

Closure lets your audience know that you are going to finish soon. However fascinating your presentation might have been, your audience's attention will perk up when they realize that you have almost concluded. Your audience might also expect to hear a final reiteration of your main message, knowing intuitively that good speakers follow the "three-part" presentation model. You can show closure by a phrase as simple as "In conclusion, . . ." or "To conclude, . . ." Or you might incorporate an emphasis marker, saying, for example, "I'd like to make one final point before I finish."

Remember, however, that if you show closure and then continue for more than a very few minutes, your audience will very quickly become irritated and lose the interest your presentation might have elicited.

Final Summary

This is point at which you "tell 'em what you told 'em." It is, in effect, a paraphrase of your initial summary, serving to remind your audience of the main points you wish them to remember.

Call to Action

Most engineering communication concludes with a call to action (also known as an action request). If your presentation's prime purpose is persuasive, your call to action is likely very clear: for example, "So, please, when the vote is called, speak up in favor of the code of ethics changes that your executive has proposed."

When your purpose is informative, descriptive, or instructive, your call to action is no longer a "request" of the same type. Often, it resembles a "personal statement" or an indication of your own hopes. You might, for example, say something like "I believe that one of the responsibilities of being a professional engineer and a member of this association is taking part in determining its direction. I believe that our decisions should be educated ones, and that the more we know about our code of ethics, the more informed our decisions will be. I hope my presentation has helped you come to the same conclusion."

Close

Occasionally, your last few words—perhaps a quip, perhaps the call to action itself—will serve to let your audience know that you have finished. Most of the time, however, it is a far better idea to let them know—quite explicitly—that your presentation is, in fact, over. You might do so by saying, "Thank you for your time today; I've appreciated the opportunity to speak with you." If you think it sufficient, you might simply say, "Thank you," nod, and smile; your audience will understand.

Make sure that your voice doesn't disappear during your close. Your audience should hear your final words as clearly as they should have heard your attention-getter. And do not disrespect your audience by turning away from them while making your close.

13–8 Paralinguistic Features

Many experts feel that at least 60 percent of what your audience assimilates and uses to assess you and your presentation does not rely on the words you use. Rather, your audience's judgment is based on its perceptions of how you speak and of what you do while you are speaking. Some authorities use the term "nonverbal features." For the CMAPP model, we call these elements paralinguistic (alongside or beyond language) features. And, although your audience might not be consciously aware of them, as an effective presenter, you have to be. Here are brief descriptions of some of the more important ones, in alphabetical order.

13–8a CONFIDENCE AND PRESENCE

Successful people tend to "appear" successful, confident people look confident, and so on. What they seem to have is what we often call "presence." (Perhaps it's synonymous with charisma.) Presence is an attribute that is difficult to define, although most of us recognize it quickly in others. As we pointed out earlier with regard to conquering fear, if you look confident, your audience will likely believe that you are confident; and, over time, you will become so. Self-confidence tends to manifest itself as the mysterious "presence"—but without supernatural content.

13–8b DRESS

We do judge people by the clothes they wear—at least initially. This might not be right, but it's true. Dress appropriately. You wouldn't wear jeans and an old sweatshirt to the investiture dinner for your new professional association president. Nor would you go on an inspection tour of a building site wearing a tux or formal gown.

Decide what's appropriate in light of your CMAPP analysis, consider the kind of image you want to present, and dress accordingly.

13–8c EYE CONTACT

In one-on-one conversations, most Americans naturally make eye contact with each other. When you're presenting, your audience members like to know that you know that they're there. Consequently, making eye contact with your audience is crucial to a successful presentation. Here are some suggestions.

- Dealing with a small audience, try to make eye contact with everyone several times during your presentation. Note, however, that some people will not give you the opportunity; don't force the issue. (Recall our comments in Chapter 1 regarding cultural preferences.)

- If the audience is large, mentally divide the room into segments and look at each several times.
- Never proceed up one row or column and down another: Your eye contact should seem random, but be comprehensive.
- Limit the duration of each contact to half a second to a second; otherwise, people feel you're staring at them, and may become embarrassed and/or angry.

13-8d FACIAL EXPRESSIONS

Although you need not match the exaggerations of "the silent screen," your presentation will be more effective if you don't remain expressionless. Just as in person-to-person conversations, allow your face to show your emotions and your intellectual energy.

13-8e GESTURES

Everyone gestures—on the phone, with other people, perhaps even when you're talking to yourself. Again, try not to exaggerate, but do let yourself gesture naturally. You can use gestures, including pointing, to emphasize your words, for example. (Note, however, that you should not point at individuals in your audience; some people will feel conspicuous and uncomfortable; some will simply consider you rude.)

Recall that gestures can have very different—sometimes very impolite—meanings in other cultures. As an example, you might be familiar with the "Hook 'em Horns" gesture often used with very positive connotation by University of Texas Longhorns fans. Longhorns fans in Europe have discovered to their dismay that the same gesture is perceived as a serious insult in Italy.

13-8f HESITATION PARTICLES

You may have noticed that some inexperienced speakers seem to say "um" or "uh" every second word. Just as unconsciously, others may say "you know" or "like" with irritating regularity. Such "fillers," as they are sometimes called, eventually draw your audience's attention away from your message. Avoiding them is, for most of us, difficult; doing so requires a conscious awareness of our words while we are speaking. Gaining that facility will make you a more effective presenter.

In most cases, it is better to pause—briefly. Your audience will usually accept such pauses as the mark of thoughtfulness. Frequent hesitation particles, however, will inevitably work against you.

13-8g MOVEMENT

Your audience wants to know that you're alive, so move around a bit. This doesn't mean that you should prowl around the podium with your hands waving all around; it simply means that you should avoid remaining motionless.

13-8h POSTURE

This relates to presence. Don't slump. Your audience prefers to see that you're actually alert throughout your presentation.

13–8i PRONUNCIATION AND ENUNCIATION

Ensure that you speak clearly and that you have mastered the pronunciation of whatever terms or names you intend to use. If you falter or err in the pronunciation of words you yourself have chosen, your audience is unlikely to request a repetition. (Indeed, an educated audience understands that a speaker who can't pronounce words in his or her own presentation likely did not rehearse that presentation, even once; this discourtesy will usually be judged harshly.) Analogously, remember that your audience has to understand every syllable—the first time you utter it. If you allow yourself to slur or mumble, or to "swallow" some syllables (as we all often do in rapid face-to-face conversation), your audience will not try to make sense of what you've muttered; they'll ignore your poor enunciation by ignoring you.

13–8j SPEED

A metronome keeps time by ticking at unvarying speed. A lack of variety in the speed with which you speak generates similar monotony, and a loss of audience interest. You should vary your speed of delivery while also varying the remaining paralinguistic features mentioned below.

13–8k TONE, PITCH, AND INTONATION

In face-to-face conversations, you automatically vary your tone, pitch, and intonation. When you ask a question, for example, you tend to end on a rising tone. When surprised, you tend to use a higher pitch. As you converse, your intonation changes to show emotion, emphasis, engagement, and so on. Those changes should not disappear when you are speaking in public.

Be careful, however, of what is often called "uptalking": ending every sentence with a rising tone, as though it were a question. The phenomenon is surprisingly common among American youth. Thus, although it might be quite "correct" in some informal contexts, it remains inappropriate for the level of discourse required for professional contexts, including presentations.

Natural variations in tone, pitch, and intonation add interest and luster to your presentation and help keep your audience attentive. If you speak with little or no variety of tone, pitch, and intonation, you will sound as though you are merely reciting something of no interest to you. Computer-generated voices often do that, but effective presenters should not.

13–8l VOLUME

Vary your volume for effect. Some highly competent presenters use everything from a stage whisper to a shout. Even if you avoid extremes, you should also avoid the monotony of unchanging volume; it is ineffective and may well put your audience to sleep.

13–9 Time Management

Pacing your presentation takes practice and experience. Ensuring that you have left enough time for all of your points is just as important as making certain that you do not run out of message halfway through your allotted time. Your audience expects you to take the time you have—not much less and certainly not much more. How professional do you think you will seem, and how much confidence do you think you will inspire, if your audience of potential clients has allotted you an hour of their meeting time, and you run out of material after 25 minutes . . . or if they're forced to "stop" you after 75 minutes? The best way to test your time management is to rehearse.

13–10 Visuals and Visual Aids

More and more, audiences expect visuals in professional presentations. From Chapter 7, you will understand that presentation visuals should

- illustrate, not overpower;
- explain, not confuse;
- enhance, not detract;
- simplify, not complicate;
- fulfill a CMAPP purpose, not merely decorate;
- be visible to all; and
- be intelligible to all.

Employing whatever today's hot technology is may make you feel your presentation is current and edgy; however, what you think of as cutting edge may, in the minds of some of your audience, already seem old school. Furthermore, it may not be wise to assume that, for your presentation, you will have access to whatever hardware and/or software you would like to use. As part of your CMAPP analysis for *Context*, verify whether you'll have what you need at your presentation venue.

Here, then, are a few comments on some types of visuals that you may want or need to incorporate into your presentations. You may think some of them antiquated. But remember, sometimes oldies can still be goodies; like Newton's laws of motion, they can still fulfill useful roles. And, analogous to Newton's third law, for every advantage to using a certain type of visual, there'll be an equal and opposite disadvantage.

13–10a HANDOUTS

Handouts remain a popular form of presentation visual. They allow your audience to take a more active part in the communication. As well, many audiences appreciate being able to take away information for future reference. You can distribute handouts before, during, or after your presentation.

Before

- Distribution will not create interference during your presentation.
- You can refer to any part of the handout at any time; however, your audience members will likely continue to examine (and perhaps rifle through) the material until their curiosity has been satisfied—not just until you want to begin.

During

- You can retain your audience's attention until you distribute the material.
- You can refer to your handouts from the moment they have been distributed, but every means of distribution will be an interruption: People will be looking to find their copy and will begin to examine what they have received, and will not pay full attention to you until they've satisfied their curiosity.

After

- You have your audience's attention throughout your presentation.
- But you cannot effectively refer to material in your handout, because your audience does not have it.

You must therefore design your handouts and select your time of distribution carefully, according to your CMAPP analysis.

13-10b PROPS

"Props" is a generic term that can refer to a variety of visuals, including, for example,

- maps—discussing the changes in weather patterns as you follow the Mississippi
- models—using a scale model to explain construction of the Niagara Escarpment locks at Lockport, New York
- people—using a person to demonstrate first aid
- photos—illustrating the CERN Large Hadron Collider in Switzerland, shown in Figure 13.1
- samples—explaining the use of a data projector by using a real one

FIGURE 13.1: Large Hadron Collider

© Pascal Boegli/Moment/Getty Images

Note that passing around a *real* example of a prop is usually ineffective. Only the person actually holding the item will be able to relate it to what you are saying; all others will have their attention divided between listening to you (as you talk about something they do not then see) and looking around for the prop to appear. In addition, remember that the same example might be too small to be effective. Imagine holding up a sample of the small fiber optic switch your company has just manufactured and expecting your audience of 50 engineers to see its details clearly enough to follow your discussion. It would not work.

13-10c OVERHEAD PROJECTOR

Once the most common means of displaying text and graphics in presentations, overhead projectors and their acetates or transparencies have been almost universally replaced by the electronic data projector (see section 13-10d). Nonetheless, companies such as 3M and Apollo continue to manufacture overhead projectors, and companies such as Amazon.com continue to sell them. In fact, in some smaller engineering firms and in some schools and colleges across the country, overhead projectors continue to be used quite effectively.

Advantages

- An overhead works well even with normal ambient lighting (with the lights on in a classroom, for example).
- You need not (and should not) turn your back on your audience: What you see on the overhead's plate is what your audience sees on the screen behind you.

- If you are standing at the overhead, you can draw your audience's attention to any item on your transparency merely by pointing at it with a pen—or even with your finger.
- Anything you can print or write on a sheet of paper can be quickly, easily, and cheaply turned into an overhead visual of fairly acceptable quality. You can photocopy or print directly onto an acetate, in black and white or in gray scale (as in the 1970s photo of author Dave Ingre's parents in Figure 13.2) or in color.

FIGURE 13.2: Photo of Dave Ingre's Parents

David Ingre

Disadvantages

- Screens usually hang vertically, but the overhead projects its image toward the screen at an angle. The result, termed keystoning, is a distortion of what is called aspect ratio, and the top of the image becomes widened.
- The further the overhead lies from the screen, the larger the image it projects. At the same time, however, the larger the image, the lower its quality. Thus, overheads are not always effective for a very large audience in a very large hall.

13-10d DATA PROJECTORS

The use of data projectors has become quite standard in the marketplace. Receiving a signal from a computer or other device, the equipment projects that digital information onto a screen.

Advantages

- Data projectors work well even in normal ambient light.
- Anything that can be made to appear on a computer's monitor can be projected onto a screen for the audience.
- Most data projectors can be connected to other input sources such as DVDs or flash drives.
- Projected material can thus include websites, slides containing text and/or graphics and/or animations, screen captures (such as the one in Figure 13.3), sound, and video.

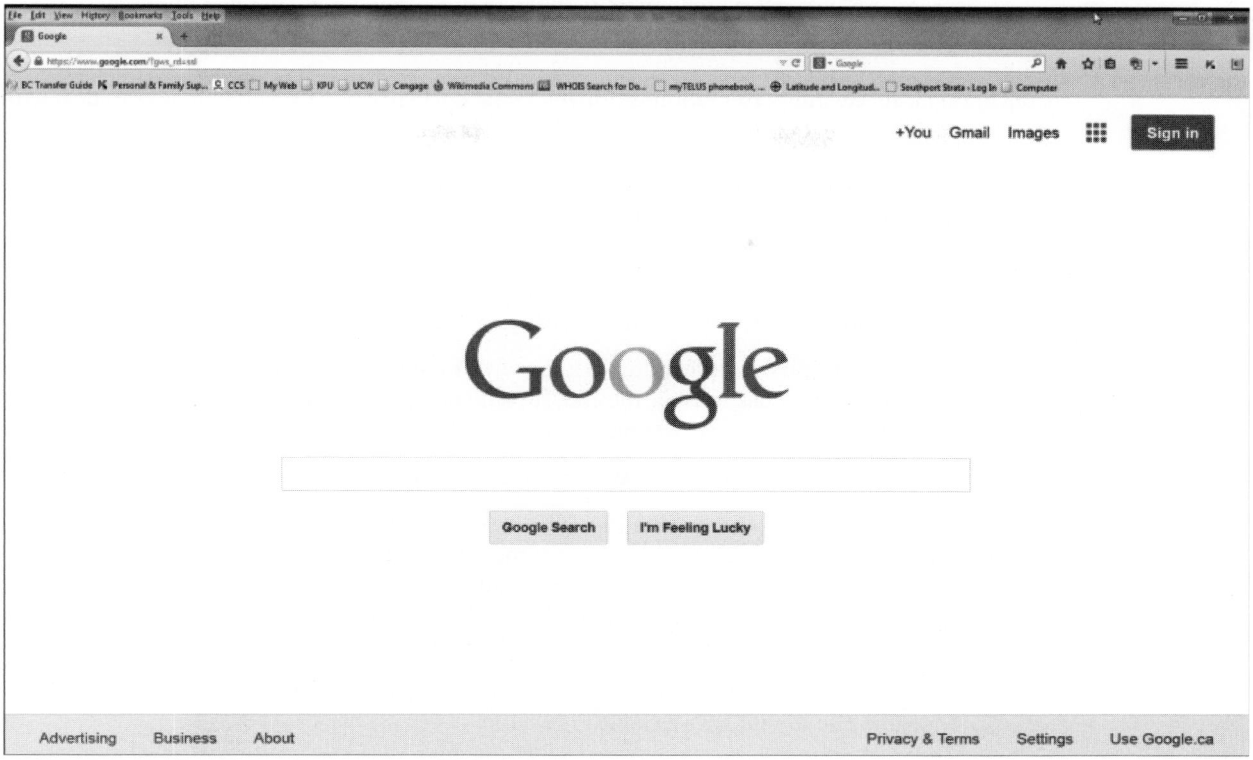

Source: Google, www.google.com

FIGURE 13.3: One of the World's Most Recognizable Websites

Disadvantages

- To avoid significant keystoning, data projectors work best when ceiling mounted, a sometimes unavailable option.
- It can be difficult to resist using the multitude of themes, colors, fonts, animations, sounds, and other options commonly offered by slide-creation software such as Microsoft PowerPoint®. However, if you don't adhere to the KISS principle when creating your slides, your audience may remember dazzling colors, amazingly diverse fonts, piercing sound effects, hopelessly cute clip art, eye-catching videos, and spectacular photos . . . rather than you or the content of your presentation.

As you surely understand, no hardware is perfect, no software is bug-free, Internet connections sometimes fail, and—surprise!—Wi-Fi might *not* be available. In fact, the likelihood of your presentation's being subjected to Murphy's Law remains too high for you not to prepare for that possibility. (If you've forgotten, Murphy's Law states that whatever can go wrong, will.) Thus, verify everything you can *before* the presentation, and make sure that you have a "Plan B," just in case. (For example, you might bring transparencies and/or handouts just in case the room's computer fails. For their "Plan C," your authors also always make sure they have a marker and a piece of chalk to use on whiteboards or blackboards when all else fails—including the copier machine needed for one of the first Plan B's handouts. Incidentally, one of the corollaries to Murphy's Law is called O'Shaughnessy's Law; it states, simply, that Murphy was an optimist.)

13–11 After the Presentation

Concluding a successful presentation is not your final challenge. Likely, you will have to deal with questions and comments and may have to justify some of your assertions. Here are some suggestions.

- Assume that you will have to explain what you say. Be ready to do so without being defensive; rather, look forward to this as an additional opportunity to reinforce your message. Welcome feedback and queries gratefully, with a nod and a smile.
- When you are asked to explain or even to justify, think of your response as a very brief impromptu presentation, and try to make use of the "three-part rule."
- When you begin, let your audience know if you want them to ask questions as you go along or to save questions for the end. If your context so dictates, inform attendees that you will not be able to answer questions at all. Whatever your decision, stick to it.
- Whatever the provocation, always remain polite. And don't let yourself be drawn into an argument.
- If you don't get any questions, don't wait more than 5 to 6 seconds. A longer expectant silence will embarrass both you and your audience. Rather, say something like, "Well, since there don't seem to be any questions at this time, I'd just like to thank you again for your attention," and consider your presentation over.
- Don't let any single questioner monopolize the floor. Remain polite, but suggest that someone else should have a chance as well.
- It is often a good idea to paraphrase a question before answering it. This indicates to your audience that you want to ensure that they have heard the question clearly, and it also allows you to "massage" the question so that you are better able to deal with it effectively.
- Many people don't like to be the first to ask a question, but will comfortably follow someone else's lead. Thus, to preclude a potentially embarrassing silence, you might wish to arrange beforehand for a "plant" to ask a question. Similarly, you could "start the ball rolling" by asking one yourself. Introduce it by saying, for example, "People often ask . . . ."

13–12 An Ethical Aside

Do the last two pieces of advice conform to ethical conduct? As was the case with Chapter 7's discussion of creating different impressions through adjusting chart axes, the answer would seem to be one of intent. If you are deliberately misleading your audience, you are likely culpable; if you are sincerely trying to help them understand, you are probably on firm ethical ground. Examine your CMAPP purpose, look closely at your own motivation, and decide where you think you stand. And remember that, as a professional engineer, you are expected to conform to a code of ethics. A 20-second ethical lapse during a Q&A—willfully cherry-picking report data that supports your main point while knowing that the entire report is more ambiguous—can undermine the many years it took you to acquire the reputation you had when you started your presentation.

CASE STUDY

13 AEL Presentation to Students

SITUATION

AEL's senior partners, Sarah Cohen and Frank Nabata, have for some years had a social relationship with Tina Trann, RAI's General Manager in Chicago, and her husband, Ben. Over supper one evening while Cohen and Nabata are in Chicago on business, Tina mentions RAI's highly positive assessment of RAI Boston's participation in Grandstone Technical Institute's Internship Program. The conversation sets Cohen and Nabata to thinking, and over the next few weeks they discuss AEL's position at length.

While they do not conclude that AEL would be able to accommodate work-term students on an internship or co-op program, they do decide to make a more organized attempt to attract bright young people into the business. Finally, they settle on asking Flavio Santini, a Senior Consultant in their Dallas office, to undertake what they whimsically dub a "speaking tour." As well as visiting each of the cities in which AEL maintains offices—Dallas, Detroit, Lansing, New York, and San Francisco—Santini is to deliver a talk at Grandstone Technical Institute in Randolph and at Ann Arbor University.

Santini has his office arrange with institutions in all those cities for him to deliver 25- to 35-minute presentations to groups of seniors in the programs of particular interest to AEL: chemical, civil, and geological engineering, and software design and development.

His preparatory work for his presentations can be seen in Figures 13.4, 13.5, 13.6, and 13.7.

FIGURE 13.4: Santini's CMAPP Analysis

Recruitment Presentation: CMAPP Analysis

Context:

- Students likely interested in available employment after graduation.
- Many may not have heard of AEL: probably necessary to establish relationship.
- Top students likely to hear several recruiters and have several options.
- Others likely there because of few possibilities: important for me to present AEL as top firm looking only for best students.
- Presentations likely to be in classrooms, small conference rooms, etc.: will probably vary from university to university.
- Technology available: data projector a possibility; BUT, prepare visuals for overhead projector—almost certainly available.
- Probably groups of 10–50 students: will vary according to institution and their "marketing."
- Length of presentations: 25–35 minutes, including question-and-answer period at end.
- Some students may expect me to be signing recruitment offers; if so, I must try to maintain their enthusiasm, but temper it with patience.
- AEL hopes to receive applications from a total of at least 50–60 students.

FIGURE 13.4: (continued)

Message

- AEL's reputation
- Outline of AEL's organization and activities
- Opportunities for after-graduation placement in the various fields: What? Where? When? Starting salaries?
- Typical career advancement opportunities at AEL
- Means to indicate interest: statement of interest forms
- "AEL is interested in *you*"

Audience/Audience Analysis

- Education level: impending graduation in technical specialty areas means close to 4 years college or equivalent education
- Age range: most likely to be 21–24
- Attitude: most likely to be enthusiastic; some skeptical; some unsure/unmotivated
- Gender: likely equal M/F split; possible preponderance of males
- Cultural background: extremely varied, but interested in "American dream"
- Economic status: most likely to be from middle- to upper-middle-class upbringing; as students, many may be having difficult time monetarily, but most likely expect to achieve at least upper-middle-class status within a few years of graduation
- Political bias: because of fields of study, more likely Republican-leaning than Democrat; thus no need to for me to explain "customer satisfaction" and "profit motive" in a large business
- Know: most likely aware of current entry-level job market in their fields; likely have idea of type of position/company they are looking for
- Need to know: what opportunities available, how to contact us, how to apply
- Want to know: whether AEL's opportunities are really of interest to them (including type of work, benefits, corporate culture, cost of living in respective cities, travel opportunities, etc.), how to find out more

Purpose

Persuasive—I want to

- obtain a sense of numbers and types of potential employees.
- present AEL as modern, energetic, ethical, profitable, exciting, etc.
- elicit interest of the *best* students.
- get students to follow up with AEL offices in respective cities.

Product

- Extemporaneous presentation with PPT slides as visuals

FIGURE 13.5: Santini's Topic, Thesis Statement, and Objectives

Topic: Working with AEL

Thesis Statement: If you are a qualified, ambitious graduate in a relevant field, you should apply to AEL for excellent entry-level opportunities.

Objectives

- Description of AEL
- What we are looking for
- Whom we are looking for
- Next steps

FIGURE 13.6: Santini's Multilevel Outline

Presentation Outline

I. Introduction	
A. Career success won't come by itself!	Visual #1: AEL logo/building photo
B. My position	
C. Relevance to you as impending graduates	
D. Why AEL is interested in you	
II. Description of AEL	
A. History	
1. Engineers in Lansing	Visual #2: photos of Cohen & Nabata
2. Growth of consulting activities	
3. Opening of offices	Visual #3: Map showing all US offices
B. Organization	Visual #4: Organization chart
1. Senior Partners and Partners	
2. Senior Associates and Associates	
3. Senior Consultants and Consultants	
4. Junior Consultants	
5. Staff Associates	
C. Locations	Reprise: Visual #3: Map
1. Dallas	
2. Detroit	
3. Lansing (Head Office)	
4. New York	
5. San Francisco	
D. Areas of endeavor	
1. Civil engineering	Visual #5: Fremlin Hall project, AAU
2. Chemical engineering	Visual #6: MicroGen Test Labs
3. Geological engineering	Visual #7: Martini Falls Dam Project
4. Software design	Visual #8: HelioTech "Vivacious" brochure cover
5. Software development	Visual #9: HelioTech Software Test Lab

FIGURE 13.6: (continued)

6. Now investigating:
    a) Environmental engineering
    b) Industrial safety
    c) Urban planning

III. What we are looking for
- A. Demonstrated attitudes
    1. Cooperation
    2. Team spirit
    3. Enthusiasm
    4. Positivity
    5. Desire to succeed
    6. Client satisfaction
- B. Qualifications
    1. Only the best of the best!
    2. Appropriate degree (see #II, D, 7 above for list)
    3. Marks
        a) At least 85% average in relevant subjects, AND
        b) At least 80% average each term throughout degree program
- C. Experience
    1. Not mandatory but an advantage
    2. Internship / Co-op is desirable if in appropriate field
    3. Extracurricular if relevant

IV. Whom we're looking for
- A. Entry-level candidates for:
    1. Dallas
    2. Detroit
    3. Lansing
    4. New York
    5. San Francisco
- B. People willing to relocate
- C. People willing to travel
- D. People looking for self-motivated work
    1. Senior firm members seek contracts
    2. Other consulting staff often work independently
- E. People able to work under little or no supervision
- F. People who take pride in their own work and want the company's name to be recognized for quality and integrity
- G. You represent the company and it represents you

V. Next steps	
A. Take brochures	Visual #10: Show sample brochure
B. Think over what I've said	
C. Send letter of application to Senior Associate in city of interest according to handout	Visual #11: List of Senior Associates
D. If no word within three weeks, follow up with phone call.	
VI. Conclusion	
A. Brief summation of points	
B. Recommendation to act immediately: competition becomes stiffer	
C. Question period	
D. Thanks and hope to see you again	Reprise Visual #1: AEL logo/building photo

FIGURE 13.7: Santini's Speaking Notes

1. a) CAREER SUCCESS ▭ #1 b) ME + YOU + AEL c) START & GROW ▭ #2 d) OFFICE OPENINGS ▭ #3	**2.** a) COMPANY ORG. CHART ▭ #4 b) CURRENT LOCATIONS ▭ #3 (2nd time) c) CIVIL ▭ #5 (Fremlin) d) CHEMICAL ▭ #6 (MicroGen)
3. a) GEOLOGICAL ▭ #7 (Martini) b) S-DESIGN ▭ #8 (Vivacious) c) S-DEV ▭ #9 (Helio) d) POTENTIAL EXPANSION	**4.** a) ATTITUDE: – Cooperation – Team – Enthusiasm – Positive – Success – Clients b) QUALIFICATIONS: – Cream – Degree – 85% topic + 80% overall
5. a) EXPERIENCE: – Advantage + Co-op + Other b) WHO: – Entry: ALL CITIES – Moves – Motivation >> << Supervision – Pride >>> Mutual Recognition	**6.** a) BROCHURES: ▭ #10 (brochure) b) CONSIDER! c) APPLY ▭ #11 (list) – Follow-up d) QUESTIONS / THANKS ▭ #1 (close)

ISSUES TO THINK ABOUT

1. What is your overall impression of Santini's CMAPP analysis (Figure 13.4)? Think in terms of the complementary attributes of CAP that you studied in Chapter 3.
2. How well do you think Santini has planned for his presentations? Justify your point of view with examples.
3. How would you have tried to deal with the problem of multiple audiences that Santini faces?
4. How accurately do you think that his *Topic* and *Thesis Statement* in Figure 13.5 represent his ideas? How might you rephrase them?
5. Look at Santini's *Presentation Outline* in Figure 13.6 and note his intended use of visuals. What would be your explanation for the lack of visuals to illustrate sections 3 and 4?
6. Thinking back to what you learned about multilevel outlines in Chapter 5, comment on his adherence to the principles of subordination, division, and parallelism. What changes (improvements?) would you suggest?
7. What does the format of Santini's *Speaking Notes* in Figure 13.7 suggest to you in terms of how he plans to use them?
8. Having studied his *Outline*, would you make any changes to those notes? What would you change and why?
9. Explain what you think you can deduce about Santini's character and professionalism in light of Figures 13.4, 13.5, 13.6, and 13.7.
10. Explain how you think his presentations might be received.

EXERCISES

13.1 Distinguish between a presentation and a speech.

13.2 Briefly explain how each of the five CMAPP components can apply to presentations.

13.3 Identify and describe five factors considered in audience analysis.

13.4 Identify the four main types of presentation delivery and indicate the principal advantages and disadvantages of each.

13.5 Identify and describe at least three strategies for dealing with stage fright.

13.6 Identify and describe the four types of *primary purpose* discussed in this chapter.

13.7 Describe the purpose and physical makeup of speaking notes, and explain the relationship of those notes to the multilevel outline.

13.8 Identify and describe at least three elements found in each of the following:

a. introductory segment
b. body
c. concluding segment

13.9 Identify five paralinguistic features discussed in this chapter, and briefly explain the importance of each to an effective presentation.

13.10 Briefly describe the advantages and disadvantages of the following visual aids:

- data projector
- props

13.11 Explain the advantages and disadvantages of distributing handouts

- before the presentation
- during the presentation
- after the presentation

13.12 As an audience participant, which of the guidelines for presenting visual aids have you seen presenters overlook most frequently? Why do you think that guideline is violated so often? What are the results?

13.13 Watch a presentation and prepare a written evaluation of the speaker's effectiveness. Comment on the delivery of the presentation as well as the effectiveness of any visual aids.

13.14 Prepare the introduction (attention-getter, self-introduction, main message, your credibility/qualifications, and audience relevance) and the closing (sense of an ending, final reiteration of message, call to action, and a polite thank-you/goodbye) to a presentation you might be asked to give to a student engineering association at your school. Your topic will be informative, describing an internship you had at a county zoning authority the previous summer. Make up details as you see fit. Keep your introduction and closing to a total of three minutes, plus or minus 10 seconds. (You will, of course, need to rehearse to keep to this strict time limit. You will also notice that the more you rehearse to keep to the time limit, the better your spoken presentation becomes.)

USEFUL WEBSITES

Note: For ease of reading, we have removed the "http://www." from most of the URLs below. However, some browsers might still require that you insert it to precede some URLs so as not to receive an error message.

■ Atissa Banuazizi, of MIT, offers a series of slides for "Creating your BE.109 Presentation." Although designed for the Laboratory Fundamentals of Biological Engineering course, the material will be useful for students in all engineering disciplines.
Visit: openwetware.org/wiki/BE.109:Creating_your_BE.109_presentation

■ Denny Coon, of the University of South Dakota, offers a PowerPoint presentation called "Effective Mechanical Engineering Presentations." Despite the title, the slides offer advice and additional Web links that will be useful to students in all engineering disciplines.
Visit: http://www.uwyo.edu/ceas/current-students/classes/meref/general_eng_presentation_tips.ppt

- Matt Napolitan, of CX Associates, has a very useful blog post called "Effective Presentations: 12 Tips for Engineers (Don't Laugh)."
Visit: buildingenergy.cx-associates.com/2013/02/effective-presentations-12-tips-for-engineers-dont-laugh/

- MindTools® offers a page with advice on "Creating Effective Presentation Visuals."
Visit: mindtools.com/pages/article/creating-presentation-visuals.htm

- Powtoon.com helpfully assesses 10 alternatives to the ubiquitous PowerPoint presentation software used so often in academia and industry.
Visit: powtoon.com/blog/10-best-powerpoint-alternatives/

- Purdue University's Online Writing Lab (OWL) offers a "Visual Rhetoric" page, with a variety of useful links regarding the creation and analysis of visuals for both documents and presentations.
Visit: owl.english.purdue.edu/owl/section/1/7

- Toastmasters International's home page offers links to information about Toastmasters and tips on successful public speaking.
Visit: toastmasters.org/

- University of Texas at Austin's McKetta Department of Chemical Engineering offers a Communication site that provides advice on "Formal Presentations."
Visit: engrcomm.che.utexas.edu/che-333t-engineering-communications/formal-presentations

- University of Wyoming's Information Technology Department offers tips similar to those in Denny Coon's PowerPoint file (see above).
Visit: http://www.uwyo.edu/infotech/services/multimedia/wyocast/tips.asp

CHAPTER 14

Seeking Employment

Seeking and securing employment is the subject of a vast library of books, from serious college texts to material promising to teach you to get the best job in the world without any effort at all. To add to the information, and to the confusion, your authors' search for "resumes" under "books" at Amazon.com produced more than 30,000 hits. Googling the same word produced some 112 *million* hits. (In both cases, of course, your results may differ.) So, as an engineering student with aspirations to work as a professional engineer, what can you expect from a short chapter in this brief text that is more than the familiar, obvious, and vague?

You will, in fact, find several recommendations and tips, all designed primarily to point you in a useful direction. How far you progress along that path will be up to you. Our own experience, along with that of many others, suggests that there is no panacea, no "magic bullet" for success; rather, there is only determination, hard work, and tenacity—and a welcome helping of plain luck.

Remember, first, that we're not talking about the kind of job you might take on over a summer to help pay for expenses during the school year. And, unless you're considering a career change away from engineering and looking to climb the corporate ladder of a company like Walmart or open your own set of Starbucks franchises, we're not speaking about what are often called "Mcjobs." (By the way, we don't believe in disparaging Mcjobs. Your parents' and older generations and many new Americans view that kind of work as an *opportunity*: To get into the workforce, you need to start *somewhere*.) Rather, this chapter is about the kind of job that you hope will at least *lead* to the engineering career you have your eyes and your heart set on.

14–1 Gambling

Engineering success usually depends on certainties: the strength of materials, the measurement of known forces, the interaction of chemical elements, and so forth. However, in the 1990s, six MIT engineering and math students turned their genius and their academic skills into the most lucrative—and legal—"card counting" business ever to rock the world of gambling. (You can read about them in Ben Mezrich's 2003 book, *Bringing Down the House*.) So what does this have to do with finding a career? Well, as the cliché goes, we've got some good news and we've got some bad news.

The bad news is that, regardless of what the self-help gurus might assert, obtaining a good position in the engineering field does involve a certain amount of luck: dealing with the right person in the right place at the right time. Unfortunately, not everyone who does everything right gets that perfect job. Notwithstanding what your self-esteem might wish to hear, sending off cover letters and resumes is, in some ways, like playing the horses or betting in the football pool. Admittedly, the more you know about the subject, the better your chances of winning something; however, unless you expect to be hired by a relative or friend who wants you regardless of anything else, the element of chance still plays an inevitable role. More about this later.

On the other hand, consider this: If you buy a lottery ticket, you might win; but if you don't buy one, you can't win. The good news, therefore, is that knowledge, skills, and determination invariably help. And, despite the growing competitiveness of the marketplace, the more of an edge you have when applying for that engineering position you want, the greater the likelihood that you, rather than someone else, will get it.

A quick look at job postings on LinkedIn shows that a pronounced majority of companies looking to hire engineers specifically require that applicants have excellent written and oral communication skills. If you can't communicate, you can't collaborate; and if you can't collaborate, you will find that most professionals won't want to work with you, no matter how brilliant you might be.

This chapter is predicated on the assumption that having better communication skills than your job applicant competitors will give you that edge in your search for rewarding employment.

14–2 The Employment Application Package

The constituents of a successful employment application package are as follows:

1. Preparation: doing your homework before you send out your application
2. Cover letter: using an effective CMAPP analysis to compose the best one for the occasion
3. Resume: tailoring your content, form, and format as needed
4. Interview: saying the right things at the right time
5. Follow-up: remembering that you're probably not the only player in the game

Each part has to work in conjunction with the others. So here is a series of recommendations.

14–3 Preparation

1. If your sole employment goal is wealth, don't try to become an engineer and then try to find a way to like the profession. That probably won't work. Rather, if you find that you enjoy what you're learning as an engineering student and that you're good at it, your next step will be to find a way to make a good living at it.

2. Don't believe that looking for a job is different from working at one: Finding the job *is* your job. After all, when you're trying to find employment, you are working for someone—yourself.

3. Don't believe that your credentials, qualifications, experience, and enthusiasm make you unique, a word that literally means "one of a kind." Notwithstanding the probable uniqueness of your DNA, you are most likely not one of a kind when it comes to future employers. However good you are at whatever you want to be paid to do, you can count on there being a great many others of equal, if not necessarily better, merit, trying to make sure that they, rather than you, get the position.

4. Don't believe the fairy tale that if you want it enough, your dream job will drop into your lap. Doing, not wanting, brings results.

5. Unless you graduate at the top of your class from a really prestigious engineering school, in a branch of engineering that's currently in extremely high demand, don't believe that employers are desperately seeking you. They are more likely to be searching for ways to keep their businesses profitable—with or without you.

6. Many employers have indicated that their hiring is based 90 percent on aptitude, and their firing based 90 percent on attitude. The moral behind that fact is that if you look for employment on the assumption that you can always "take the course again" or submit a "rewrite" for better marks, or that a potential employer has a responsibility to help you obtain what you want, you are likely to be cast aside before you are even hired.

7. Unless you're forced to seek absolutely any work at all—in which case, refer to the third paragraph at the beginning of this chapter—think about why you are applying. Look at your application as a product, and do a thorough analysis of your context, message, audience, and purpose. You might be surprised.

8. A 2013 article by Nancy Collamer in *Forbes* magazine (http://www.forbes.com/sites/nextavenue/2013/08/12/6-ways-to-crack-the-hidden-job-market/) states that "It seems almost a cruel joke being played on people desperate to find work: Most jobs getting filled these days aren't even advertised. Instead, they're typically part of the hidden job market—those millions of openings that never get formally posted. It now accounts for up to 80% of hires, according to some estimates."

   So, network. In the engineering fields, most good career positions will not be obtained through newspaper or Web notices. Rather, people know people who know people who know people. Therefore, get to know people. Since the first edition of this book was published, the helpful online platform called LinkedIn (sometimes referred to as *social media for business*) has made getting to "know" people who know people remarkably easy. Author Bob Basil was once extolling the virtues of LinkedIn to professionals in the construction field, showing an industry-related "help wanted" ad on that site. It turned out that the new hire who viewed that ad was sitting right in front of him, next to the happy field manager who had given that person the job.

a. Use personal contacts, libraries, and Internet resources to obtain the names of people who work in companies that operate in the discipline and industry that interest you.

b. Contact them and ask for a bit of their time, stressing that you are not seeking a job—you are trying to find out more about that particular branch of engineering and what it values. Most people will be quite accommodating if you specify that you want no more than half an hour—and then stick to it.

c. Thus, when you do apply for a job, you are likely to have—and be able to demonstrate—greater familiarity with your specialty. That knowledge gives you an added edge.

9. Read ads carefully. Notwithstanding the previous points, it is likely that you will at times respond to an advertised position. If so, treat the text of that ad as a combination of explicit and encoded information. Pay attention to what the ad says and to what it probably implies. For example, if it says, "No phone calls, please," it is likely that your deciding to call anyway will be treated not as a display of initiative but as an unwillingness to accept direction. Similarly, if the ad asks for specific remuneration expectations, the employer may be trying to find out whether you are already sufficiently familiar with the field to know typical pay and benefits.

10. Judge yourself. If you aren't thoroughly familiar with all the terminology in the ad, perhaps you'd best not apply. If an ad mentions Clojure and XSLT programming, the Kepner-Tregoe process, or Monte Carlo simulations for molecular modeling, the potential employer presumes that qualified candidates know and use that jargon.

11. Don't procrastinate. Murphy's Law implies that things left to the last minute will fail.

a. If you're in a co-op, internship, or similar program, expend the effort necessary to let that program work for you, by fortifying relationships with your colleagues there and investigating career possibilities actively.

b. Several months before you expect to graduate, start investigating the job market in the field that interests you.

14-4 Cover Letters

You'll recall from Chapter 10 that much persuasive communication targets the intellect rather than the emotions. That's what a cover letter should do.

14-4a RATIONALE

Why bother using a cover letter? After all, isn't the resume the important thing? The rationale for taking the trouble to create an effective cover letter is as follows:

1. Analogous to a report's transmittal letter, which introduces both you and the attached report, a cover letter introduces you and your resume.

2. It is your opportunity to use the letter's content, form, and format to your advantage in selling your resume and yourself.

3. It is an expected convention; thus, its absence would be jarring.

14-4b CONTENT

An effective cover letter should incorporate well-reasoned points to press your case.

1. Refer precisely to the position you are applying for. For example, a reference to "your recent ad in the local paper" is worse than vague; alluding to "Competition #99-A343, described on page F8 of the Careers Section in the April 17, 2015, edition of the Seattle Post Intelligencer" does two things: It allows the reader to know what you are talking about, and it suggests to the potential employer that you will be specific and precise in your work—both considered commendable qualities in an engineer.
2. Avoid superfluous information. For example, beginning with "Let me introduce myself. My name is <your name>, and I would like to apply for . . ." leaves your reader thinking, "I know that from the letterhead; don't waste my time." Putting serious effort into your initial CMAPP analysis for your cover letter will help you ensure it is precise, accessible, and concise—similarly laudable attributes for an engineer.
3. In his 1961 inaugural address, President Kennedy made what became a defining motto for a generation: "And so, my fellow Americans: Ask not what your country can do for you—ask what you can do for your country." It's worth remembering his words when you're looking for a position: Focus explicitly on what you can do for the company, not what the company can do for you. Remember: Employers are quite rightly looking to their own benefit, not to yours.
4. Reiterate any specific points in your resume that you feel should be of particular interest. While bearing brevity and KISS in mind, don't expect your audience to be as interested in you as you are.
5. Include a clear call to action. Again, think of your CMAPP analysis, and consider the language you are using. A very weak (though unfortunately common) action request is, "If you have any questions, please don't hesitate to call me at . . . ." In effect, you are telling your reader *not* to call you unless he or she has questions—even, for example, if the audience were already considering offering you an interview. Remember, if your audience finds it advantageous to do so, he or she will interpret your words quite literally and will act on what you wrote rather than on what you probably meant. In short, ask to get together for an interview to discuss how you might be of value to the company.

14-4c FORM

If you can obtain next-to-no information about your specific audience, how can you determine the appropriate levels of discourse and technicality? Your CMAPP analysis will help you decide. For example, many engineering positions will be highly technical and specialized; you should probably assume, therefore, that your audience has certain technical expectations. If the context is more generalist, that factor, too, should influence the construction of your letter. In all cases, you should be careful to keep your writing as clear, concise, and precise as possible.

14-4d FORMAT

Beyond a requirement for professional appearance, a cover letter need follow no prescribed format. You should, though, follow certain guidelines.

1. Do not exceed a single page.
2. Use letterhead stationery—all good word processors now enable you to quickly and easily produce your own—rather than simply putting a *return address* at the top of the page.

3. Your letterhead should include your first name and surname, your mailing address, any phone numbers you think would be useful to your audience, and your email address. If you maintain a website, include it only if you think it reflects well on you, and if its content would be relevant to the employer. You might want to remember that a potential employer is likely to look for your Web presence—from blogs to Facebook to LinkedIn to Twitter.

4. Make effective use of appropriate document visuals as outlined in Chapter 7.

5. As much as possible, follow the business letter conventions discussed in Chapter 12. If you don't have enough information to identify your audience appropriately, you might consider the *simplified* letter format, since it requires neither an honorific nor a salutation.

14-5 Resumes

A resume (sometimes spelled with accents, "résumé") is occasionally called a *curriculum vitae* (abbreviated as "c.v."), Latin for "life's path." Irrespective of terminology, you might consider your resume

- a short analytical summary of your salable credentials and qualifications
- a short analytical report on the relevant aspects of your background
- a short persuasive document targeting the intellect, and functioning as a complementary attachment to your cover letter
- introduced by its cover letter, a short, external, probably solicited, proposal for assisting the organization to which you are applying

Your resume is, in effect, all of the above. Furthermore, it is one of the most important documents you will ever create: It is a reflection of you that you want examined by people from whom you hope to obtain substantial long-term benefit.

14-5a MESSAGE

A common mistake of many people looking for employment is to carefully construct a resume, and then use it when applying for a variety of positions. Doing so, however, violates the principles of good CMAPP analysis. Each application necessarily involves a different context and a different audience; the individual resume's message, therefore, must be specific to that context and audience. Although your background necessarily remains intact, as might the basic elements of your CMAPP product—the resume—you must tailor your document. This means you must create a "new" resume each time you submit an application: You will want to highlight certain experience, stress particular accomplishments, use the document's organization and format to convey specific impressions, and so on.

Professional Objective

Very common at the beginning of a resume is an elliptical clause or a single sentence indicating your professional objective. An example might be the following:

> **Professional Objective:** To apply my engineering skills and academic experience in a large, well-respected firm, which offers the potential for career advancement to senior management ranks.

Should you include one? One school of thought still advocates including a professional objective, maintaining that it demonstrates forethought and long-term commitment. Another current perspective, however, advises against including one, pointing out the following:

- Why should the employer really be interested in *your* objective? The company has its own.
- Should you be interviewed, and should the matter be of interest, the manager would more likely prefer to discuss the issue directly.
- If your stated objective happens not to match that of the employer (to you, a poorly known audience), it will likely work against your candidacy.

Nonetheless, since there is an element of gambling involved here, too, you should conduct the most accurate CMAPP analysis you can and decide in which basket to place this particular egg.

As many younger engineers tend to stay in one job for a shorter period of time than do older generations of their professional colleagues, an "objective" statement can indicate to a human resources professional or subject-matter expert at the company, where on your professional development "path" you are and what explicitly you hope to accomplish at the company. Including both a short- and long-term objective might usefully identify this path for your audience.

14-5b AUDIENCE

Although the situation may well vary, you are likely to have both an *initial* and a *second* audience when you submit a resume to an organization.

Initial

Your initial audience is probably going to be a personnel or administration officer. Though this, too, might jar your self-esteem, remember that this person's customary role is not to choose you from among the likely numerous applications received. Rather, it is to find reasons (indicated or implied by your resume and/or your cover letter) to *exclude* your application. Thus, your resume and letter must be persuasive enough to convince this primary audience not to relegate your application in the "never mind" pile.

Second

Your second audience is, quite possibly, a "line manager"—one who will have authority over the position in question. That individual's role likely includes reviewing the initial audience's "inclusions" (but not the excluded applications), and deciding which few candidates are most suitable for interviews. Thus, your resume must persuade your second audience as well.

14-5c PURPOSE

Another frequent but mistaken assumption is that your resume's purpose is to get you a job. It won't. Its purpose is to obtain an interview; *that* will be your opportunity to convince your audience to offer you the position—based in large part, you will recall, on their conclusions regarding your *aptitude*.

14-5d PRODUCT

You should consider several CMAPP issues in particular.

Form

Brevity, precision, and concision are crucial. Avoid wasting words and space. For example, using headings of "Name" and "Address" to introduce your name and address is counterproductive: Your audience recognizes the information.

Choose action verbs in preference to weaker "roundabout" constructions: "apply" not "make application," "ensure" not "make certain that," "guided" not "looked to the guidance of," "created" not "established the creation of," and so forth.

Resumes make use of phrases or elliptical clauses rather than complete sentences. Remember, however, to observe the principle of parallelism (see Chapter 5) rigorously; otherwise, your text is liable to lose cohesion, coherence, and consistency.

The levels of discourse and technicality you should use will depend on your context and your audience; your choice should derive from your CMAPP analysis.

Finally, typographical, grammatical, or structural flaws reflect badly on your professionalism and lessen, if not altogether destroy, your credibility and thus your audience's interest in you.

Format

The appearance of your resume reflects your abilities. Make effective use of level heads. Commonly, these appear as though in the left-hand column of a table, complemented by the "information" that appears in a wider right-hand column. (Examine Figure 14.1 for an example.) Remember accessibility: You want your audience to extract salient items without having to pore over every word.

14-5e LENGTH

How long should your resume be? There are three common schools of thought:

1. Your resume should never exceed a single page. The rationale is that the audience will not be inclined to take the time or trouble to read more than one page. After all, the resume will merely identify candidates for the next step in the selection process. This point of view is valid.
2. Regardless of the number of pages, your resume should include everything relevant to the application. The rationale is that the audience will be trying to exclude applications, and is likely to do so if the sparse contents of a single page do not provide enough information to permit a confident decision. Your audience will inevitably exclude a resume of uncertain value rather than take the trouble to contact you for further information. This point of view is also valid.
3. If your relevant content will definitely require more than a single page, use the first one as a summary of salient points, title it as such, and follow it with however many pages you need to provide brief and concise amplification. Here, the idea is that you may be able to satisfy both of the previous audience types. However, some audiences may react negatively, seeing your choice as an attempt to please all the people all the time.

Notwithstanding the last two points, if a company specifies that your resume not exceed one page, obey.

Which of the three would we recommend? We can't say, because we don't know your audience. And, since you're not likely to either, you're back to "gambling." No, life isn't fair.

14-5f TYPE

Broadly speaking, you can classify resumes as follows:

- **Chronological**: to reflect when events occurred (see chronological organization of reports in Chapter 8)
- **Functional**: to arrange information according to its function (see topical organization of reports in Chapter 8)
- **Electronic**: to submit your resume by fax, by email, or over the Web, although in a variety of formats

Which type should you use? First, of course, try to conduct the most accurate CMAPP analysis your information permits. Should a potential employer state or imply a preference to see your experience over time, you should likely choose chronological; should the intended audience somehow suggest you show your skills rather than your experience over time, you might be better off with the functional; and, of course, if the company you're targeting requests electronic submission, you're fine except that you *may* now need to decide whether to submit as an email attachment or within an email, in a particular word processor's format, in scannable format, in ASCII, or in HTML. We hope you're still in a gaming mood, because here, too, you'll be trying to play the odds.

FIGURE 14.1: A One-Page Reverse-Chronological Resume

Rick Vanderluin, a senior employee with the Hartford, Connecticut, marketing firm of Avery and San Angelo, heard through the grapevine that RAI might be looking to create a marketing department. He decided to apply to Celine Roberts, RAI's company manager. Here is the one-page curriculum vitae he attached.

Richard (Rick) Vanderluin

1151 St. James Court Hartford, CT 06101
(860) 555-4438 Fax: (860) 555-7878 Email: vanderl@rite.com

Academic MBA (2004) — University of Connecticut
BA in Marketing (2002) — University of Massachusetts

Career

2013 to Present — Senior Associate: Avery and San Angelo (Hartford, CT)

- Report to Senior Partner.
- Supervise staff of six.
- Develop promotional strategies for several large clients, including Loews, Safeway, and Bank of America.
- Assess and recommend new target opportunities: generated $2.5 million in new business in 2015.

2010–2013 — Director of Marketing: Rawlinson Hotel (Hartford, CT)

- Reported to General Manager, set up Hotel's Marketing department.
- Supervised marketing staff of two.
- Developed new marketing strategies: increased Hotel revenue by 27%.
- Secured profitable linkages with out-of-town business and government, including Hartford Chamber of Commerce, American Dentistry Association, and Connecticut Agricultural Experiment Station.
- Actively pursued convention and meeting business: in 2011–2013, secured a minimum of five association AGMs.

2007–2010 — Sales Director: FastService Hotels Ltd. (Farmington, CT)

- Exercised full line responsibility for all company sales staff in Farmington, Burlington, Manchester, and Marlborough.
- Coordinated corporate-sponsored functions for all FastService Hotels.
- Actively solicited tourism business: received FastService's Most Successful Manager award in 2009 and 2010.

2004–2007 — Sales Associate: Hartford Motors (Hartford, CT)

- Reported to Sales Director.
- Recommended and developed marketing campaigns, increasing sales by 13% in 2006.
- Sold new and used Cadillacs, Lincolns, and Hummers.

References Available on request.

14-5g CHRONOLOGICAL RESUMES

Since the ordering of employment information is normally from most recent to earliest, this type of resume is often referred to as being in reverse chronological order. (An example is Rick Vanderluin's c.v. in Figure 14.1.) Skills and accomplishments are integrated within each employment "segment." This is viewed as a "traditional" format and is likely still the most common—and thus the most commonly expected on the part of many audiences.

The chronological resume is particularly effective for illustrating an improving career path. On the other hand, it tends to draw attention to any lack of career progress, if you are just beginning, for example, and to any chronological gaps, if you were unemployed for a protracted period while caring for a family.

14-5h FUNCTIONAL RESUMES

The functional resume focuses on your skills and accomplishments. You organize your background information around the practical benefits you can offer your potential employer. Functional resumes are very effective for highlighting *what you can do* rather than *when you did it*.

Note, however, that functional resumes remain somewhat less common than chronological ones. Consequently, some audiences may question them, suspecting that they are designed to obscure unflattering information. In making your choice, therefore, you must perform as accurate a CMAPP analysis as possible—and then cast your dice. Figure 14.2 shows Vanderluin's resume reworked as a functional resume. Note that every skill on each list—Marketing, Sales, and Administrative—is "proved" by precise examples.

14-5i DIGITAL RESUMES

Most American employers now want chronological or functional resumes submitted electronically, often as an attachment to an email sent to a company's human resources office. The reason: Digital resumes are easier to store and to search. Some resume pundits will claim that an effective digital resume is entirely different from a good "paper-based" one, in content, form, and format. On the other hand, other authorities contend that an electronic resume is, in effect, merely a chronological or functional resume formatted to fit a digital template. The main difference between most paper and digital resumes, in our view, is that the latter permits you to include hyperlinks that quickly and easily give your resume another dimension, allowing readers to find out more about companies and activities you've listed.

If you are looking to file your resume electronically, however, you should bear in mind a few generalities.

- Your wording will remain of crucial importance: Not only must it persuade your audience, but it must be effective in allowing the electronic search engine the employer uses to flag important terms. Remember that no *person* is likely to actually read all resumes in the data bank; rather, a search engine will cull for those that match the criteria (the words) chosen by the employer.
- Don't assume that your ability to simply file your resume electronically necessarily gives you a real advantage. Just as with more traditional job searches, you will be competing with many other skilled candidates. Networking remains perhaps *the* advantage.

Scannable Resumes

Scannable resumes are designed to be printed or faxed and then scanned by your potential employer. To economize their search for employees, many companies use scanners and specialized software programs to screen employees. Your authors note, though, that in recent years the use of scanners has declined, though they are still common.

Richard (Rick) Vanderluin

1151 St. James Court Hartford, CT 06101
(860) 555-4438 Fax: (860) 555-7878 Email: vanderl@rite.com

Academic	MBA (2004) BA in Marketing (2002)	University of Connecticut University of Massachusetts

Marketing Skills

- Developed promotional strategies for several large clients of the marketing firm of Avery and San Angelo, including Loews, Safeway, and Bank of America.
- Assessed and recommended new target opportunities: generated $2.5 million in new business for the marketing firm of Avery and San Angelo.
- Developed new marketing strategies that increased Rawlinson Hotel revenue by 27%.
- Secured profitable linkages with out-of-town business and government, including Hartford Chamber of Commerce, American Dentistry Association, and Connecticut Agricultural Experiment Station.
- Recommended and developed marketing campaigns for Hartford Motors, increasing sales by 13% in 2006.

Sales Skills

- For Rawlinson Hotel, secured a minimum of five association annual conventions in each of two years.
- Actively solicited tourism business for FastService Hotels, receiving the Most Successful Manager award in 2009 and 2010.
- Sold new and used Cadillacs, Lincolns, and Hummers.

Administrative Skills

- Set up and managed Rawlinson Hotel's marketing department.
- Coordinated corporate-sponsored functions for all FastService Hotels.
- Exercised full line responsibility for staff of up to six sales and marketing professionals.

References Available on request.

FIGURE 14.2: A Functional Resume

Companies scan the resumes they receive and store the information using keywords. When a position opens in the company, a computer looks up all the resumes that include their targeted keywords. For example, a company looking for a mechanical engineer might search for keywords such as "design," "erection sequences," "injection molding," and so on, and even "American Society of Mechanical Engineers."

A keyword summary can be included in your resume to highlight keywords. Keywords should be nouns rather than verbs or adjectives, and should comprise roughly 25 words that can include job titles, skills, software programs, and selected jargon for your field. Many authorities recommend that your list of keyword descriptors appear immediately following your name and address on your resume. After your keyword summary, your resume should continue with the standard parts and sections.

When submitting a scannable resume, follow these guidelines:

- Send only originals on white paper with black ink.
- Limit visual effects to capitalization and bold headings. Avoid italics, underlining, boxes, columns, or shaded areas because they sometimes confuse the software.
- Use only 10 to 14 point font sizes.
- Mail flat resumes. Do not fold or staple pages. Creases can cause scanner problems, and staples have to be removed.

Figure 14.3 shows Rick Vanderluin's resume prepared in a scannable format.

Richard (Rick) Vanderluin

1151 St. James Court Hartford, CT 06101
(860) 555-4438 Fax: 555-7878 Email: vanderl@rite.com

KEYWORDS

Senior Associate, Director of Marketing, Sales Director, Supervisor, Promotional Strategy Development, New Business Development, Revenue Growth, Advertising Industry, Hotel Industry, Automotive Industry, Leadership, Project Management, Team Building, MBA, University of Connecticut

EXPERIENCE

Senior Associate, Avery and San Angelo (Hartford, CT)

2013 to Present

Report to Senior Partner.

Supervise staff of six.

Develop promotional strategies for several large clients, including Loews, Safeway, and Bank of America.

Assess and recommend new target opportunities: generated $2.5 million in new business in 2010.

Director of Marketing, Rawlinson Hotel (Hartford, CT)

2010–2013

Reporting to General Manager, set up Hotel's marketing department.

Supervised marketing staff of two.

Developed new marketing strategies: increased Hotel revenue by 27%.

Secured profitable linkages with out-of-town business and government, including Hartford Chamber of Commerce, American Dentistry Association, and Connecticut Agricultural Experiment Station.

Actively pursued convention and meeting business: from 2003–2005, secured a minimum of five association AGMs.

Sales Director, FastService Hotels Ltd. (Farmington, CT)

2007–2010

Exercised full-line responsibility for all company sales staff.

Coordinated corporate-sponsored functions for all FastService Hotels.

Actively solicited tourism business: received FastService's Most Successful Manager award in 2009 and 2010.

Sales Associate, Hartford Motors (Hartford, CT)

2004–2007

Recommended and developed marketing campaigns, increasing sales by 13% in 2006.

Sold new and used Cadillacs, Lincolns, and Hummers.

EDUCATION

MBA, 2004

University of Connecticut

BA in Marketing, 2002

University of Massachusetts

FIGURE 14.3: Scannable Resume

14–5j ENTRY-LEVEL RESUMES

If you're creating a resume while you're still a student, you may well have relatively little employment background. You may, though, have other relevant items to list, such as more details regarding your education, or volunteer activities you've undertaken.

To examine the situation of someone with little career experience, contrast the chronological and functional resumes of Rosalind Greene, the Ann Arbor University student whose transcript appeared as Figure 8.1 in Chapter 8's discussion of reports. Her resumes are shown in Figures 14.4 and 14.5. As well, some of the websites listed at the end of this chapter offer advice on resume creation for people who are starting out.

Your authors have found that with students or young professionals in particular, a systematic interview process with a friend or colleague helps "generate" skills or accomplishments that might

Rosalind Rebecca Greene

44702 Leslie Lane Plymouth, MI 48170
(734) 555-9873 Cell: (734) 555-1837 Email: rgreene@britenet.com

Education	
2015	BIS (Bachelor of Information Science): Ann Arbor University Major: Systems Design Minor: Political Science Overall GPA: 3.47
2011	High School Graduation: Plymouth Salem High School (Canton, MI)
Employment	
Fall 2011–present	Senior Server (part-time): La Champagne Restaurant, Livonia, MI • Provided table service to clientele in a four-star downtown restaurant • Maintained own table and tip accounts for restaurant management • Supervised work of two other part-time servers
April–September, 2010	Supervising Server (full-time): La Champagne Restaurant, Livonia, MI • Trained all new regular servers • Supervised work of four servers and three bussers • Provided table service to clientele • Maintained own accounts and oversaw those of the four servers and three bussers
2005–2010	Server (part-time): La Champagne Restaurant, Livonia, MI
Volunteer Work	
May–September, 2008–Present	Lifeguard (A.L.A. certified): Camp Harnessy, Belleville, MI • Every Sunday, held full shift responsibility for life guarding at Olympic-size pool at children's day camp • Supervised swimming activities of some fifty children, aged 6–8 • Submitted shift reports to Camp management • In July, 2009, received Lifesaver Hero award after resuscitating and providing first aid to a 10-year-old who had fallen and knocked himself unconscious against the diving board support.
Hobbies	Competitive swimming, track and field
References	Available upon request

FIGURE 14.4: Rosalind Greene's Chronological Resume

Rosalind Rebecca Greene

44702 Leslie Lane Plymouth, MI 48170

(734) 555-9873 Cell: (734) 555-1837 Email: rgreene@britenet.com

Education	
2015	BIS (Bachelor of Information Science): Ann Arbor University Major: Systems Design Minor: Political Science Overall GPA: 3.47
2011	High School Graduation: Plymouth Salem High School (Canton, MI)
Training and Supervisory Skills	Over a period of four years, trained a total of fourteen servers at La Champagne, one of Livonia, MI's four-star restaurants Regularly supervised the activities of four servers and three bussers Supervised the swimming activities and held responsibility for the pool safety of up to fifty children, aged 6–8, at Camp Harnessy in Belleville, MI
Organizational and Financial Skills	Working as Server, Senior Server, and Supervising Server at La Champagne, maintained my own table and tipping accounts and oversaw those of other servers and bussers
Interpersonal and Communication Skills	Dealt regularly with clientele at one of Toronto's four-star restaurants As lifeguard at Camp Harnessy children's camp in Livonia, provided regular shift reports to Camp management Dealt effectively with up to fifty children during their swim times
Awards	As ALA-certified Lifeguard at Camp Harnessy in Livonia in July 2009, received Life-saver Hero award after resuscitating and providing first aid to a 10-year-old who had fallen and knocked himself unconscious against the diving board support.
Hobbies	Competitive swimming, track and field
References	Available upon request

FIGURE 14.5: Rosalind Greene's Functional Resume

have otherwise been overlooked. Here is why: You might not always know what sets you apart, what is special about you. If you have been training young athletes at your local YWCA since you were a sophomore in high school, winning a lot of games along the way, you might forget to include that fact in your resume because it comes from the sports arena, so to speak, rather than an academic or workplace setting. Yet it shows several important qualities: leadership, conflict resolution, time management, goal-setting, and teaching ability. All of these are transferable, or portable, skills that are sought in many sectors, including engineering.

To demonstrate portable skills, we often ask our students to prepare a grid that charts what they might have accomplished in various "arenas of action." Common ones include school and work, of course, but also volunteering, sports, and the arts; you might also have been involved in sizable projects in a condominium association, a neighborhood group, or a religious organization.

Common transferable skills often sought in engineering fields include communication and linguistic skills—knowing more than one language can be a big plus in a company that has an international project portfolio, as well as organizational, teamwork, leadership, and teaching skills.

With your friend or colleague, go through each "arena of action" and answer the question, "What have I accomplished as a communicator (as an example) in school, at various jobs, as a volunteer, in my neighborhood, in sports or the arts, or at my church, mosque or temple?" You may very well be surprised by what turns up.

Common proofs of accomplishment include raises and promotions at work, good grades at school, write-ups in local or online media, awards or scholarships, and publications (as author or co-author) in traditional or digital media.

14–5k ELECTRONIC RESUMES

An online or hypertext resume is an HTML document that can be viewed on the Internet. This type of resume generally uses links to other screens, which expand on information that might be omitted in a one-page resume. For example, as part of work experience, you might provide a link to detailed job descriptions, drawings or blueprints, or other items from a portfolio. Online resumes can have distinct advantages for candidates in the engineering fields—provided your primary audience has indicated a willingness to *accept* this form. Once again, specific advice for the creation of Web documents varies greatly, even to the point of being contradictory at times. So, once more, conduct a CMAPP analysis and try to create the most professional product you can.

Author Bob Basil's resume—http://basil.ca/resume—on his home website is a clear example of how hyperlinks deepen one's self-presentation without lengthening the document itself.

Online Forms

Note that some employers specify that you should fill out an online form instead of submitting a resume of *any* kind. That way, they will have the same kind of information from all candidates, organized in the same way, allowing for much easier and faster extraction of data and, therefore, initial selection of candidates. As well, such a database can also be used later to look for suitable candidates to replace those who have left. If you are applying for a position and are told to fill in an online form, follow Nike's advice and "just do it." Trying to show your individuality and initiative by doing something else will almost inevitably backfire.

14–5l CONCLUSION

For engineers accustomed to working with standardized, verifiable ideas, all these inconsistencies can be troubling. Nonetheless, everyone does seem to agree that the professionalism necessary for you to obtain the interview you want is as necessary for digital as for "snail mail" submissions. Several of the websites provided at the end of this chapter will offer advice—occasionally conflicting, unfortunately—on how best to develop and write a resume intended for electronic submission.

14–5m ETHICS

It is undeniable that some people have submitted successful applications that contain deliberate and inaccurate exaggerations. While your resume must be persuasive, and while you should try to word it in a way that casts a favorable light on you and on your activities, we strongly recommend you not veer away from the truth. You will recall our earlier comments regarding aptitude and attitude. Consider how you would be viewed if, for example, your resume specified that you were an expert user of the UNIX operating system, and, after a few weeks' employment, your lack of expertise became obvious. As the proverb states, "The truth will out." In most cases, that is what does happen. Therefore, don't compromise your integrity by untruthfully exaggerating your worth.

14–6 Interviews

A profitable way to look at your participation in an employment interview is as a series of rapid impromptu presentations (see Chapter 13). Therefore, consider the following.

14-6a INTERVIEW TIPS

1. Your CMAPP analysis, part of your preparation before submitting your application, is crucial.
2. Your overall purpose at this stage is to convince your audience to hire you rather than someone else.
3. Each time your audience asks a question, or asks you to play a role in a scenario, you benefit from the impromptu presentation skills you have learned.
4. You should make effective use of paralinguistic features and of the "three-part rule for perfect presentations."
5. Your audience will likely want to see how well you can think on your feet (even though you're probably seated).
6. Before you submitted your application, you would have had the luxury of broad preparation: At the very least, make sure you have a good idea of the company's main functions and goals.
7. This same preparation, coupled with on-the-spot audience analysis, should allow you to ask intelligent, relevant questions of your own—an opportunity commonly afforded applicants. (In fact, you should have prepared a few such questions before coming to the interview.) What you ask will depend on your context and audience. Nonetheless:
   a. Avoid seeming flip: What you consider humorous, your audience may find cavalier or disrespectful; thus, for example, the question, "So did I do OK?" might be asked with a smile but received without one.
   b. There may be a fine line between laudable self-assurance and apparent arrogance or conceit. Unless you are negotiating for a very senior position, concluding your interview by asking about vacation times is not likely to sit well with your audience.
   c. If you have been successful in doing your homework about the organization, and if you're sure of the accuracy of your information, you can ask about company operations. Make sure, however, that what you mention is public knowledge: This is not the time to try to impress the audience with your ability to ferret out confidential information on them. Similarly, "name dropping" is unlikely to impress your audience. In both cases, you're courting the consequences of the aptitude/attitude issue mentioned earlier.
   d. It is acceptable to ask how long it might be—in "ballpark" terms—before they make their decision, and whether there is anything else you can offer that might help them reach it. Be careful of connotation here: You should not be asking whether you can now "correct" something you think you did poorly during the interview; you are diplomatically offering to help them. Furthermore, note our later advice under *Follow-up*.

14-6b COMMONLY ASKED INTERVIEW QUESTIONS

Your interviewer(s), of course, may not ask all—or even any—of the questions mentioned in this section; further, note that the rubrics are possibilities rather than definitive classifications. You should also notice that these questions seem to presuppose a candidate (that's you) seeking an entry-level position, rather than one who has been working as a professional engineer for some time.

Education

- What did you major in? Why?
- Which courses did you like best? Least? Why?
- What motivated you to seek a college education?

Experience

- What do you know about our company?
- Why do you want to work for us?
- What aspects of the job you're applying for appeal to you most?
- What kind of work did you do in your last job?
- What were your responsibilities?
- Describe a typical day on your last (or present) job.
- What was the most difficult problem you encountered on your last (or present) job and how did you handle it?
- What did you like best about your previous position(s)?
- What did you like least? Why?
- Why did you leave (or why do you want to leave) your current job?

Human Relations

- How do you get along with others in work situations?
- What kind of people do you enjoy working with?
- What kind of people do you find difficult to work with? Why?
- Have you worked previously in teams?
- How do you get along with others in a team environment? What makes you think so?

Goals

- What are your career goals?
- Why did you choose this particular field?

Self-Concept

- What are your greatest strengths? Why do you think so?
- What are your weaknesses? Why do you think so?
- What qualities do you need to strengthen?
- What could you tell us about yourself?
- Why do you think you are qualified for this job?
- How do you spend your leisure time?
- What do you consider to be your chief accomplishment in each position you've held?
- What have your supervisors complimented you on? Criticized you for?

14-6c LEGAL ISSUES REGARDING QUESTIONS

Your audience is there to assess you—not only to find out about your likely ability to do the job, but probably to decide if they would like to work with you. Thus, as recommended in Chapter 1's discussion of ethics in engineering communication, you should be honest and forthright.

With certain exceptions, current American employment law prohibits questions about your sex, age, religion, or ethnic background. Such questions include the following:

- Are you married/living common-law/single/divorced/widowed?
- Do you have children?
- Do you plan to have children?
- What is your date of birth?
- Have you ever been arrested?
- Where were you born?
- Where does your husband/wife/father/mother work?
- Are you pregnant?
- Do you belong to a religious organization? Which one?
- Do you rent or own your home?
- What is your maiden name?
- Do you have a girlfriend/boyfriend?

If your interviewer does ask questions like these, consider politely indicating that you feel the question is inappropriate or tactfully deflecting it. Remember, however, that it is possible the interviewer has no interest in the *information*; rather, he or she wants to see how you react to the unexpected and/or awkward. Thus, you might pause, then diplomatically indicate that you don't quite understand how the question relates to the job involved, and ask if they could explain that before you try to answer.

If you feel that you are being unfairly pressed, consider whether you are seeing a reflection of the working environment. Ask yourself if you would really like to be a part of such an organization. Remember that unethical interview questions may reflect how the employer treats employees.

14-6d FOLLOW-UP

Some authorities specify a standard follow-up to your interview: a short letter, for example, addressed to at least one of the interviewers, expressing your appreciation for their time, and reiterating your interest in the position. That *might* be an excellent strategy. It shows your continued involvement, stresses your interest, and may well be what distinguishes you from other candidates.

Unfortunately, however, some audiences react to such communications negatively: They see them as meddling or as an inappropriate attempt to influence their decision, or as an indication that you do not recognize the value of *their* time. Therefore, you might wish to conclude your interview by asking whether you could take the liberty of following up within the next few days. But, unless the response is a clear and definitive "yes," sending a follow-up communication will probably not be to your advantage. Remember that your peace of mind, however important to you, is probably not the first priority in their working lives. Take even an implied "no" as the final answer—for the time being.

Notwithstanding the above, some professionals recommend you begin an interview by offering your business (or personal) card to the interviewers—and requesting theirs in return. Then, two or three days later, having obtained their addresses from their business cards, you send them a *very* brief email that says, for example, "I just wanted to say thank-you for the opportunity of talking with you on [*put in the date*] about the [*put in the job title*] position."

But make sure you do not allow that kind of reasonable follow-up to become pestering. Unless your interviewers have clearly indicated otherwise, they are unlikely to welcome repeated queries about their decision. There might well be opportunities with this firm in the future. Remember that your goal is, always, to foster and/or maintain relations as a communicator. Don't bug people.

14–7 A Final Word about LinkedIn

We discussed this social media site in Chapter 6. It is a very useful platform to network with professionals in your field. It is also a fine way, as indicated before in this chapter, to find work.

There are two career-forming advantages of having a LinkedIn profile. First, its content is completely under your control, and, second, it has a powerful algorithm that works beautifully with search engines so that someone trying to look you up will get a profile of your own composition at, or at least usually very near, the top of the search engine's rankings. (Reason number two, by the way, is why you should compose—and frequently revise—your profile very carefully.)

A November 9, 2014, story by Natasha Singer in the *New York Times*, called "A Reference List You Didn't Write" (nytimes.com/2014/11/09/technology/on-linkedin-a-reference-list-you-didnt-write.html), reveals how the site can work against you, too. Human resources workers had been using LinkedIn's "professional" suite of products to look up former colleagues of applicants and contact them—without the applicants' permission, essentially to "dig up dirt" and shrink their pile of resumes. "The LinkedIn service in question was called 'Reference Search.' It was available only to premium account holders, who paid a monthly fee. An employer or recruiter could use it to generate a list of people in its own network who worked at the same company at the same time as a job candidate. It also allowed premium members to use the site's messaging system to contact people who appeared on those lists, without notifying a job candidate."

After a great deal of public outcry, LinkedIn discontinued the Reference Search feature in June 2015; nevertheless, terms of service and privacy statements are always changing on social media platforms, so be sure to keep up with how LinkedIn and other social networks share private information.

CASE STUDY

14 Cover Letters from AAU Students to AEL

SITUATION

Wally Strong, a fourth-year electrical engineering student at AAU, and George Nagmara, an AAU civil engineering student, heard Flavio Santini's presentation on employment with AEL. (See Chapter 13's case study.) Both men decide to apply and they prepare cover letters; these appear in Figures 14.6 and 14.7, respectively.

1225 Davenport Road Ann Arbor, MI 48103
555-0908 Cell: 555-3435
03/03/20__
Employment Manager, AEL
342 Center Street West
Lansing, MI 48980

Dear Employment Manager,

Hello. My name is Wally Strong and I have been going to Ann Arbor U for a while in engineering. I'd like to apply for any of the positions that your person talked about when he visited where I was at school here.

My studies were completed when I finished my forth year at AAU. I took a lot of different courses but really majored in engineering, most of my grades were pretty good but I did get a couple of not so good ones too, but I'm sure what I learned in engineering can be applied practically to whatever kind of engineering you do there in Lansing, where I've always wanted to live. That's the reason why I'm applying to you rather than trying to get a good offer closer to home here.

I'm sure you'll be able to make really good use of all my engineering and communication skills because I know you do a lot of consulting and my specialty is in electrical engineering, I hear in the papers that you guys are involved with laying some new electrical grids in the east. Yes I do try to keep up with the news, that's another reason why you'll probably want me.

I'm not married and don't have any responsibility which would make it easy for me to move whenever you'd need me to start. (Actually, I have a camper van and I don't have a lot of stuff, so I can move myself pretty easily and save you the money.)

I sure hope you'll at least interview me for this job because I'm sure that when you see my credentials and background in detail (I can send a copy right away if you want to see them before you speak to me) you'll think I'm worth it.

Thanks for taking the time to read this and think of me. If you have any questions at all just call me anytime, even on my cell at 555-3435, though that's probably long distance, hope you don't mind.

Thanks again, it'll be really great to meet you!

Wally Strong

Wallace (Wally) Strong

FIGURE 14.6: Wally Strong's Cover Letter

ISSUES TO THINK ABOUT

1. Strong and Nagmara attended the same presentation and created cover letters addressed to the same company. What can you say about the two letters in terms of the following CMAPP components?
   a. **Audience:**
      i) Examination of sources to determine a specific audience
      ii) Response to what the audience knows, needs to know, and wants to know
   b. **Message:**
      i) Content of the letter to fit the context and the audience
      ii) Language appropriate to the context and the audience
   c. **Purpose:** Clear indication to the audience of the purpose of the letter
   d. **Product:**
      i) Credibility deriving from the quality of the product
      ii) Impression of author as potential employee
      iii) Specific format issues

George Nagmara

March 6, 2015

Mr. Frank Nabata
Senior Partner
Accelerated Enterprises Ltd.
342 Center Street West
Lansing, MI 48980

Dear Mr. Nabata:

I am writing in the light of Mr. Flavio Santini's February 3, 2015, presentation to students at Ann Arbor University (AAU).

You will note from my enclosed curriculum vitae that I am currently completing my M.Sc. degree in Civil Engineering here at AAU, for which I expect to maintain my current GPA of 4.00. You will likewise note that I returned to university to undertake my graduate degree after a hiatus of three years, during which I was employed as an Engineer Trainee at the firm of Lather and Brown in Detroit, my home town. Mr. William Brown has kindly agreed to offer me a reference, should you wish to speak to him. He may be reached at his office at (734) 555-1435.

I would be keenly interested in the possibility of a civil engineering career with Accelerated Enterprises. I expect to be in Lansing from May 2 through May 12, following my thesis defense here in Ann Arbor. Therefore, I would be grateful if you could spare me an hour of your time during that period, so that we might discuss how my education and experience could be of benefit to Accelerated Enterprises.

Could you perhaps have your office contact me as soon as possible at the address, phone, fax, or email shown below, so that we might arrange a time convenient to you?

Thank you for your consideration. I look forward to meeting with you.

Yours truly,

George Nagmara

George Nagmara
Encl.

478 Blair Drive Ann Arbor, MI 48103
Tel: (734) 555-0907 Fax: (734) 555-1421 Email: nagmara@aau.edu

FIGURE 14.7: George Nagmara's Cover Letter to AEL

2. What do you think may have been Strong's principal error or errors?
3. Why do you think that Strong did not attach a resume?
4. Why do you think Nagmara did?
5. How would you describe Strong's action request?
6. How does it differ from Nagmara's?
7. How would you formulate one for Strong?
8. How would you account for the differences between the two letters?
9. How do you think Frank Nabata is likely to respond to each letter?

EXERCISES

14.1 What are the five elements of the employment application package discussed in this chapter?

14.2 On what attributes do many employers seem to base their hiring and firing decisions?

14.3 Briefly explain the value of networking before you apply for jobs.

14.4 What are the principal reasons for using a cover letter with your resume?

14.5 How do you decide on the appropriate level of technicality for your cover letter?

14.6 What are the principal format aspects that you should consider for your cover letter?

14.7 What is another term for a resume?

14.8 A resume could be considered a combination of which four CMAPP products?

14.9 How long should a resume be and why?

14.10 What is the principal advantage and what is the main possible disadvantage of including a statement of professional objective?

14.11 What are the two basic types of resumes you're likely to create for an audience to read, what are their principal differences, and what are their principal advantages and disadvantages?

14.12 Create an effective resume that might accompany a revised cover letter from Wally Strong (see Figure 14.6). Use the type of resume shown in Figure 14.1.

14.13 Create the resume that George Nagmara might have enclosed with his cover letter (Figure 14.7). Use the type of resume shown in Figure 14.2.

14.14 Contact at least two people from different companies or organizations who are authoritatively involved in the hiring process. Ask them individually for a few minutes of their time to help you in your project. Ask them to tell you what are the most important things they look for in a cover letter, a resume, and an interview. In a short oral or written analytical report (perhaps to your class), compare and contrast the answers you received and draw conclusions from them.

14.15 a. Select an advertisement from a newspaper or website regarding an engineering position that would interest you.

b. Research the company and position, and then create a cover letter and resume that you might use to apply.

c. Discuss the cover letter and resume in a short evaluative report.

14.16 How would you alter your resume if you were submitting it

a. in a digital format?

b. as a scannable document?

c. as a page on a website?

USEFUL WEBSITES

Note: For ease of reading, we have removed the "http://www." from most of the URLs below. However, some browsers might still require that you insert it to precede some URLs so as not to receive an error message.

■ Accent Resume Writing's site features practical guidance for resume writing as well as examples and critiques of resumes.
Visit: accent-resume-writing.com.

■ CareerLab's Cover-letters.com site features an extensive sampling of letters for the job search.
Visit: cover-letters.com

■ Contract Job Hunter offers a useful "Guide to Effective Resume Writing."
Visit: cjhunter.com/cew/resume_writing_guide.html

■ Edmonds Community College in Lynnwood, Washington, maintains an online Career Action Center that includes advice on "Creating Your Technical Résumé (CIS Students)." Though targeting one field, the information is readily applicable to most branches of engineering.
Visit: edcc.edu/careeractioncenter/jobsearch/resume/technical-resume.html

■ JobStar Central offers career information such as guides for specific careers and online assessment tests. It also provides tips and advice for writing successful resumes and cover letters.
Visit: jobstar.org

■ Monster.com remains one of the largest electronic clearing houses for potential employers and job seekers alike, both in the United States and around the world. This site offers a wealth of potentially useful information and advice.
Visit: monster.com

■ Purdue University's Online Writing Lab offers a Job Search Writing page. Its advice and links are more than worth the time to investigate.
Visit: owl.english.purdue.edu/owl/section/6

■ TopUSAJobs.com uses the tag line "Where Engineering Job Seekers Find the Top Jobs in Engineering." It lists a great variety of available engineering-related positions.
Visit: engineering.jobs.topusajobs.com

CHAPTER 15

The Importance of Language Use

This textbook has been concerned with engineering communication here and now. Underpinning every chapter has been our conviction that you cannot communicate effectively without using the language effectively. And, like it or not, effective language is largely based on the effective application of what we often call grammar.

15–1 Overview

The passages in Figure 15.1 are extracted from a grammar compendium written by author David Ingre a number of years ago. As you read them, we'd ask you to recall Chapter 1's definition of correct language: what the majority of educated, native speakers tend to use in a particular context. For you, that context will most often require a level of discourse appropriate to a professional engineer, and that implies making appropriate use of English grammar.

FIGURE 15.1: Passages

It would seem that the majority of North American English speakers either disliked the little grammar they did study (and thus forgot it) or managed to skip all the grammar classes they should have attended. In fact, recent generations of students apparently escaped all exposure to traditional grammar. Having averted possible boredom at the time, they now lack a profitable skill.

Why do so many people believe they have always disliked grammar?

I suspect you did not want to learn it for two reasons. First, you felt that you already spoke the language quite well, thank you, and it was a total waste of time to memorize speech patterns that no one you knew ever practiced. Second, you were convinced that it was appallingly artificial: You were supposed to study arcane terms and obscure paradigms, all of which reminded you precisely of what you detested about the Latin that you were afraid you might also be forced to learn.

Time, of course, has gone by. So it is likely that life's little surprises have now all but extinguished the passion of your antipathy to grammar. You have recognized that the world of business has certain expectations, one of which is an accepted written standard. Allowing your language use to fall short of that standard tends to curtail your success.

Many people tend to think of grammar as the rules for correct language. For a very long time, in fact, written grammars were what we now call *prescriptive* (the equivalent of thou shalt = *do it!*) and *proscriptive* (meaning thou shalt not = *don't do it!*). Thus, for every possible utterance, someone would be able to say with self-righteous confidence, "that's right" or "that's wrong." And, consequently, generations of native speakers of English learned that there was a rule book to which they were supposed to defer.

However, not that long ago (in terms of language), perhaps less than a hundred years, people started to look at language differently. Grammar, they began to say, is—or should be—more like a set of observations: It is a detailed description of how the language actually works. That is, grammar rules are really a way of writing down the way that people use the language. This type of grammar is, not surprisingly, called *descriptive*, and has become fairly standard.

In this chapter, we try to offer a number of succinct observations about the way we use the kind of English you're going to need as an engineer in America today.

15-2 Parts of Speech

Every word in a message has a use. Understanding word usage will help you communicate more clearly and effectively. Familiarity with the parts of speech will help you choose the best word at the right time.

15-2a NOUNS

A **noun** is a word used to name people, places, or things. Nouns are the largest group of words in the language, and they are used more frequently than any other part of speech. Almost every sentence you read, write, speak, or hear contains at least one noun. Most sentences contain more than one noun.

Categories of Nouns

It is helpful to sort all nouns into one of two very broad categories: proper nouns and common nouns.

Proper Nouns A proper noun names a specific person, place, or thing. Proper nouns are always distinguished by capital letters.

SPECIFIC PEOPLE	SPECIFIC PLACES	SPECIFIC THINGS
Meera	Seattle	Bic pens
Mr. Yukimura	Tibet	Canon copiers

Common Nouns A common noun is a word that identifies a person, place, or thing in a general way.

baseball fan	movie-goer	assets	goodwill
boy	table	joy	team

Common nouns can be compound ("editor-in-chief," "vice president," "son-in-law," "board of directors").

Noun Plurals

Most noun plurals may be formed by using one of three rules.

Rule 1. Add *s* to the end of most nouns.

	SINGULAR	PLURAL
Common Nouns	pamphlet	pamphlets
	song	songs
Proper Nouns	Chang	the Changs
	Corvette	Corvettes
	New Yorker	New Yorkers
Abbreviations	CPA	CPAs
Numbers	1800	the 1800s

Rule 2. Add *es* to any singular noun that ends in *s, x, z, sh,* or *ch.*

SINGULAR	PLURAL	SINGULAR	PLURAL
lens	lenses	tax	taxes
Lopez	the Lopezes	bush	bushes
Lynch	the Lynches	wrench	wrenches

Rule 3. Add an *s* to form the plural of any noun ending in *y* when the *y* follows a **vowel** (*a, e, i, o,* or *u*). With the exception of proper nouns, to form the plural of nouns that have a **consonant** (all letters except vowels) before the final *y*, change the final *y* to *i* and then add *es.*

	SINGULAR	PLURAL
Nouns ending in a vowel + y	delay	delays
	key	keys
	relay	relays
Nouns ending in a consonant + y	city	cities
	territory	territories

Foreign Words. Because the *s* ending is standard for English plurals, plurals that do not end in *s* may sound odd. English claims a considerable number of such words, mostly borrowed from Latin and Greek.

SINGULAR	PLURAL	SINGULAR	PLURAL
medium	media or mediums	alumnus	alumni
crisis	crises	matrix	matrices

Nouns Ending in *o*. Nouns that end in *o* form their plurals in one of two ways. Many simply add *s* to form their plurals. Others add *es* to form their plurals.

SINGULAR	PLURAL	SINGULAR	PLURAL
ratio	ratios	radio	radios
potato	potatoes	veto	vetoes

Compound Nouns. Compound nouns may be spelled as separate words; in this case, the most important word is made plural.

SINGULAR	PLURAL
editor-in-chief	editors-in-chief
side scan sonar	side scan sonars

Compound nouns may be joined by hyphens; in this case, the base form is made plural.

SINGULAR	PLURAL
brother-in-law	brothers-in-law

Compound nouns may be spelled as one word; in this case, the plural is formed by adding *s* or *es* or changing the *y* to *i* and adding *es* (depending on the word ending).

SINGULAR	PLURAL
letterhead	letterheads
housefly	houseflies

Proper Names. When forming plurals, treat a proper name like any other noun with this exception: Add only *s* to all proper nouns that end in *y*; ignore the "change *y* to *i* . . ." rule with proper names. In the examples shown below, the word "the" is inserted before the plurals to simulate real-life use.

SINGULAR	PLURAL
John Haggerty	the Haggertys
Rosemary Portera	the Porteras

One-Form Nouns. Some nouns have only one form. Depending on the noun, that one form may be either always plural or always singular.

ALWAYS PLURAL	ALWAYS SINGULAR
thanks	news
scissors	headquarters

A **collective noun**, such as "tribe" or "jury," represents a group that usually acts as a single unit.

The *jury* eats in the cafeteria at noon.

However, when we want to emphasize the idea of the "members" acting "individually," the noun takes a plural verb.

The *committee* appear unable to agree on a decision.

Count Nouns and Noncount Nouns

In general, we assign nouns to one of two categories, depending on *how* we "count" them: **count nouns** and **noncount nouns** (often called *countables* and *noncountables*).

Count nouns are those which we normally turn into plurals when we wish to designate "more than one" of them. This includes the vast majority of nouns referring to tangibles: for example, "book," "bridge," "microwave," "person," "computer"—and even multiword nouns such as "ball bearing," "mine shaft," "valve assembly," and so forth.

Noncount nouns are those which we do not normally turn into plurals in the same way, since the language tends to see them as somehow "different." In this category are many nouns that refer to abstract qualities. We do not, for example, talk of "honesties" or "courages" or "cowardices." Noncount nouns also include things like "thunder" and "lightning": Instead of "two thunders" or "three lightnings," we would refer to "two crashes of thunder" or "three flashes of lightning."

"Mixed-use" nouns can function as both count or noncount nouns, depending on what we mean. For example, if you are a chemist, you might discuss different *salts*—sodium salts and potassium salts, for instance; you might refer to *sugars*—natural versus synthetic, for instance. In these cases, "salt" and "sugar" are **count nouns**. However, when we think of "salt" or "sugar" in terms of condiments, we "pluralize" differently—*grains* of salt or *cubes* of sugar, for example.

Note, also, that "search" is a count noun. Thus we would say, "I conducted five detailed *searches* on those websites." However, "research" is a noncount noun: We might say, "I conducted over three hours of detailed *research* on those websites."

Finally, note that formal language makes the following distinctions:

WITH COUNT NOUNS	WITH NONCOUNT NOUNS
Many	***Much***
We noticed *many* inaccuracies in their report.	We did not have *much* difficulty completing the task.
N.B.: "a lot of" is used in less formal speech: We noticed *a lot of* inaccuracies.	N.B.: "a lot of" is used in less formal speech for noncount nouns, too: We did not have *a lot of* difficulty.
A ***number*** of	An ***amount*** of
We noticed a large *number of* inaccuracies in their report.	We discovered a large *amount* of detritus below the ridge.

WITH COUNT NOUNS	WITH NONCOUNT NOUNS
Few / fewer / fewest	***Little / less / least***
The new software caused *few* problems for staff.	The new software caused *little* confusion among staff.
It caused *fewer* problems than the old version did.	It caused *less* confusion than the old version did.
It caused the *fewest* problems of all the applications.	It caused the *least* confusion of all the applications.

Possessive Nouns

Possessive nouns show possession. To form the possessive, add an apostrophe plus *s* to all singular nouns, both common and proper.

Man	+	's	=	one man's opinion
Mr. Ross	+	's	=	Mr. Ross's district

Add only an apostrophe to any plural noun if it ends in *s*.

executives	+	'	=	three executives' goals
district attorneys	+	'	=	the district attorneys' ideas

Irregular plural nouns (such as "men," "women," "children," and "alumni") and some compound nouns are examples of plural forms that do not end in *s*. For these exceptions, add an apostrophe plus *s* to form their possessives; in other words, apply the rule for singular nouns.

women	+	's	=	both women's investments
brothers-in-law	+	's	=	my two brothers-in-law's kitchens

15–2b PRONOUNS

Pronouns are convenient substitutes for nouns, and they help to communicate the nominative, objective, and possessive forms to listeners and readers.

Personal Pronouns

A personal pronoun substitutes for a noun that refers to a specific person or thing.

Nominative Case. A nominative case pronoun ("I," "we," "you," "he," "she," "it," "who," "whoever"), sometimes referred to as a subjective case pronoun, may be used as a subject or a predicate nominative. A predicate nominative is a noun or pronoun that refers to the subject and follows a form of the verb "to be" ("am," "is," "are").

Carla and I voted for Josh.
It is she who received all the attention. (predicate nominative)

Objective Case. An objective case pronoun ("me," "us," "you," "him," "her," "it," "them," "whom," "whomever") may be used as a direct or indirect object of a transitive verb, which is a verb that denotes action and needs an object. A direct object is a noun or pronoun directly affected by the action of the verb. (They chose *me*.) An indirect object is a noun or pronoun that receives the verb's action. (They gave *me* a gift.) An objective case pronoun may also be used as the object of a preposition.

Please send *them* by express mail.
Ned bought *her* a burrito.
Give it to *whomever* you see first.

Possessive Case. A pronoun that indicates ownership or possession is a possessive case pronoun ("my," "mine," "our," "ours," "your," "yours," "his," "her," "hers," "its," "their," "theirs," "whose"). Unlike nouns, pronouns do not need an apostrophe to signal possession.

Those folders are *ours*.
The fancy clothes are *hers*.
My going to the party surprised Jose.

Reflexive Pronouns

A **reflexive pronoun**, which ends in "self" or "selves," refers to a noun or pronoun that appears earlier in a sentence.

We found *ourselves* reminiscing at the reunion. (The reflexive pronoun "ourselves" refers to "we.")
You, *yourself*, designed that assembly.

Interrogative Pronouns

An **interrogative pronoun** begins a question that normally leads to a noun response. Interrogative pronouns are "who," "whose," "whom," "which," and "what."

Who is in your office?	*Whom* do you want to call you?
Whose are these?	*Which* of those are important?

Demonstrative Pronouns

A **demonstrative pronoun** is used to "point to" a specific person, place, or thing. The four demonstrative pronouns are "this," "that," "these," and "those."

Do you prefer *this* monitor or *that*?
These books should be moved next to *those*.

15-2c VERBS

A verb expresses action, a state of being, or a condition of the subject of the sentence. No sentence is complete without a verb, and some sentences have more than one verb.

Types of Verbs

Every sentence must have a verb in order to be complete. Verbs are either action or linking verbs. Linking verbs, also called copula verbs, include state-of-being verbs and condition verbs.

Action Verbs. **Action verbs** help to create strong, effective sentences. Action verbs may take objects and indirect objects.

Mr. Gomez *teaches* me Finance 102.
Gabrielle *wrote* legibly.

State-of-Being Linking Verbs. **State-of-being verbs**, sometimes called *to be* verbs, do not have objects or indirect objects; instead, these verbs have predicate nominatives and predicate adjectives. The verb "to be" has many different forms to denote the present, past, or future state of being.

The new president *is* Mr. Jongg. (The predicate nominative is "Mr. Jongg.")
The old software programs *were* expensive and inefficient. (The predicate adjectives are "expensive" and "inefficient.")

Condition Linking Verbs. A **condition verb** does not have an object or an indirect object. Instead, it connects an adjective to the subject. Condition linking verbs either refer to a condition or appeal to the senses.

The assistant *appears* cooperative.
The health food *tastes* delicious.

Transitive and Intransitive Verbs

A **transitive verb** is a verb that must have a direct object to complete the meaning of a sentence.

Clark *suggested*. (Incomplete—What did he suggest?)
Clark *suggested* a profitable method. (Complete)

An **intransitive verb** is a verb that does not need a direct object to complete the meaning of a sentence.

The recruits *laughed*.
The error *weighed* heavily on his conscience.
He *will be* treasurer. ("Treasurer" is a predicate nominative, not an object, and "will be" is a copula verb.)

Verb Tenses

There are six verb tenses in English; they indicate the time an action takes place. These six tenses are categorized into two groups, simple and perfect.

Simple Tenses. The simple tenses are called present, past, and future. A present tense verb expresses present occurrences (what is happening now).

Computer services *sell* information.
Georgette *is teaching* a course in merchandising.

A **past tense verb** expresses action recently completed.

Restless, the Commander *walked* all night.
Tammy *was visiting* her bed-ridden father.

A **future tense verb** expresses action or condition yet to come. The future tense is formed by placing the helping verb (sometimes called the auxiliary verb) "will" before the main verb (sometimes called the principal verb).

I *will vote* on election day.
The accountants *will be consulting* with their clients.

Perfect Tenses. A perfect tense verb describes the action of the main verb in relation to a specific time period in the past, from the past to the present, or in the future. The three perfect tenses are present perfect, past perfect, and future perfect. Form the perfect tense by preceding the past participle form of the main verb with either "have," "has," or "had."

A **present perfect tense verb** indicates continuous action from the past to the present. "Has" or "have" precedes the past participle form of the main verb.

Francois *has voted* in every election since 1986.
They *have been jogging* every day since the beginning of the month.

A **past perfect tense verb** indicates action that began in the past and continued to the more recent past when it was completed. "Had" precedes the past participle form of the main verb.

Francois *had voted* in every election until last week.
They *had been jogging* every day until this past Monday.

A **future perfect tense verb** indicates action that will be completed at a specific point in the future. "Will have" precedes the past participle form of the main verb.

Including next year, Francois *will have voted* in every election since 1986.
By next Tuesday, they *will have been jogging* every day for a month.

Active and Passive Voice

Voice indicates whether the subject is doing the action or receiving the action of a verb. **Active voice** means that the subject of a sentence is doing the action.

Gianni *completed* his report using his computer.
Alberta *rode* her bike to work and back.
The young sprinter *won* the race.

Passive voice means that the subject of a sentence is receiving the action. The passive voice is formed with the past participle and a form of the verb "to be."

The report *was completed* by Gianni.
The bike *was ridden* to work and back by Alberta.
The race *was won* by the young sprinter.

Verbals

A **verbal** is a verb form used as a noun, adjective, or adverb. The three verbals are infinitives, gerunds, and participles.

Infinitive. An **infinitive** is a verb form that functions as a noun, an adjective, or an adverb, but not as a verb. An infinitive is formed by placing the word "to" in front of a present tense verb.

To run like the wind is Jaime's dream. (noun)
Her desire *to become* principal is noble. (adjective)
Racine International was founded *to promote* world peace, and its mission remains the same today. (adverb)

Gerund. A gerund is an *-ing* verb form that functions as a noun. Gerunds may be used in a phrase that contains the gerund, an object, and its modifiers.

Avoiding the awful truth was her tendency. (subject)
Su Yung's career is *refurbishing* boats. (predicate nominative)
Every Friday, they love *swimming* at the YMCA. (direct object)
Telly's habit, *falling* off to sleep, gets him into trouble in class, especially on test days. (appositive, defined in section 15.4)
His talent for *guessing* someone's age is uncanny. (object of preposition)

Participle. A **participle** is a verb form that can be used either as an adjective or as part of a verb phrase. The present participle is always formed by adding *ing*. The past participle is usually formed by adding *d* or *ed* to the present tense of a regular verb, or it may have an irregular form. The perfect participle always functions as an adjective and always is formed by combining "having" with the past participle of the verb.

Casey has a *snoring* dog on her front porch. (present participle)
There were six *launched* satellites that summer. (past participle)
This is a *broken* arrow. (irregular past participle)
Having understood the problem, she was able to find a solution. (perfect participle)

15-2d ADJECTIVES

An adjective is a word that describes or limits nouns or noun substitutes (pronouns, gerund phrases, and infinitive phrases). Adjectives answer questions about nouns.

Which one? *this* proposal, *those* appointments
How *many? six* calls, *few* tourists
What kind? *ambitious* student, *creative* teacher
She is reading a *suspenseful* book. (The adjective "suspenseful" describes "book.")
Casey's stylish suit was perfect for *Dot's* wedding. (The adjectives "Casey's" and "stylish" describe "suit"; "Dot's" describes "wedding.")

Articles

Although classified as adjectives, the words "the," "a," and "an" are also called **articles**. "The" denotes a specific noun or pronoun. "A" or "an" denotes a nonspecific noun or pronoun.

Place the article "the" before a noun to designate that the noun is specific, not general.

the person (a specific person)
the toy (a specific toy)

Place the article "a" before a noun that begins with a consonant sound to designate that the noun is general, not specific.

a person (a nonspecific person)
a toy (a nonspecific toy)

Place the article "an" before a noun that begins with the sound of a vowel.

an honorable leader
an attractive child

Nouns and Pronouns Used as Adjectives

Nouns or pronouns that precede and modify other nouns and answer questions such as "which one" or "what kind" are used as adjectives.

Luis had four *theater* tickets. (Usually a noun, "theater" serves as an adjective describing the kind of tickets.)
Did you see *my mathematics* assignment? (The pronoun "my" and the noun "mathematics" are used as adjectives to identify which assignment.)

Proper Adjectives

Proper nouns that precede and modify other nouns serve as **proper adjectives**. Begin proper adjectives with capitals.

Burton is proud of his *New York* accent.
Our family thoroughly enjoys *Thanksgiving* dinner.
Marta has lost her favorite *Ann Arbor University* sweatshirt.

Compound Adjectives

A **compound adjective** is two or more hyphenated words that precede and modify nouns.

The *well-known* mystery writer is signing copies of his book.
Vivian is selling *long-term* health care insurance policies.
This *fast-acting* medicine will lessen your pain very quickly.

Comparison of Regular Adjectives

Adjectives have three degrees for comparison: the positive degree, the comparative degree, and the superlative degree.

Use the **positive degree** to describe one item.

Ryan is an *efficient* worker.
That item is in a *big* carton.

Also use the positive degree to express equality.

He is as *big* as you.

Use the **comparative degree** to describe two items.

The box is a *bigger* carton than the first one.
Ryan is *less efficient* than Hsinchen.

Use the **superlative degree** to describe three or more items.

The box is the *biggest* carton of the three.
Ryan is the *least efficient* of the new employees.

To create the comparative degree of regular adjectives:

- For most monosyllabic adjectives, add *er* to the positive form (sometimes doubling its final letter):

  bold → bolder
  big → bigger
  fine → finer

- Precede most polysyllabic adjectives with "more" or "less," depending on your meaning:

  effective → more effective
  reasonable → more reasonable
  complex → less complex
  determined → less determined

To create the superlative degree of regular adjectives:

- For most monosyllabic adjectives, add *est* to the positive form (sometimes doubling its final letter):

  close → closest
  rare → rarest

- Precede most polysyllabic adjectives with "most" or "least," depending on your meaning:

  judgmental → most judgmental
  relaxed → most relaxed
  complicated → least complicated
  measurable → least measurable

Comparison of Irregular Adjectives

A few frequently used adjectives do not form their comparisons in the usual manner (adding *er* or the word "more" or *est* or the word "most").

POSITIVE	COMPARATIVE	SUPERLATIVE
good book	better book	best book
bad result	worse result	worst result
little amount	less amount	least amount
many reports	more reports	most reports
much laughter	more laughter	most laughter

Absolute Adjectives

Some adjectives cannot be compared because they do not have degrees; they are already at the maximum level of their potential. These adjectives are referred to as **absolute adjectives**. Some examples are "immaculate," "perfect," "square," "round," "complete," "excellent," and "unique." When you use these words in your sentences, use them alone or precede them with the terms "more nearly" or "most nearly."

The food at Tim's restaurant is *excellent*.
Your yard is *more nearly square* than your neighbor's.
This typeface is *unique* to the economics textbook.

15–2e ADVERBS

An **adverb** is a word that modifies an action verb, an adjective, or another adverb. Most adverbs end in *ly*. An adverb answers the questions "how," "when," "where," "how often," or "to what extent."

He wrote the paper *correctly*. (how)
He wrote the report *yesterday*. (when)
He wrote the report *here*. (where)
He wrote the report *twice*. (how often)
He wrote the report *very* quickly. (to what extent)

Modifying Action Verbs

Adverbs modify action verbs but not linking verbs. Linking verbs are modified by adjectives, as in "She *appears* happy" or "George *is* intelligent." Action verbs, on the other hand, require adverbs.

She gave it to me *gladly*.
The dog sat up and begged *just once*.

Modifying Adjectives

An adverb that modifies an adjective usually answers the question "to what extent."

The cookies are *very* good. (The adverb "very" modifies the adjective "good.")
That new project is *tremendously* complex. (The adverb "tremendously" modifies the adjective "complex.")

Modifying Other Adverbs

An adverb also can answer the question "to what extent" about another adverb in a sentence.

The grammar school pupil did her work *too* quickly. (The adverb "too" modifies the adverb "quickly.")
We purchased the printer *very* recently. (The adverb "very" modifies the adverb "recently.")

Accompanying Verb Phrases

Because adverbs modify action verbs and verb phrases that include action verbs, adverbs such as "never" or "always" frequently appear in the middle of verb phrases.

Ned is *always* writing email messages.
I have *never* seen such an amazing use of cinematography before.
That has *already* been ordered.

Conjunctive Adverbs

A special group of adverbs called **conjunctive adverbs** includes words such as "therefore," "moreover," "however," "nevertheless," and "furthermore." A conjunctive adverb is a transitional word that joins two independent but related sentences.

They remained at work late; *therefore*, they were able to complete the project.
She works after school as a docent; *moreover*, she waits tables on the weekend.

Comparison of Adverbs

Like adjectives, adverbs have three degrees of comparison: positive, comparative, and superlative. Adverbs usually show their comparative form by adding *er* or the words "more" or "less" to the simple form (positive degree). They show their superlative form by adding *est* or the words "most" or "least" to the simple form.

POSITIVE	COMPARATIVE	SUPERLATIVE
arrived late	arrived later than	she arrived latest of all
clearly written	more clearly written	most clearly written

15–2f PREPOSITIONS

A **preposition** is a word that usually indicates direction, position, or time. A preposition is linked to a noun or noun substitute to form a phrase.

She walked *into* the classroom. (direction)
She stood *behind* the open gate. (position)
She left work *before* lunch. (time)

The examples below are some of the most commonly used prepositions.

about	at	by	like	toward
above	before	concerning	of	under
across	behind	during	off	until
after	below	except	on	up
against	beneath	for	out	upon
along	beside	from	over	with
around	beyond	into	to	without

The Role of the Preposition

Prepositions introduce phrases called **prepositional phrases**. A prepositional phrase begins with a preposition and ends with a noun or noun substitute that functions as the object of the preposition. In addition, one or more adjectives that modify the object may appear in a prepositional phrase.

Place the carton *behind the tall cabinet*. (The preposition is "behind"; the object is the noun "cabinet"; "the" and "tall" are modifiers.)

Gary believes that learning a spreadsheet software program is *beyond him*. (The preposition is "beyond"; the object is the pronoun "him.")

Prepositional Phrases Used as Adjectives

Prepositional phrases may be used to modify nouns and noun substitutes in sentences. They can have the same function as adjectives and answer questions such as "what kind" or "which one" about the words they modify.

Robert is *among those here*. (The prepositional phrase "among those here" modifies the noun "Robert.")

They, *without a doubt*, are the most considerate people I have ever met. (The prepositional phrase "without a doubt" modifies the pronoun "they.")

Prepositional Phrases Used as Adverbs

Prepositional phrases may be used to modify action verbs, adjectives, or adverbs. Prepositional phrases can have the same function as adverbs and answer questions such as "when," "where," "why," "how," or "to what extent" about the words they modify.

> *After lunch*, Maria filed the papers. (when)
> Ms. Torres is very knowledgeable *about the subject*. (how)

15-2g CONJUNCTIONS

A **conjunction** is a word that joins two or more words, phrases, or clauses. There are three types of conjunctions: coordinate conjunctions, correlative conjunctions, and subordinate conjunctions.

Coordinate Conjunctions

A **coordinate conjunction** joins words, phrases, and clauses of equal grammatical rank. **Equal grammatical rank** means that the connected elements are the same part of speech. For example, the connected elements may be nouns, verbs, prepositional phrases, or independent clauses. The coordinate conjunctions are "for," "and," "nor," "but," "so," "or," and "yet."

> Leo is studying computer science, *for* he plans to be a systems analyst. (The conjunction "for" joins two independent clauses.)
> The teacher *and* the principal spoke. (The conjunction "and" joins two nouns.)
> Tien wanted to attend the workshop, *but* she couldn't spare the time. (The conjunction "but" joins two clauses.)
> Philippa says she loves to travel, *yet* she has never been on an airplane. (The conjunction "yet" joins two clauses.)

Correlative Conjunctions

A **correlative conjunction**, like a coordinate conjunction, is a word that connects words, phrases, and clauses of equal grammatical rank. Correlative conjunctions differ from coordinate conjunctions because they are always used in pairs for emphasis.

> *Both* Greg *and* Barbara applied for the teaching position.
> *Neither* Greg *nor* Barbara applied for the teaching position.
> *Not only* Greg *but also* Barbara applied for the teaching position.

Subordinate Conjunctions

A **subordinate conjunction** joins elements of unequal grammatical rank. It is primarily used to connect dependent clauses with independent clauses.

> *Although* we couldn't attend, we sent a donation.
> Jules and Beth will visit *provided* they are allowed.

15-2h INTERJECTIONS

An **interjection** is a word or expression that has no grammatical relationship with other words in a sentence. An interjection is primarily used to express strong emotion; therefore, it is often followed by an exclamation point.

> *Hey*, get your coffee cup off my monitor!
> Your idea is sure to work. *Super!*

15–3 Sentence Parts and Sentence Structure

A **sentence** is a group of related words that contains a subject and a predicate and expresses a complete thought. When forming sentences, the parts of speech are arranged into subjects and predicates.

15–3a THE SUBJECT IN A SENTENCE

A **subject** is either the person who is speaking, the person who is spoken to, or the person, place, or thing that is spoken about.

Simple Subject

The **simple subject** is the main word in the complete subject that specifically names the topic of the sentence. The simple subject of a sentence is never in a prepositional phrase.

> *John* writes articles.
> *John*, the young journalist, has written articles.

Complete Subject

The **complete subject** includes the simple subject plus all the other words in the sentence that are not part of the complete predicate.

> *John* writes articles.
> *John*, *the young journalist*, has written articles.

Compound Subject

A **compound subject** is two or more simple subjects joined by conjunctions such as "and," "or," "nor," "not," "only/but also," and "both/and."

> *John and Sally* work for our company.
> *Not only Jorge but also Svetlana* will attend the career fair tomorrow.

When two nouns in a subject refer to one person, place, or thing, the article "the" (or "a") is omitted before the second noun.

> The *teacher* and *counselor* is my friend.

When two nouns in a subject refer to two people, places, or things, the article "the" (or "a") is placed before each noun.

> *The teacher* and *the counselor* are my friends.

15–3b THE PREDICATE IN A SENTENCE

The discussion of a predicate is divided into three brief parts: the simple predicate, the complete predicate, and a compound predicate.

Simple Predicate

The **simple predicate** is the verb in the complete predicate.

> John *writes* articles.
> John, the young journalist, *has written* articles.
> Marisol *walked* the dog around the block.

Complete Predicate

The **complete predicate** is everything in the sentence said by, to, or about the subject; it always includes the main verb of the sentence. Whatever is not included in the complete subject of a sentence belongs in the complete predicate.

> John *writes articles.*
> John, the young journalist, *has written articles.*
> Marisol *walked the dog around the block.*

Compound Predicate

A **compound predicate** consists of two or more verbs with the same subject. The verbs are connected by conjunctions such as "and," "or," "nor," "not only/but also," and "both/and."

> John and Sally *discussed* the matter and *concluded* that our actions were incorrect.
> The engineer not only *complained* but also *refused* to finish the project.

Objects and Subject Complements

Objects and **subject complements** help complete the thought expressed by a subject and simple predicate.

Objects

An **object** is a noun, pronoun, clause, or phrase that functions as a noun. It may be direct or indirect.

A **direct object** helps complete the meaning of a sentence by receiving the action of the verb. In fact, only action verbs can take direct objects. Direct objects answer the questions "what" or "whom" raised by the subject and its predicate.

> Louis closed the *door*. (Louis closed *what*?)
> The boy lost his *mother*. (The boy lost *whom*?)

An **indirect object** receives the action that the verb makes on the direct object; you cannot have an indirect object without a direct object. Neither the direct object nor the indirect object can be part of a prepositional phrase.

The indirect object usually answers the question "to whom is this action being directed." You can locate the indirect object by inverting the sentence and using the passive voice, and adding "to."

> Michiko gave *Thomas* the candy bar. (The candy bar was given by Michiko to Thomas.)
> Nancy brought the *twins* broccoli with cheese. (Broccoli with cheese was brought to the twins by Nancy.)

Subject Complements

A **subject complement** is either a noun or pronoun that renames the subject or an adjective that describes the subject. In either case, it always follows a state-of-being or linking verb (such as "am," "is," "are," "was," "were," "has been," "seems," "appears," "feels," "sounds," "looks," and "tastes").

> Petersmeyer is an honest *banker*. (The noun "banker" renames "Petersmeyer.")
> Her writing appears *magical*. (The adjective "magical" describes "writing.")

15-3c CLAUSES, PHRASES, AND FRAGMENTS

A **clause** is a group of words with a subject and a predicate; a **phrase** is a group of words with no subject or predicate.

Clauses

A clause is labeled **independent** if it can stand alone as a complete sentence.

> *One of our sales managers has developed an excellent training manual*, which we plan to use in all future training sessions.

A clause is labeled **dependent** if it cannot stand alone as a complete sentence.

> One of our sales managers has developed an excellent training manual, *which we plan to use in all future training sessions.*

Phrases

A **phrase** is a group of related words that does not contain both a subject and a predicate.

A **verb phrase** is a group of words that functions as one verb.

> Frederico *will be finished* when we call him.
> The IBC Corporation *has been supplying* us with these products.

A **prepositional phrase** is a group of words that begins with a preposition and ends with a noun or a noun substitute.

> Place both cartons *on the desk.*
> The boxes *in the office* belong *to him.*

Phrases add detail, interest, variety, and power to your writing. Compare the examples below.

> Horace writes.
> An avid storyteller, Horace writes shocking, turn-of-the-century ghost tales for impressionable teenagers.

Fragments

A **fragment** is an incomplete sentence that may or may not have meaning. Fragments that have meaning in context ("Yours truly,") can be used in business messages. However, do not use fragments that have no meaning.

> **Fragment:** Sam, the vice president's brother.
> **Sentence:** Sam, the vice president's brother, got a hefty raise.
> **Fragment:** Because the beds were uncomfortable.
> **Sentence:** Because the beds were uncomfortable, Goldilocks slept on the floor.
> **Fragment:** As soon as I receive a raise.
> **Sentence:** I will plan my vacation as soon as I receive a raise.

15-3d SENTENCE STRUCTURES

Your communication will be more stimulating if you vary the types of sentences you write. There are four basic **sentence structures**, which are classified by the number and type of clauses they have.

Recall that the two types of clauses are independent (main or principal) and dependent (subordinate). As an effective workplace communicator, you can put emphasis on an idea by placing it in an independent clause, or you can take emphasis off of it by placing it in a dependent clause.

The Simple Sentence

A **simple sentence** contains one independent clause and no dependent clauses. There may be any number of phrases in a simple sentence. Especially in workplace writing, a simple sentence can clearly and directly present an idea because there are no distracting dependent clauses. However, if overused, too many simple sentences can sound monotonous or abrupt.

Michael Jackson sang. (simple sentence)
Michael Jackson and the Beatles sang. (simple sentence with compound subject)
Michael Jackson sang and danced. (simple sentence with compound predicate)
Michael Jackson, one of the world's best-known celebrities, sang and danced. (simple sentence with various phrases)

The Compound Sentence

A **compound sentence** contains two or more independent clauses and no dependent clauses. In other words, two main ideas share equal importance. Note in these examples that the two independent clauses are joined by a coordinating conjunction, a conjunctive adverb, or a semicolon.

Mr. Feinstein is the founder, and he was the first president of FSI. (coordinating conjunction)
Are you going to the farmer's market, or are you going to the grocery store? (coordinating conjunction)
It's getting late; however, I am glad to stay here and finish this project. (conjunctive adverb)
Erin loves to ride horses; Connor loves to draw horses. (semicolon)

The Complex Sentence

A **complex sentence** contains one independent clause and one or more dependent clauses. In this structure, one or more ideas are dependent upon the main idea. Use dependent clauses to de-emphasize less important or negative ideas or to provide detail and support to the main clause.

Although it is important to proofread, many people feel they do not have the time.
Dan, who cannot swim, hates wading in Lake Waldo because it appears polluted.
You should understand that Karen and Tim are happily married.

The Compound-Complex Sentence

A **compound-complex sentence** contains two or more independent clauses and one or more dependent clauses. This structure offers the technical writer a variety of ways to present ideas and emphasize or de-emphasize details. Because this structure can become long and complicated, be careful how you use it in engineering communications. In the examples below, the independent clauses are in **bold** and the dependent clauses are in *italics*.

Since Noni left the folders on the desk, **her assistant decided to finish up**, and **he did a good job**, *even though he was dead tired.*
Sierra and Casey, *who are cousins*, **play together** often; **however, their fathers**, *who are brothers*, **don't see enough of each other** *because they both travel so much.*

15–3e SUBJECT-VERB AGREEMENT

Good communicators make sure that their subjects and verbs always agree ("he walks," "they walk"). Grammatical errors in subject-verb agreement offend the receiver and label the person who erred as a careless writer or speaker.

Number

Third-person singular pronouns and singular nouns require a singular verb that ends in *s* when the present tense is used. Third-person plural pronouns and plural nouns require a plural verb that does not end in *s* when the present tense is used.

> *Joy telephones* her parents daily. (singular)
> *Ari drives* to his client's warehouse every Monday. (singular)
> Joy's *parents telephone* her daily. (plural)
> The *musicians record* their music when they have a chance. (plural)

Inverted Sentences

If a sentence is **inverted** (predicate precedes subject), putting the sentence in normal order will help you check subject-verb agreement.

Inverted Order:	In the recruit's many strengths *lies* her *admiration*.
Normal Order:	Her *admiration lies* in the recruit's many strengths.
Inverted Order:	In the box *are* two *bags* of apples.
Normal Order:	Two *bags* of apples *are* in the box.

Intervening Phrases

Intervening words do not affect subject-verb agreement and should be ignored when making subjects and verbs agree. Note these examples with the intervening words in **bold** and the subjects and verbs in *italics*.

> The *manager* **of the sports teams** *is traveling* to New Orleans.
> *A professor*, **rather than the college administrators**, *represents* the institution at the convention.
> My *assistants*, **along with the company comptroller**, *work* overtime on this project.

A Number, The Number

When used as a subject, the expression "a number" is considered to be plural and needs a plural verb.

> *A number* of inquiries *come* to our office each day.
> There *are a number* of tourists at our concert.

When used as a subject, the expression "the number" is considered to be singular and needs a singular verb.

> *The number* of attorneys in Philadelphia *is* on the rise.

Names of Companies

Names of companies are usually considered singular. Although a firm's name may end in *s* or include more than one individual's name, it is still one business.

> *Gordon, Rodriguez, and Ramirez is representing* the plaintiff.

Amounts

An amount that is plural in form takes a singular verb if the amount is considered to be one item.

> *One hundred dollars is* a generous wedding gift.

An amount that is plural in form takes a plural verb if the amount is considered to be more than one item.

Fifty-three dollar bills are in my wallet.

Compound Subjects Joined by "And"

Because errors in subject-verb agreement commonly occur with compound subjects, take a careful look at some special guidelines. Usually a compound subject joined by "and" is plural and requires a plural verb.

Mei-ling and Yuan are visiting their parents in Wuxi.

Sometimes compound subjects are treated as one item and require a singular verb.

Peanut butter and jelly is popular in the grammar school.

If "each," "every," or "many a" precedes a compound noun, always use a singular verb.

Many a homeowner and investor *has supported* this tax increase.

Compound Subjects Joined by "Or/Nor"

When a compound subject is joined by "or," "nor," "either/or," or "neither/nor," the verb agrees with the subject that is closer to the verb.

Tracey or *Hal seems* to be well qualified for the position.
Either George or *his sisters are* catering the buffet.
Neither the supervisors nor *the security guard has seen* the criminal.

Pronoun-Antecedent Agreement

The noun or noun phrase that is replaced by the pronoun is called the **antecedent** of the pronoun. The pronoun must agree with its antecedent in person, number, and gender.

- *Person*. Use a first-person pronoun to represent the persons speaking ("I", "we"). Use a second-person pronoun to represent the persons spoken to ("you"). Use a third-person pronoun to represent the persons spoken about ("he," "she," "it," "they").
- *Number*. Use a singular pronoun ("he," "she") to refer to an antecedent that is a singular noun. Use a plural pronoun ("they") to refer to an antecedent that is a plural noun.
- *Gender*. Use a masculine pronoun ("his") to refer to an antecedent that is a masculine noun. Use a feminine pronoun ("her") to refer to an antecedent that is a feminine noun. Use a gender-neutral pronoun (such as "it") to refer to an antecedent that is a gender-neutral noun (such as "table").

John encouraged *his* staff.
Anyone can state *his* or *her* opinion on the matter.

Third-Person Pronoun Agreement

While writers do not have many problems matching first- and second-person pronouns with their antecedents, they do, on occasion, find that third-person pronouns present problems in gender and number.

The gender of the antecedent in a sentence is not always obvious. For example, nouns such as "manager," "nurse," "astronaut," "president," "systems analyst," or "worker" could apply to either gender. Here are two alternative solutions:

1. Use both masculine and feminine pronouns to agree with an antecedent if its gender is unknown.

A *doctor* tends to *his or her* patients without favoritism.

2. Change the antecedent to a plural form and use the gender-neutral plural pronoun "their."

The *astronauts* cooperate 100 percent with *their* peers at NASA.

A problem may arise when applying the number-agreement principle to a collective noun ("jury," "panel," "committee"). You must first determine whether the group is acting as a unit or individually.

The *committee* submitted *its* report. (acting as a single unit)
The *police* were given *their* assignments. (acting as individuals)

Compound Antecedents

A **compound antecedent** is an antecedent that consists of two or more elements. Agreement in number may present a problem if an antecedent is compound. To eliminate errors when this occurs, follow these three principles:

1. When two or more elements are connected by "and," use a plural pronoun to refer to the antecedent.
2. If two or more elements of a compound antecedent are joined by "or/nor," "either/or," and "neither/nor," (a) use a singular pronoun if all elements are singular; or (b) use a plural pronoun if all elements are plural.
3. If elements are connected by "or/nor," "either/or," or "neither/nor" and one part of the antecedent is singular and the other is plural, the pronoun must agree with the part that is closer to the verb. If applicable, place the plural item last and use a plural verb and pronoun.

The manager and the administrator planned *their* itinerary.
Faye or Tom can work on *her or his* papers now.
Neither Lars nor Hal has completed *his* book report.
The trainees or their supervisors will finish *their* statistical computations.
Neither the men nor the women plan to share *their* profits on the sale.
Neither the boxers nor the manager expressed *his* (or *her*) opinion.
Either the engineers or the architect gave *her* (or *him*) suggestions for renovation.
Either the architect or the engineers gave *their* suggestions for renovation.

Indefinite Pronoun Agreement

An **indefinite pronoun** refers in general terms to people, places, and things. Some pronouns in this category are always singular, such as "one," "each," "every," "anybody," and "anything."

Every auditor had an opportunity to ask *his or her* questions.
Each of the data operators is concerned about *his* or *her* job.

Other indefinite pronouns are always plural, such as “many,” “few,” “both,” and “several.”

Many will hand in *their* questionnaires.
Few accountants receive *their* CPAs.

Some indefinite pronouns, such as “all,” “any,” “some,” “more,” and “most,” can be either singular or plural depending on the noun or object of the preposition that follows them.

Singular: *Most of the* report had *its* spelling checked.
Plural: *Most of the reports* have *their* spelling checked.

Parallel Construction

Another kind of agreement is parallel construction. A construction that is not parallel will have a conjunction that joins unmatched elements. An adverb may be joined to a prepositional phrase, or a verb phrase may be joined to a noun. Constructions that are not parallel are ungrammatical.

Incorrect: Customers want *not only* good service *but also* to be treated with courtesy. (A correlative conjunction joins a noun with an infinitive verb phrase, which is a nonparallel construction.)
Correct: Customers want *not only* good service *but also* courtesy.
OR
Customers want *not only* to receive good service *but also* to be treated with courtesy.
Incorrect: The expert works cleverly *and* with speed. (A coordinate conjunction joins an adverb with a prepositional phrase, which is not a parallel construction.)
Correct: The expert works cleverly *and* speedily.
Incorrect: Jack is responsible for washing, ironing, and to fold the clothes. (an unparallel series)
Correct: Jack is responsible for washing, ironing, and folding the clothes.
Incorrect: His territory is larger than the Brainerd Realty Company. (confusing comparison)
Correct: His territory is larger than the Brainerd Realty Company’s territory.

15-4 Punctuation

For readers to interpret your ideas and inquiries precisely as you intend, you need to use correct punctuation in every message you write. Punctuation tells your readers where one thought ends and the next begins. Punctuation also clarifies and adds emphasis.

Punctuation includes external marks such as periods, question marks, and exclamation points. Punctuation also includes internal marks such as commas, semicolons, colons, quotation marks, parentheses, dashes, apostrophes, and hyphens.

15-4a THE PERIOD

A **period** can be used to indicate the end of a sentence, to indicate the end of an abbreviation, and to accompany an enumeration.

Periods at the End of Sentences

A period is used at the end of a declarative sentence, a mild command, an indirect question, and a courteous request. A **declarative sentence** makes a statement.

Gloria and Ralph are upgrading their software programs.
The choir members will sing in Italy during the holiday season.

A **mild command** is a stern request from the writer to the reader.

You should watch your step or you will fall.
Return the defective hard disk to the plant today.
Please pay attention to this section.

An **indirect question** is a statement that contains a reference to a question.

They inquired how your parents are feeling since their accident.
The judge asked if the prosecutor had any more questions for the witness.

A **courteous request** is a polite way to ask for action on the part of the reader; it is not, in fact, a question.

May I have an interview when convenient.
Would you be kind enough to revise the proposal and return the corrected copy.

Periods with Abbreviations

Periods are placed after many commonly used abbreviations to indicate that the words are shortened forms of longer words.

Mr. (Mister)	Jr. (Junior)	Dr. (Doctor)
Ltd. (Limited)	Inc. (Incorporated)	Sr. (Senior)

Periods in Enumerations

When numbers or letters are used in a vertical list, periods are placed after each number or letter.

Your child will need the following items for the outing:

1. one change of clothing
2. bathing suit, swim cap, sandals, and towel
3. snack money
4. sunscreen lotion and bug repellent

15-4b THE QUESTION MARK

A **question mark** is used after a direct question and after each part in a series of questions. The response may be a single word, or it may be one or more sentences.

Question Marks after Direct Questions

Use a question mark after a sentence that asks a direct question.

Do you agree that summer seems to pass more quickly than winter?
Have you considered relocating to find suitable employment?

Question Marks in a Series

Occasionally a series of questions may help your writing. For emphasis, follow each segment in the series with a question mark.

Were all the votes counted? all the winners notified? all the losers contacted?
Did she apply to Temple University? to Boston College? to the Univeristy of Miami?

15-4c THE EXCLAMATION POINT

An exclamation point is a punctuation mark that follows a word, a group of words, or a sentence. When that group of words shows excitement, urgency, anger, for example, the exclamation point conveys the strong emotion intended by the writer. In engineering writing, use exclamation marks very sparingly if at all.

Quick! Here's an opportunity to make money!
I'll never do that again!

15-4d THE COMMA

External punctuation marks tell the reader whether a sentence is a statement, question, or exclamation. Internal punctuation marks clarify the message intended by the writer. Of all the internal punctuation marks discussed here, the comma is without a doubt the most frequently used and misused.

Commas are used with introductory elements, independent clauses, nonessential elements, direct addresses, numbers, abbreviations, and repeated words. Commas also are inserted in a series and between adjectives. In addition, commas can indicate the omission of words and can enhance clarity in sentences.

With Introductory Elements in Sentences

Insert a comma after an introductory word, phrase, or clause.

Meanwhile, I will begin the next phase of the project. (introductory word)
Therefore, I wish to announce my candidacy. (introductory word)
In the long run, the cutback will be beneficial. (introductory phrase)
Because we have no record of the sale, we cannot help you. (introductory clause)
Although he was not present, his influence was evident. (introductory clause)

With Independent Clauses in Compound Sentences

When independent clauses with the same subject in a compound sentence are joined by a coordinate conjunction such as "for," "and," "nor," "but," "or," or "yet," precede the conjunction with a comma.

I will go to the hockey game on Friday, or I will babysit for my niece.
We thought he was guilty at first, but now we have changed our minds.

When each independent clause in a compound sentence has fewer than four words, no comma is needed.

Yoshi spoke and they responded.
I rode but he walked.

With Nonessential Elements

Nonessential elements are set off from the rest of a sentence with commas. Examples of nonessential elements are interrupting expressions, nonrestrictive elements, and appositives. Nonessential elements include information that may be interesting but is not necessary to the meaning or the structure of a sentence.

To determine if the information is essential, temporarily omit it. If the functional meaning of the sentence stays the same, set off the nonessential word, phrase, or clause with commas.

Interrupting Expression

Any expression that is nonessential and interrupts the flow of a sentence is set off with commas.

The most interesting part of the movie, *I believe*, is the ex-wife's entrance.
He should, *on the other hand*, separate the items in the box.

Nonrestrictive Element

A nonrestrictive phrase or clause adds information that is not essential to the meaning of the sentence.

Jeffrey Chang, *who graduated from Loyola*, is my neighbor.
We plan to order Part 643, *which Steve recommended*.

A phrase or clause that is essential to the meaning of a sentence is called a **restrictive phrase**, or a **restrictive clause**, and is not set off with commas.

Ask the nurse *who was on duty that night*.
The man who *was just hired* is part of my team.

Appositive

An appositive is a noun or noun substitute that renames and refers to a preceding noun. Appositives provide additional information that is not necessary to the meaning of a sentence. They are set off from the rest of the sentence with commas.

The paper contained the forecasts for the next quarter, *July through September*.
Ruby Muñoz, *the councilwoman*, is soliciting suggestions to bring up in council.

With Direct Address

To personalize a message, a writer may use direct address by mentioning the reader's first or last name in the beginning, middle, or end of a sentence. Because the name is not needed to convey the meaning of the sentence, it is set off with commas.

Dr. Oakes, you have been exceedingly helpful to my family.
Have I told you, *Gwen*, that we appreciate your purchase?

In a Series

Use a comma to separate three or more items in a series of words, phrases, or clauses. Although some experts omit the comma before a conjunction in a series, we recommend that you include the comma to avoid confusion.

> Evan's college essay was *thoughtful, humorous, and brief.*
> I will be *going to the movies, to the mall, or to my grandparents' home* Saturday evening.
> *Wake up early, prepare and serve breakfast, and take the children to the school bus.*

Between Adjectives

Use a comma between two adjectives that modify the same noun when the coordinate conjunction "and" is omitted. If the word "and" wouldn't make sense between the adjectives, do not insert a comma.

> The *short, thin* teenager envied the *tall, husky* football players.
> Janet's *royal blue* suit is inappropriate attire for a job interview.

With Omission of Words

Occasionally, a writer may omit words that are understood by the reader. Inserting a comma at the point of omission provides clarity.

> The treasurer is Johnetta; the *secretary, Garth*; and the *vice president, Warren*. (The word "is" is omitted twice in the sentence; commas are inserted at the points of omission.)

In Numbers

Use commas to indicate a whole number in units of three whether in money or items.

> $2,468
> 34,235 hot dogs
> 526,230 pins

For Clarity

Occasionally, a sentence requires a comma solely to ensure clarity.

> **Not Clear:** Shortly after the teacher left the classroom.
> **Clear:** Shortly after, the teacher left the classroom.

With Abbreviations

Writers who use abbreviations such as "etc.," "Jr.," "Sr.," and "Inc." should be familiar with the following comma rules concerning these abbreviations.

Rule 1. In a series, insert a comma before "etc." Insert a comma after "etc." when it appears in the middle of a sentence.

> We will be taking camping clothes: shorts, boots, swimwear, *etc.*
> The computer setup consisted of a CPU, a monitor, a printer, *etc.*, and cost $1,500.

Rule 2. Generally, place a comma before "Jr.," "Sr.," and "Inc." when the abbreviations appear in a name. Also insert commas after the abbreviations in the middle of a sentence.

> Harry Larkin, *Jr.*, was elected to the presidency.
> Able, *Inc.*, is owned by a conglomerate in New York.

15-4e THE SEMICOLON

A **semicolon** is a form of punctuation used to denote a pause. Semicolons are stronger than commas but weaker than periods.

Between Independent Clauses

Use a semicolon between two related independent clauses instead of using a comma and a coordinate conjunction.

> George is studying economics; his brother Dave is majoring in accounting.
> Elaine will attend the July convention; she then will vacation in Lofton.

Before Conjunctive Adverbs

Use a semicolon before a conjunctive adverb ("moreover," "nevertheless," "however," "consequently") that joins two independent clauses. Conjunctive adverbs, which function as transitional expressions, introduce the second clause.

> His report is too long; *therefore*, he cannot submit it until he revises it.
> The voice mail system can be easy; *however*, it tends to confuse some callers.

In a Series

Use a semicolon before expressions such as "for example (e.g.)," "that is (i.e.)," and "for instance" when they introduce a list of examples.

> You can attend some interesting functions; *for example*, an art show, a dance performance, or a special film screening.
> They must follow smart money management principles; *that is*, save part of their income, make purchases they can afford, and avoid buying inferior goods.

In Compound Sentences

Use a semicolon before a coordinate conjunction in a compound sentence when either or both of the clauses have internal commas and the sentence might be misread if a comma is inserted before the conjunction.

> I requested a return call, information about a particular check, and a phone number; instead, I received a reference to the wrong check and an incorrect phone number.
> On Wednesday, March 12, 2000, the group will meet; but Flora will not officiate unless she has recovered from her illness.

In a Series Containing Commas

Use semicolons to separate items in a series when an item or items contain commas.

> The mortgage company has branches in Newport, Rhode Island; Atlanta, Georgia; and Chicago, Illinois.
> Acklin, the chairman; Ikuko, the secretary; and Maria, the treasurer, were there.

15-4f THE COLON

A **colon** is a form of punctuation that directs the reader's attention to the material that follows it. The material that follows the colon completes or explains the information that precedes the colon.

Before a Series

Use a colon when the words "the following," "as follows," or "are these" are near the end or at the end of a sentence that introduces a series of items.

> Each person will need the following at the meeting: a computer, a printer, a set of instructions, and a writing tablet.

Before a List

Use a colon before a vertical, itemized list. As with a series, the words "the following," "as follows," or "are these" may precede the colon.

> Your instructions for Sunday are these:
>
> 1. Open the office at 9 a.m.
> 2. Check Saturday's mail, and call me if Pinder's check arrived.
> 3. Answer the telephone until noon.

Before a Long Quotation

Use a colon to introduce a long quotation of more than two lines.

> Chien remarked: "When I think of my home in Beijing, I can just picture the people riding their bicycles to work early in the morning and returning from work late in the evening. I now live in the center of busy Seattle, and that image has become one of my fondest memories."

Between Special Independent Clauses

Use a colon instead of a semicolon to separate two independent clauses when the second clause explains the first.

> Lucia is a skilled artist: She won an award for sketching animals.
> Here is one way to improve your sense of humor: Recall experiences that seemed serious at the time, and realize how funny they actually were.

After a Salutation

When using mixed punctuation in a letter, use a colon after a salutation.

> Dear Sir: Dear Dr. Santiago: Dear Ms. Linden:

In Time Designations

Use a colon between the hour and the minutes when the time is expressed in numerals.

> Let's meet at 11:30 a.m. in the lobby of the office building.

15-4g THE DASH

A **dash** (created in some word-processing software by keying two unspaced hyphens) is an informal punctuation mark. It is used with appositives or other nonessential elements that contain commas, before a summarizing statement, with a sudden change of thought, or before a detailed listing.

Note that many writers identify four distinct punctuation marks: em dash, en dash, hyphen (see the entry below), and minus sign. For brief explanations, see the last three entries under *Useful Websites* at the end of this chapter. Creating these punctuation marks in your document depends on your software. Microsoft Word, for example, allows for a variety of special characters through its *Insert/Symbol* feature.

With Nonessential Elements

For emphasis, use a dash to set off appositives and other nonessential elements from the rest of the sentence. Some of the nonessential elements may have internal commas.

> The stockbroker's office—newly equipped, nicely decorated, and spacious—is perfect for the hospitality reception.
> Paolo's new car—a Toyota—is equipped with power locks and a sunroof.

Before a Summarizing Statement

Use a dash after a listing at the beginning of a sentence that is followed by a summarizing statement. Summarizing statements often begin with the words "all" or "these."

> A nurturing manner, a love of people, and an unselfish attitude—these are three traits school counselors need.
> Precision in grammar and facts—both are necessary for effective writing and speaking.

With a Sudden Change of Thought

Use a dash to indicate a sudden change of thought or a sudden break in a sentence.

> Here is the perfect suit for work—and it's on sale, too!
> "Then we both agree that—oh no, now what's wrong?" asked Amy with a troubled look on her face.

Before a Detailed Listing

Use a dash to set off a listing or an explanation that provides details or examples.

> The restaurant features exotic desserts—Polynesian pudding, Hawaiian coconut sherbet, and Samoan almond supreme cake.
> Do your graduates have employable skills—excellent oral communications, keyboarding at least 70 wpm, and desktop publishing experience?

15-4h THE HYPHEN

A hyphen is a punctuation mark used after some prefixes and in forming some compound words.

After Prefixes

Use a hyphen after prefixes in some words. If you are unsure whether a word needs a hyphen, consult a dictionary.

ex-president	pro-American
de-emphasize	co-coordinator

In Compound Words

Use a hyphen in some compound words. In the English language some compound words are written as one word, others are written as two words, and others are hyphenated.

up-to-date reports	self-confident speaker
well-informed reporter	two-year-old child
Abe's mother-in-law	one-half the members

Some compound adjectives, such as "up to date," "well informed," and "two year old," are hyphenated if they precede the noun they modify, but they are not hyphenated if they follow the noun.

The report is up to date.
Our up-to-date equipment improves productivity.

15-4i QUOTATION MARKS

Quotation marks indicate a direct quotation, a definition, nonstandard English, a word or phrase used in an unusual way, or a title.

With Direct Quotations

When stating someone's exact words, enclose the words within opening and closing quotation marks.

Betty exclaimed, "It's getting late; let's go!"
"We'll leave now," answered Jeff. "We don't want to miss the train."

Within Quotations

Use single quotation marks to enclose a quotation within a quotation.

Amanda stated, "They listened to the president when he said, 'Our competition is getting ahead of us.'"

With Other Punctuation Marks

When placing ending quotation marks, follow these guidelines.

Place periods and commas within ending quotation marks.

"I concur," said the investor, "with your suggestion."

Place semicolons and colons outside ending quotation marks.

His best lecture is called "Psychoanalysis in the 1990s"; have you heard it?
This is the "beauty of San Diego": ideal temperatures and clear skies.

Place question marks and exclamation points inside the ending quotation marks when the quoted material is a question or an exclamation.

She shouted, "Watch out!"
He replied, "What's happening?"

Place question marks and exclamation points outside the ending quotation marks when the sentence, but not the quoted material, is a question or an exclamation.

Did Lydia actually say, "I will attend the seminar"?
What a deplorable situation; he's just "goofing off"!

With Definitions and Nonstandard English

Use quotation marks to designate a term that is defined in the same sentence in which the term appears.

A "couch potato" is someone who watches television all day and all evening.

Use quotation marks to enclose slang words or expressions.

He referred to his car as a "dumb bunny."

With Titles

Use quotation marks to enclose the titles of parts of whole works such as magazine articles and chapters. Also use quotation marks to enclose titles of lectures, songs, sermons, and short poems.

I read the article "The New Subcompact Cars" in *Consumer's Digest*.

15-4j PARENTHESES

A **parenthesis** is used in pairs to set off nonessential words, phrases, or clauses. The pair is called **parentheses**. Parentheses also are used with monetary designations, abbreviations that follow names, references and directions, and numerals and letters accompanying a list.

With Nonessential Elements

De-emphasize nonessential elements by placing them in parentheses. When the items in parentheses appear at the end of a sentence, place the external punctuation mark after the ending parenthesis.

A high percentage of the alumni (73 percent of those surveyed) opposed changing the name of the college.
We received a visit from our ex-president (1997–1998).

When an item in parentheses is a complete sentence, capitalize the first word and end the item with an internal punctuation mark.

Luis and Ramona relocated to Brooklyn last month. (Didn't you meet them in San Juan?)

When a dependent clause is followed by an item or items within parentheses, place the comma after the ending parenthesis.

When they arrive at the airport (around 7 p.m.), George will meet them and drive them to their hotel.

With Monetary Designations and Abbreviations

Used primarily in legal documents, parentheses can enclose a numerical designation ($500) following a verbal designation of money.

Mr. Chin has deposited the sum of five hundred dollars ($500) in your Swiss bank account.

In addition, parentheses are used with abbreviations that follow names.

The Association for Business Communication (ABC) had selected Clifford Chung as its Interim Executive Director.

With References and Directions

Use parentheses to set off both references and directions to minimize their importance in a sentence.

You may consult the appendix (page 345) for the correct format.
This trip (see the enclosed brochure) is a once-in-a-lifetime opportunity.

With Numerals and Letters Accompanying a List

When numerals or letters are used to list items in a sentence, parentheses may be used to enclose the numerals or letters.

> Please include (a) your date of birth, (b) your social security number, and (c) your password.

15-4k THE APOSTROPHE

The apostrophe is used primarily to indicate the omission of one or more letters or numbers to show a contraction, to indicate possession in nouns and indefinite pronouns, to denote time and money, or to avoid confusion in some plurals.

Contractions

Although sometimes considered overly informal, contractions are accepted in today's business world by many communicators. We recommend, however, using contractions sparingly. To indicate a contraction, insert an apostrophe in the space where the missing letter or letters belong.

don't (do not)	didn't (did not)	we'll (we will)
won't (will not)	doesn't (does not)	I'm (I am)
you're (you are)	aren't (are not)	hasn't (has not)
haven't (have not)	I'd (I had or I would)	shouldn't (should not)

To indicate an omission in a number, insert an apostrophe in the space where the missing number or numbers belong.

> Martin graduated in '99. (1999)
> The reunion had been planned for November of this year but was rescheduled for '02. (2002)

Possession

Apostrophes are used in the possessive case in nouns.

> The *boy's* suit needs pressing. (singular possessive)
> The *boys'* suits need pressing. (plural possessive)

Add an apostrophe plus *s* to an indefinite pronoun such as "someone" or "everyone" to show possession. In compound words, add the apostrophe to the last word to indicate possession.

> *Someone's* monitor has been left on.
> Although we had not asked her to, the server came with the water pitcher and refilled *everyone's* glass.
> My *brother-in-law's* education prepared him for his career as a lawyer and human rights activist.

Time and Money

Add an apostrophe or apostrophe plus *s* to "dollar," "day," "week," "month," and "year" to indicate each word's relationship with the noun that follows it.

> A *week's* salary is needed to pay the rent, but two *weeks'* salary is needed for the car payment and the insurance bill.
> Buy 10 *dollars'* worth of produce at the farmer's market.
> Phoenix is only an *hour's* drive from here.

Plurals

Add an apostrophe plus *s* to lowercase letters and to some abbreviations to form the plural.

We sometimes find it difficult to distinguish her *a's* from her *o's*.
Be sure to cross your *t's* and dot your *i's*.
Do not include so many *etc.'s* in your listings.

15–5 Abbreviations, Capitals, and Numbers

The basic rules for abbreviation, capitalization, and number usage may be called elements of writing style. Writers who are concerned about these three aspects of their workplace writing will minimize the number of distractions in a message and bring consistency to their writing.

15–5a ABBREVIATIONS

An abbreviation is a shortened form of a word or a group of words. Shortened forms should be used sparingly in business letters because they sometimes obscure the writer's meaning and they also present an informality that may offend the reader.

Shortened forms that apply to workplace writing include courtesy titles, "Jr." and "Sr." designations, and initials; professional titles and academic degrees; addresses and states; and names of companies, organizations, and government departments. In addition, abbreviations such as "a.m.," "p.m.," "Co.," "Inc.," "Corp.," and "Ltd." appear in engineering communications. Notice that although many abbreviations are followed by periods, some abbreviations are not.

Courtesy Titles and Family Designations

Abbreviate a personal title that precedes a person's name.

Messrs. White and Rome represent our firm at the negotiations. (The title "Messrs." is the plural of the title "Mr.")
We will interview *Ms.* Violeta Ruiz. ("Ms." is a title for a woman that omits reference to marital status; it does not have a full-length form. "Ms." is not an abbreviation for "Miss" or "Mrs.")

Abbreviate family designations, such as "junior" and "senior," that appear after a person's name. Commas usually set off the family designations.

Carl Brockman, *Jr.*, is the first speaker on the program.

Sometimes people use an initial to indicate the first letter of their first name or middle name.

I. H. Roth uses his first and middle initials, not his first name.

Gladys *S.* Blackwood insists that her middle initial appear on all correspondence.

Professional Titles

Some professional titles are abbreviated in business writing.

Dr. Sergio Silva is an internist in private practice.
The company lawyer, Sonia Ramos, *Esq.*, has an office on the 11th floor. (The title "Esq." is set off with commas.)

Academic and Professional Degrees

Abbreviate academic and professional degrees that follow a person's name.

Luisa Barnes, *Ed.D.*	Letitia Anderson, *M.D.*
Steven Joffe, *Ph.D.*	Edwin Jeifreys, *D.D.S.*

Addresses

In business correspondence, do not abbreviate words such as "street," "avenue," "boulevard," "road," "north," "south," "east," and "west." However, do abbreviate compass designations after street names.

Our new address is 123 South Main Street.
The meeting will take place at 4 Spring Boulevard.
Our president lives at 1605 Bird Lane *NW*.

States

Two-letter postal abbreviations appear in all capital letters without punctuation. Use these abbreviations with the appropriate five- or nine-digit ZIP codes in your correspondence.

Two-letter postal abbreviations are used in full addresses within the text of a letter but are not used when a state name appears in a sentence by itself.

Please send to Ms. Lucy Sands, 1004 Clemens Avenue, Roslyn, *PA* 19001-4356.
The cell phone will have to be shipped directly to Pennsylvania.

Companies, Organizations, and Government Departments

You may abbreviate the names of some well-known companies and organizations if the institutions themselves use the abbreviations. This policy also applies to U.S. government departments.

ABC (American Broadcasting Company)	AMA (American Medical Association)
FBI (Federal Bureau of Investigation)	IBM (International Business Machines)

Company, Incorporated, Corporation, Limited

The abbreviations "Co.," "Inc.," "Corp.," or "Ltd." may be used in a company name if the company uses it as part of its official name.

Our accountant previously worked for Mobil Oil *Corp.*
The British firm Lourdes, *Ltd.*, distributes this product.

Do not abbreviate "company," "incorporated," "corporation," or "limited" when it appears in lowercase letters in a sentence.

One firm has *incorporated* into the other.
She now owns her own software development *company*.

Expressions of Time

The abbreviations "a.m." and "p.m." may be used to designate time when they accompany numerals.

The next meeting is called for 8 *a.m.* on Tuesday, but I don't think I can make it at that time.
I stayed late at work—I was there until 8:30 *p.m.*

Familiar Workplace Abbreviations

Here are more examples of abbreviations, some of which tend to be used in informal workplace communications such as memos.

ASAP (as soon as possible)	CEO (chief executive officer)
C.O.D., c.o.d., COD (cash on delivery)	EST (Eastern Standard Time)
FYI (for your information)	GNP (Gross National Product)
P.O. Box (Post Office Box)	vs. (versus)

Miscellaneous Abbreviations

Some abbreviations used in statistical documents should not be used in business letters.

mfg. (manufacturing)	reg. (registered)
pd. (paid)	whlse. (wholesale)

Other abbreviations such as "No." (number) and "Acct." (account) may be used in technical documents when they are followed by numerals.

Please refer to check *No.* 654.
This information pertains to *Acct.* 6J843.

Units of Measure

The following abbreviations, though not acceptable in standard correspondence, are widely used in technical documents.

mph (miles per hour)	in. (inches)
oz. (ounce)	ft. (feet)
lb. (pound)	kg. (kilogram)
cm. (centimeter)	yd. (yard)

Days and Months

In lists and forms, the abbreviations for days and for months are acceptable. They are not acceptable in general correspondence.

Mon. (Monday)	Tues. (Tuesday)
Wed. (Wednesday)	Thurs. (Thursday)
Fri. (Friday)	Sat. (Saturday)
Sun. (Sunday)	Jan. (January)
Feb. (February)	Mar. (March)
Apr. (April)	Aug. (August)
Sept. (September)	Oct. (October)
Nov. (November)	Dec. (December)

15-5b CAPITALIZATION

A capital letter is used in the first word of a sentence, quotation, salutation, complimentary close, and outline. Further, capitalize titles of persons, written works, and proper nouns.

First Word of a Sentence or a Quotation

To indicate the beginning of a sentence, capitalize the first letter of the first word.

The tax collector is at the door.
When did this problem begin?

When a complete sentence that states a rule or emphasizes a statement is preceded by a colon, capitalize the first letter of the first word.

It is a perfect beach day: *The* sun is out, the breeze is warm, and the temperature is balmy.

Capitalize the first word of a direct quotation.

He said, "*Let* me help you perform the end-of-month audit."

Do not capitalize the second part of an interrupted direct quotation.

"We should congratulate Gail," James stated, "*on* her recent promotion."

Salutation and Complimentary Close

In a letter, capitalize the first letter of the first word, the person's title, and the proper name in a salutation. Also capitalize the first word in a complimentary close.

Dear Ms. Morales   Yours truly   Sincerely

Titles of Persons

Capitalize professional titles that precede proper names.

Dr. Nancy Musi   *Governor* Louis Ramos

Capitalize professional titles that do not precede proper names but that refer to specific, well-known individuals.

The *President* is concerned with the uprising in Europe. (refers to the President of the United States)

Generally, do not capitalize a job title that follows a name.

Tanya Blank is a junior engineer for our company.

Titles of Written Works

Capitalize all words in report headings and the titles of books, magazines, newspapers, articles, movies, television programs, songs, poems, reports, and chapters except for the following:

- The articles "the," "a," or "an"
- Short conjunctions (three or fewer letters)
- Short prepositions—including the word "to" in an infinitive

U.S. News and World Report (magazine)
How to Succeed in Business Without Really Trying (movie)

Capitalize an article, a short conjunction, or a short preposition when it is the first or last word in a heading or title.

The Far Pavilions
As You Like It

Proper Nouns and I

Capitalize the names of specific people, places, and things. Also, always capitalize the pronoun "I" wherever it appears in a sentence.

Names of People. Capitalize all proper names and nicknames.

Yoko Tanaka is a professor at the university.

Capitalize all titles of family members when the titles are used as proper nouns and are not preceded by a possessive noun or pronoun. Do not capitalize titles for family members, however, if they are preceded by a possessive noun or pronoun.

Let's visit *Grandmother* this morning.
Are you accompanying *Mother* and *Father* on the trip?
My *grandfather* started this business.

Names of Places. Capitalize the names of streets, parks, buildings, bodies of water, cities, states, and countries.

He lives at 106 *Green Street*. (street)
I have not visited *Jackson, Mississippi*. (city and state)

Names of Things. Capitalize the proper names of historical events, companies, documents, organizations, institutions, government departments, periods in history, course titles, and automobiles.

She is a veteran of *World War II*. (historical event)
Walt is a systems analyst for the *General Electric Company*. (company)
Have you studied the *Constitution* of the United States? (document)
I graduate from *Furness Junior High School* in 1995. (institution)
Glenda enrolled in *Physics 103*. (course title)
Suzanne's new car is a *Ford Taurus*. (automobile)

Capitalize some adjectives that are derived from proper nouns.

Several excellent *Spanish* students are enrolled in my class.
Do I detect a *Bostonian* accent?

Capitalize most nouns that precede numbers or letters.

Flight 643	Chapter VI	Vitamin C
Chart 6J	Invoice 1675	Check 563

Exceptions to this guideline include "line," "paragraph," "verse," "size," "page," and "note" when they precede numbers or letters.

line 4	paragraph 2	verse 16-5
size 10	page 24	note 14

Commercial Products

Do not capitalize common nouns that refer to, but are not part of, a proper noun.

Bic pen	Maytag dishwasher	Breyers ice cream

Points of the Compass

Capitalize compass points ("north," "south," "east," and "west") when they refer to a geographical area or a definite region. Do not capitalize compass points, however, when they indicate a direction or a nonspecific location.

The corporate office is in the *South*.
Travel *east* to the river and then drive *south* to the farm.

Months, Days, and Holidays

Capitalize the months of the year, the days of the week, and the names of holidays.

In *December*, we are having a company party on a *Monday* or *Tuesday*.
Where are you having your *Thanksgiving* dinner—at your house or at your aunt's house?

Seasons of the Year

Do not capitalize "summer," "fall (autumn)," "winter," or "spring" unless a specific designation accompanies the season.

Our *Spring* Blockbuster Sale begins March 21.
Old Man *Winter* is just around the corner.
After this icy *winter*, we are looking forward to spring.
The leaves turn beautiful colors in the *fall*.

Nationalities, Races, Religions, and Languages

Capitalize the names of nationalities, races, religions, and languages.

Many *Mexican* tourists visit San Diego.
Black History Month attracts noted *African-American* speakers.
Students learned about *Buddhism*, *Christianity*, and *Judaism* in Comparative Religion 101.
Her job at the World Bank requires her to learn *French*, *German*, *Italian*, and *Russian*.

Deities

Capitalize nouns that refer to a deity.

Allah   Christ   God   Jehovah   Vishnu

Academic Degrees

Because academic degrees such as Doctor of Education and Doctor of Philosophy are capitalized, also capitalize their abbreviations.

Leonard, a consultant, has *Ed.D.* printed on his business stationery and business cards.

15–5c NUMBER EXPRESSIONS

Because numbers are used in most engineering communications, writers must present them accurately and clearly to the reader. In correspondence, writers commonly refer to quantities, dollar amounts, percentages, dates, addresses, time, invoice numbers, research data and findings, etc. In reports and proposals, tables, charts, and graphs frequently accompany statistics. Numbers generally are written in word style in more formal and literary communications. Numeral style generally is used for routine correspondence.

Nine-and-Under / Ten-and-Over Rule

Write quantities of nine and under in words.

Mail *three* copies of the proposal to us.
We rented a *four*-bedroom house in the mountains.

Write quantities of 10 and over in numerals.

Would you buy *25* yellow-lined writing tablets for them?
Charles received *16* inquiries the first day of the session.

One exception to the nine-and-under/10-and-over rule involves indefinite or approximate numbers. Use words to express these numbers in a sentence.

Several *thousand* people attended the concert.

Use words for numbers in a sentence that includes two or more related numbers, all nine and under. If the numbers are all 10 and over, use numerals.

Daniel has written *five* articles, *one* anthology, and *three* textbook chapters.

When two or more related numbers are included in a sentence—some of which are 9 and under and some of which are 10 and over—use numerals for all numbers.

Bring *5* copies of the report, *2* copies of the names, and *25* copies of the newsletter.
Our inventory list of paint shows *18* cans of white, *24* cans of eggshell, and *9* cans of light blue.

Consecutive Numbers

When two related numbers appear next to each other in a sentence, write the shorter number in words and the other in numerals.

Ms. Chan received *160 two*-inch samples.
Oscar brought *twelve 36*-inch pieces to the classroom.

Consecutive Unrelated Numbers

If two unrelated numbers appear next to each other in a sentence, separate them with a comma to avoid confusion.

In *1997*, *18* of the girls made the All-State Team.

Numbers to Begin Sentences

Use words to express a number at the beginning of a sentence. If a number is very long, rewrite the sentence.

Eighty-one questionnaires were returned.
A total of *5,243* employees applied for the new health-care benefit. (Rewrite the sentence to avoid spelling out *5,243* at the beginning of a sentence.)

Numbers in Dates

When the day follows the month, express the day in numerals.

Kim's presentation is March *26*.

Use ordinals (*rd* or *th*) with the day when the day precedes the month and when the month and the year are omitted. Write out the ordinal ("first") or use numerals ("1st") if the month is omitted.

Your letter of the *26th* arrived today.

Numbers in Addresses

In ordinary text, use numerals to express house and building numbers except for the number *one*.

One East Grayson Place *6743* North Market Road

Use words for streets numbered first through tenth and numerals with ordinals for streets numbered 11th and over.

210 West *Fifth* Avenue 634 South *21st* Street

Numbers with Money

Write sums of one dollar or more in numerals preceded by a dollar sign ($).

Our total expenses are *$5.00* for the program and *$3.50* for a soda and a snack.

For sums less than one dollar, use numerals followed by the word "cents."

The small tablet costs *75 cents*.

In a series of amounts in the same sentence, use a consistent format.

Budget *$57.00* for the book, *$3.50* for the pens, and *$0.99* for the paper clips.

Write approximate amounts in words.

A *few hundred* dollars should cover the cost of the trip.

Use a combination of words and numerals to express very large amounts of money.

They won a *$20 million* state lottery last Tuesday.

Numbers with Percentages, Decimals, and Fractions

Use numerals followed by the word "percent" to express percentages.

The department store is offering a 40 *percent* discount.

Always express decimals in numerals. A zero placed at the left of a decimal point helps prevent the reader from overlooking the decimal point.

0.364 0.457 0.064

Express simple fractions in words.

We will need *three-quarters* of an hour to travel.

Express mixed numbers in either a fraction or a decimal unless they appear first in a sentence.

The job will take 2.5 hours to complete.
Two and one-half pounds of coffee are enough for the group.

Numbers with Time

Use numerals before "a.m." and "p.m.," but use words before "o'clock." To express the time on the hour, omit the colon and two zeros.

> One session begins at *9 a.m.*; the other begins at *1 p.m.*
> A *ten o'clock* meeting could extend past noon.

USEFUL WEBSITES

Note: For ease of reading, we have removed the "http://www." from the URLs below. However, some browsers might still require that you insert it to precede some URLs so as not to receive an error message.

■ Grammarly, a company that sells language-related software, also offers a great deal of free content regarding language use.
Visit: grammarly.com/handbook/

■ Jack Lynch, English Department faculty at Rutgers University, offers an excellent online "Guide to Grammar and Style." It is brief and eclectic (look it up!), but most of its advice and most of its Web links are relevant to engineering students.
Visit: andromeda.rutgers.edu/~jlynch/Writing/index.html

■ Jon Gingerich, an American journalist, wrote a useful column called "20 Common Grammar Mistakes That (Almost) Everyone Makes."
Visit: litreactor.com/columns/20-common-grammar-mistakes-that-almost-everyone-gets-wrong?

■ OWL (The Online Writing Lab of Purdue University) offers a "General Writing Resources" page that provides a wealth of information, advice, and exercises regarding writing and grammar.
Visit: owl.english.purdue.edu/owl/section/1/

■ OWL also provides a "Writing in Engineering" page of resources.
Visit: owl.english.purdue.edu/owl/section/4/19/

■ Stack Exchange's "English Language and Usage" page gives a brief explanation of how to use the em dash, the en dash, the hyphen, and the minus sign.
Visit: english.stackexchange.com/questions/2116/when-should-i-use-an-em-dash-an-en-dash-and-a-hyphen

The Chicago Manual of Style Online also provides brief guidelines for the use of the em dash, en dash, and hyphen. Note, though, that the complete Chicago Manual of Style is not available for free.
Visit: http://www.chicagomanualofstyle.org/home.html

■ The Webgrammar®, a division of Judy Vorfeld's Office Support Services, is maintained as a kind of public service. It offers a variety of pages and links that will help in improving students' grammar.
Visit: webgrammar.com/

■ Wikipedia's entries on the use of the dash punctuation mark provides a competent explanation of the difference between the em dash and the en dash.
Visit: en.wikipedia.org/wiki/Dash#En_dash_versus_em_dash

INDEX

B

C

D

E

O

P

T